# The PHYSICAL WORLδ

# QUANTUM PHYSICS OF MATTER

Edited by Alan Durrant

# I₀P

**Institute of Physics Publishing**
**Bristol and Philadelphia**
**in association with**

The Open
University

## *The Physical World* Course Team

| | |
|---|---|
| Course Team Chair | Robert Lambourne |
| Academic Editors | John Bolton, Alan Durrant, Robert Lambourne, Joy Manners, Andrew Norton |
| Authors | David Broadhurst, Derek Capper, Dan Dubin, Tony Evans, Ian Halliday, Carole Haswell, Keith Higgins, Keith Hodgkinson, Mark Jones, Sally Jordan, Ray Mackintosh, David Martin, John Perring, Michael de Podesta, Sean Ryan, Ian Saunders, Richard Skelding, Tony Sudbery, Stan Zochowski |
| Consultants | Alan Cayless, Melvyn Davies, Graham Farmelo, Stuart Freake, Gloria Medina, Kerry Parker, Alice Peasgood, Graham Read, Russell Stannard, Chris Wigglesworth |
| Course Managers | Gillian Knight, Michael Watkins |
| Course Secretaries | Tracey Moore, Tracey Woodcraft |
| BBC | Deborah Cohen, Tessa Coombs, Steve Evanson, Lisa Hinton, Michael Peet, Jane Roberts |
| Editors | Gerry Bearman, Rebecca Graham, Ian Nuttall, Peter Twomey |
| Graphic Designers | Javid Ahmad, Mandy Anton, Steve Best, Sue Dobson, Sarah Hofton, Jennifer Nockles, Pam Owen, Andrew Whitehead |
| Centre for Educational Software staff | Geoff Austin, Andrew Bertie, Canan Blake, Jane Bromley, Philip Butcher, Chris Denham, Nicky Heath, Will Rawes, Jon Rosewell, Andy Sutton, Fiona Thomson, Rufus Wondre |
| Course Assessor | Roger Blin-Stoyle |
| Picture Researcher | Deana Plummer |

The Course Team wishes to thank the following for their contributions to this book: Stan Zochowski, Ian Saunders, Ray Mackintosh, Derek Capper and John Bolton. The book made use of material originally prepared for the S272 Course Team by John Bolton, Shelagh Ross and Stuart Freake. For the multimedia package *Electrons in solids*, thanks are due to Will Rawes. *Nucleons in nuclei* and *Quarks* were originally made for S103.

The Open University, Walton Hall, Milton Keynes MK7 6AA

First published 2000

Written, edited, designed and typeset by the Open University.

Published by Institute of Physics Publishing, wholly owned by The Institute of Physics, London.
IoP Publishing, Dirac House, Temple Back, Bristol BS1 6BE, UK.

US Office: Institute of Physics Publishing, The Public Ledger Building, Suite 1035, 150 South Independence Mall West, Philadelphia, PA 19106, USA.

Printed and bound in the United Kingdom by the Alden Group, Oxford.

ISBN 0 7503 0721 8

*Library of Congress Cataloging-in-Publication Data are available.*

This text forms part of an Open University course, S207 *The Physical World*. The complete list of texts that make up this course can be found on the back cover. Details of this and other Open University courses can be obtained from the Course Reservations Centre, PO Box 724, The Open University, Milton Keynes MK7 6ZS, United Kingdom: tel. +44 (0) 1908 653231; e-mail ces-gen@open.ac.uk

Alternatively, you may visit the Open University website at http://www.open.ac.uk where you can learn more about the wide range of courses and packs offered at all levels by the Open University.

To purchase other books in the series *The Physical World*, contact IoP Publishing, Dirac House, Temple Back, Bristol BS1 6BE, UK: tel. +44 (0) 117 925 1942, fax +44 (0) 117 930 1186; website http://www.iop.org

1.1
s207book8i1.1

# QUANTUM PHYSICS OF MATTER

# Introduction

Much of the physical world can be understood on the basis of classical physics. The principles of Newton's dynamics, the elements of kinetic theory and thermodynamics and the laws of electromagnetism were established long ago. Engineers of earlier centuries built bridges and dams, and the steam engines that drove the industrial revolution, on the basis of measurements, models and empirical laws, all within the framework of classical physics.

Despite all this, the development of quantum physics has undermined the very foundations of the classical world and given us a new and much more successful way of understanding matter at the microscopic level. This sometimes creates the impression that quantum physics has its impact only on our view of the microscopic world: electrons, atoms and photons, etc. After all, bridges still stay up (usually), and engineers continue to design them without reference to wavefunctions or Heisenberg's uncertainty principle. Does quantum physics have anything to say about the macroscopic world?

Consider some simple questions: Why is steel hard and chewing gum soft? Why is it that copper conducts electricity but glass does not? Why does the Sun emit a large proportion of its electromagnetic radiation in the optical region of the spectrum? These are examples of questions that refer to matter on a macroscopic scale, but which classical physics cannot answer.

Chapters 1 and 2 of this book answer these, and other, questions by showing how quantum physics can be applied to macroscopic collections of microscopic particles, such as the electrons in a copper wire or the solar photons that warm us. Chapter 2 will also show how this understanding has lead to the technological revolution that has given us transistors, computers and other solid-state devices. Quantum physics is at the heart of the design and engineering of these devices. No team of Victorian engineers, however brilliant their classical expertise might be, could have designed and built a microchip or a solar cell!

Chapters 3 and 4 turn the spotlight back on the microscopic world. Using quantum mechanics we can investigate structures smaller than that of the atom, and answer a whole new range of fundamental questions: What holds the atomic nucleus together? What are protons made of? What are the truly fundamental constituents of matter? Answering these questions also has its pay-off at the macroscopic level: you will see why radium glows in the dark and where the Sun gets its energy from.

As you study this book you should bear in mind that many of the topics that are discussed involve quite advanced applications of quantum physics. Accordingly we do not always give full details of arguments nor give derivations of all results, since this would require a level of treatment outside the scope of *The Physical World*. It is more important that you know the physical laws and principles from which results and conclusions are obtained, and it is this, rather than the details of the derivations, that we have emphasized in this book.

Open University students should be aware that the video for this book (*Video 8 The Incredible Shrinking Chip*) is associated with Chapter 2. There is a prompt to view it at the end of that chapter.

# Chapter 1 Quantum gases

## 1 What is a quantum gas?

Quantum mechanics was invented to describe the behaviour of microscopic particles — atoms, electrons and atomic nuclei, etc. Macroscopic systems are normally described very well by classical physics. Indeed we would be very surprised if we could see the weird quantum world of wavefunctions and discrete energy levels, etc., operating on a macroscopic scale.

Well, it does happen sometimes. Liquid helium ($^4_2$He) exhibits superfluid behaviour at temperatures below 2.17 K, and some metals and ceramics become superconductors at low temperatures. These phenomena are direct manifestations of quantum mechanics in systems of macroscopic size. However, the liquid and solid phases are inherently complex, with large interactions between the constituent particles, and so the macroscopic quantum effects are not seen in their purest form. The ideal medium in which to study macroscopic quantum behaviour would be a gas where interactions between the particles play only a secondary role. However, you know from an earlier book in this series, *Classical physics of matter* (*CPM*), that classical statistical physics is able to account for most of the macroscopic properties of gases, particularly those that relate to the translational motion of the atoms or molecules that compose a gas. Despite this, gases of atoms can provide dramatic examples of quantum behaviour.

Paul Dirac (1902–84).

In 1995, a group of American physicists cooled a vapour of the alkali metal rubidium to very near absolute zero. When the vapour was cooled to a temperature of $1.7 \times 10^{-7}$ K something remarkable happened. The vapour condensed, not to a liquid or a solid, but to a phase of matter called a *Bose–Einstein condensate*. In this phase of matter, the sample is still a gas of very low density, but it is no longer a collection of individual rubidium atoms and cannot be understood in terms of classical physics. It is best thought of as a single quantum system with the entire gas represented by a single wavefunction. Bose–Einstein condensates can exhibit quantum effects on a macroscopic scale. Figure 1.1 shows interference fringes produced when two Bose–Einstein condensates of sodium vapour move into one another. The fringes appear in the region where the condensates overlap and their wavefunctions interfere. The dark fringes represent positions where the interference of the wavefunctions is constructive and the density of the overlapped condensates is large.

A Bose–Einstein condensate is a recently discovered example of a *quantum gas* — a gas for which the classical statistical description is inadequate and quantum physics must be taken into account.

Enrico Fermi (1901–54).

Quantum gases occur in many different forms, not all of them as dramatic in their properties as the Bose–Einstein condensate. In fact, the basic quantum physics of gases was worked out in the 1920s by Bose, Einstein, Fermi, Dirac, Pauli and others. Much of the motivation of that early work was the need for theoretical models that could describe such diverse phenomena as the spectrum of radiation emitted by the Sun and the conduction of electricity in metals. These are phenomena that you might not associate with the concept of a gas.

In physics, the concept of a gas is a very general one. A gas is any collection of particles that interact with one another very weakly or not at all, so that each particle can move freely within some specified region. The particles of a gas do not have to

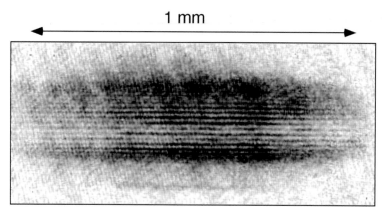

1 mm

**Figure 1.1** Interference fringes produced by the superposition of two Bose–Einstein condensates of sodium vapour. The fringe pattern is reminiscent of a two-slit diffraction pattern in optics but the fringes represent spatially periodic variations in the density of the gas due to interference of the wavefunctions.

be atoms or molecules. Photons are particles of electromagnetic radiation. The heat radiation cooking a dish in your oven is a *gas of photons*. More surprisingly perhaps, we can treat the free electrons that are responsible for the conduction of electricity in a metal as a *gas of electrons*, despite the fact that electrons would be expected to interact powerfully with one another by electrostatic Coulomb repulsion. Gases of photons and electrons do not normally exist under conditions where a classical theory can be applied. They are essentially quantum gases, and a quantum treatment is needed right from the start.

The main aim of this chapter is to introduce the quantum theory of gases and apply it to gases of molecules, photons and free electrons in metals. You will see that there are two main ways in which quantum theory enters the description of the translational motion of gases:

- The translational energy of a particle confined to a box is *quantized*.
- Identical particles are completely *indistinguishable* from one another.

Quantum physics also plays a central role in describing the rotational and vibrational motion of molecules, but in this chapter we are concerned with *translational motion* only.

You are familiar with the principle of energy quantization from the way in which electrons occupy discrete energy levels in atoms. However, the idea of indistinguishability may not mean much to you at this stage. We shall discuss it fully in this chapter.

The organization of this chapter is as follows:

In Section 2 we revisit the *Maxwell–Boltzmann energy distribution* of molecules in an ideal gas, introduced in *Classical physics of matter (CPM)*, looking carefully at its quantum-physical foundations and the limits of its validity. The classical derivation of this distribution was based on the idea of dividing up the possible positions and velocities of a molecule into 'bins' called *phase cells*. Each phase cell is supposed to specify a particular position and velocity to within a certain precision. This precision could, in principle at least, be as small as we wished. Having studied quantum theory in *Quantum physics: an introduction (QPI)* you will now appreciate that we have to abandon this idea, because the uncertainty principle ($\Delta x \, \Delta p_x \geq \hbar/2$) puts a limit on the precision to which position and velocity can be simultaneously specified. However, losing the idea of phase cells does not turn out to be a problem for us, because quantum physics itself comes to our rescue by providing bins of a different and more fundamental kind. The translational energy of a particle confined to a box is quantized, i.e. the particle can exist in *quantum states* corresponding to certain *allowed values of the translational energy*. Thus the discrete translational quantum states are the 'bins' we must use in a quantum-mechanical derivation of the Maxwell–Boltzmann energy distribution. You will see that when we do this, the discreteness of the allowed energies imposes a restriction on the validity of the Maxwell–Boltzmann energy distribution.

Quantum theory imposes another restriction, of different origin, on the validity of the Maxwell–Boltzmann energy distribution. There is an assumption in the classical treatment of gases that is very easy to overlook, although Boltzmann himself was aware of it. It was assumed that identical molecules can be distinguished from one another, i.e. it was assumed that molecules can, in the mind's eye at least, be unambiguously labelled A, B, C, etc. Quantum physics forbids the labelling of identical particles. In quantum physics, identical particles have no individual identity; they are completely *indistinguishable* from one another and cannot be unambiguously labelled. The indistinguishability of identical particles is the subject of Section 3.

One consequence of indistinguishability is the classification of particles into two types called *bosons* and *fermions* according to their intrinsic spin angular momentum. For example, photons (spin 1) are bosons but electrons (spin 1/2) are fermions. Fermion particles obey the Pauli exclusion principle while bosons do not. Thus a quantum gas of fermions is expected to have different properties from a quantum gas of bosons.

More generally, indistinguishability and the exclusion principle have profound effects on the properties of matter at all levels. By allowing only one electron to occupy any quantum state, the exclusion principle determines the electronic structure of atoms and the arrangement of the elements in the Periodic Table, as discussed in *QPI*. In Sections 4 and 5 of this chapter you will see that the exclusion principle also accounts for the dramatic differences in the behaviours of gases of photons (which are bosons) and gases of electrons (which are fermions). The Maxwell–Boltzmann distribution fails completely to describe these quantum gases.

Section 4 gives a detailed study of photon gases. Knowing that photons are bosons, we are able to derive the distribution of photon energies in a photon gas. Using this we can predict, for example, how the energy of sunlight is distributed among the different wavelengths of the spectrum, and explain why the total power radiated by a star is proportional to the fourth power of its surface temperature. We shall even see why the uniform and isotropic background radiation that bathes the entire Universe, the 3 K microwave radiation, is so cold.

Section 5 looks at free-electron gases in metals. Electrons are fermions, and the main feature here is the Pauli exclusion principle, forcing the free electrons to spread themselves out, with no more than one electron per quantum state, across a wide range of translational energies. You will see that this is one reason why metals are so hard.

# 2 Quantum theory of a gas of distinguishable molecules

This section introduces the quantum treatment of the translational motion of distinguishable gas molecules, looking carefully at the ways in which the molecules in an ideal gas can be distributed among the quantum states that correspond to the allowed values of the translational kinetic energy. These quantum states are the quantum-mechanical bins among which the molecules are distributed. Thus we shall rederive the Maxwell–Boltzmann energy distribution and reveal the limitations imposed on its validity by energy quantization.

Throughout this section we shall retain the classical notion that the molecules are distinguishable from one another and can be labelled, just as in the classical statistical theory. Thus the treatment in this section is partly quantum-mechanical and partly classical. The fully quantum-mechanical treatment, which includes the effects of indistinguishability and the exclusion principle, will be given in Section 3. You will see in that section that these principles impose further, more stringent, restrictions on the validity of the Maxwell–Boltzmann distribution.

It is useful to begin this section with a review of the main results of the classical statistical theory of ideal gases.

## 2.1 Review of the classical statistical theory of an ideal gas

We shall use the terms *molecule* and *molecular gases* in the broad sense to include atoms and monatomic gases such as helium.

You have seen (*CPM*) that the molecules of an ideal gas are assumed to be in rapid motion, continually colliding with one another and with the walls of the container. Regardless of how the gas is introduced into the container, the collisions soon establish a state of thermal equilibrium between the gas and the container walls. Because there is a huge number of molecules in a gas, an exact analysis of every collision of each molecule is beyond our grasp, and so a statistical approach is used. In this picture, the velocity of each molecule changes by a random amount with each collision and is unknown to us at any instant. However, it is possible to use statistical methods to calculate the relative numbers of molecules moving with different speeds at any instant. This results in the *Maxwell–Boltzmann speed distribution* which agrees very well with experiments such as the rotating-drum experiment.

The Maxwell–Boltzmann speed distribution corresponds to a particular distribution of molecular translational energies called the *Maxwell–Boltzmann energy distribution*. You saw in *CPM* that this distribution can be written as

$$g(E) = \frac{2}{\sqrt{\pi}} \left( \frac{1}{kT} \right)^{3/2} \sqrt{E} e^{-E/kT}.$$

Here, $E$ denotes the translational energy of a molecule (i.e. $E = E_{\text{trans}} = mv^2/2$, where $m$ and $v$ are the mass and speed of a molecule), $k$ is Boltzmann's constant and $T$ is the absolute temperature. The quantity $g(E)$ is the Maxwell–Boltzmann energy distribution function, and $g(E)\,\Delta E$ is the probability of finding a given molecule in the small range of translational energies between $E$ and $E + \Delta E$.

For the purposes of this book it is actually more convenient to consider the function

$$G(E) = Ng(E)$$

where $N$ is the total number of molecules in the gas. $G(E)$ is also called the Maxwell–Boltzmann energy distribution, and is said to be *normalized* to the total number of molecules in the gas — a description that will be made clearer below. Thus we have

$$G(E) = \frac{2N}{\sqrt{\pi}} \left( \frac{1}{kT} \right)^{3/2} \sqrt{E} e^{-E/kT} \tag{1.1}$$

We interpret $G(E) \Delta E$ as the *average number* of molecules with translational energies in the small range between $E$ and $E + \Delta E$. We will be mainly concerned with gases containing an immense number of molecules, of order $10^{23}$ in a typical sample, so fluctuations around this average value can be neglected. For this reason, the word 'average' is sometimes omitted and we interpret $G(E) \Delta E$ as the *number* of molecules with translational energies in the small range between $E$ and $E + \Delta E$. Because $G(E) \Delta E$ is just a number, without units and $\Delta E$ has units of energy, $G(E)$ must have the units of reciprocal energy, i.e. $J^{-1}$. We say that $G(E)$ itself gives the number of molecules per unit energy interval.

Note that the Maxwell–Boltzmann energy distribution depends only on $N$ and the temperature $T$. Figure 1.2 shows a graph of $G(E)$ against $E$. There is a peak followed by a long tail. In fact, the peak is located at $kT/2$ and its height is inversely proportional to temperature. As the temperature rises the peak moves to higher energies and becomes less pronounced — the curves flatten out. Figure 1.3 shows the distribution functions $G(E)$ for an ideal gas at temperatures of 200 K, 300 K and 400 K.

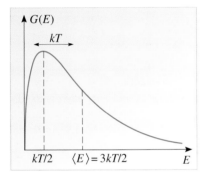

**Figure 1.2**   The Maxwell–Boltzmann energy distribution, $G(E)$. The average translational energy of a molecule, $\langle E \rangle = 3kT/2$, is shown. The energy spread of the distribution, of the order of $kT$, is also indicated.

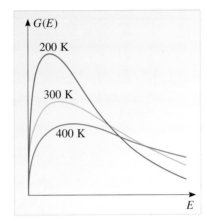

**Figure 1.3**   The Maxwell–Boltzmann energy distribution in an ideal gas at 200 K, 300 K and 400 K. The total number of molecules is the same in each case.

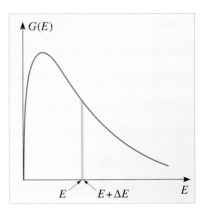

**Figure 1.4**   The area of the narrow column gives the number of molecules with translational energies between $E$ and $E + \Delta E$.

The number of molecules with translational energies in a narrow range between $E$ and $E + \Delta E$, given by $G(E) \Delta E$, is approximately equal to the area of the narrow column in Figure 1.4. The number of molecules within a wider range of energies, between say $E = E_A$ and $E = E_B$, is equal to the area under that part of the graph of $G(E)$ between the vertical lines at $E_A$ and $E_B$, as indicated in Figure 1.5. This area can be calculated mathematically. It is given by the definite integral

$$\int_{E_A}^{E_B} G(E)\, dE$$

which represents the sum of the areas of all strips, starting with a strip at $E_A$ and ending with one at $E_B$, in the limit of very narrow strips. The integral can be evaluated by standard calculus methods often with the help of tables of standard integrals, or by using computer packages. You will *not* be required to evaluate integrals. We shall always give the value of an integral where needed.

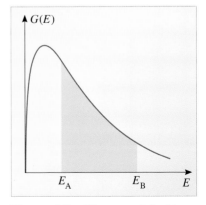

**Figure 1.5**   The area of the blue region gives the number of molecules with translational energies between $E_A$ and $E_B$.

The total number $N$ of molecules in the gas is represented by the area under the entire graph of $G(E)$. We can express this entire area as a definite integral by putting $E_A = 0$ and letting $E_B$ go to $\infty$, and so, using the integral notation, we write

$$\int_0^\infty G(E)\,dE = N. \qquad (1.2)$$

This equation is the mathematical statement of the fact that $G(E)$ is *normalized* to $N$. Thus in Figure 1.3, where the total number $N$ of molecules in the gas is the same for all three temperatures, the total area under each graph (when each graph is extended along the whole energy axis) is the same and equal to $N$.

The Maxwell–Boltzmann energy distribution, Equation 1.1, can be used to calculate average values of quantities that depend on $E$. One of the most useful quantities to know is the average translational energy of a gas molecule, $\langle E \rangle$. Let's see how this can be calculated.

We know that the number of molecules in a narrow energy range between $E$ and $E + \Delta E$ is given by $G(E)\,\Delta E$. All these molecules have translational energy $E$. The total translational energy of molecules in this group is therefore $G(E)\,\Delta E \times E$, and the total translational energy $U$ of the whole gas is the sum of the quantities $G(E)E\,\Delta E$ over all energy intervals $\Delta E$. This sum can be expressed as a definite integral, which can be evaluated to give $U = 3NkT/2$. The average energy of a gas molecule, $\langle E \rangle$, is found by dividing this total energy by $N$. Thus

$$\langle E \rangle = U/N = 3kT/2. \qquad (1.3)$$

This average energy is indicated in Figure 1.2. You can think of it as an average translational energy of $kT/2$ for each of the three translational degrees of freedom of a molecule — an example of the equipartition of energy theorem.

We can use Equation 1.3 to relate the root mean square (rms) speed of molecules, $v_{rms} = \langle v^2 \rangle^{1/2}$, to the temperature $T$. We first express the average translational energy as $\langle E \rangle = \langle mv^2/2 \rangle = m \langle v^2 \rangle /2$. Thus we have

$$m\langle v^2 \rangle/2 = 3kT/2.$$

Rearranging this result gives

$$v_{rms} = \langle v^2 \rangle^{1/2} = \sqrt{\frac{3kT}{m}}. \qquad (1.4)$$

The energy value $kT$, appearing in Equations 1.3 and 1.4, is such a useful parameter that it has acquired a colloquial name, the **typical thermal energy per particle**. You can see from Equation 1.3 and Figure 1.2 that the typical molecular energy and the spread of molecular energies are both of the order of $kT$. It is useful to remember that at room temperature $kT \approx 0.025$ eV, or 1 eV/40.

We now turn to a brief discussion of the origins of the Maxwell–Boltzmann energy distribution function, $G(E)$, from the point of view of classical statistical mechanics.

Using Equation 1.1, we can express $G(E)\,\Delta E$, the average number of molecules with translational energies in a small range from $E$ and $E + \Delta E$, as

$$G(E)\,\Delta E = \text{constant} \times (\sqrt{E}\,\Delta E) \times (Ne^{-E/kT}). \qquad (1.5)$$

Notice that apart from a constant (determined by normalization), this quantity is a product of two energy-dependent factors, $\sqrt{E}\,\Delta E$ and $Ne^{-E/kT}$. Each of these factors has a simple physical interpretation.

Note that the symbol $U$ here is used to denote the total translational energy $U_{trans}$. We are not including other possible forms of internal energy such as rotational or vibrational energy.

The factor $\sqrt{E}$ in the first bracket describes how the density of phase cells increases with energy (this was shown in *CPM*), and so $\sqrt{E}\,\Delta E$ is proportional to the number of phase cells in the energy range between $E$ and $E + \Delta E$. The factor $e^{-E/kT}$ in the second bracket is the *Boltzmann factor*, which is proportional to the probability of finding any given molecule in a particular phase cell of energy $E$. Thus the factor $Ne^{-E/kT}$ is proportional to the average number of molecules in each phase cell.

With this interpretation of the two factors in brackets, Equation 1.5 can be expressed in words as

$$\begin{pmatrix} \text{average number of molecules} \\ \text{with energies in the range} \\ \text{from } E \text{ to } E + \Delta E \end{pmatrix} \propto \begin{pmatrix} \text{number of phase cells} \\ \text{within the energy range} \\ \text{from } E \text{ to } E + \Delta E \end{pmatrix} \times \begin{pmatrix} \text{average number of molecules} \\ \text{in each phase cell} \end{pmatrix}.$$

Apart from the normalization constant, the above statement of Equation 1.5 embodies the physical interpretation of the Maxwell–Boltzmann energy distribution (Equation 1.1) according to classical statistical mechanics. The function $G(E)$ and the separate factors $\sqrt{E}$ and $e^{-E/kT}$ are shown in Figure 1.6. You can see that the Boltzmann factor $e^{-E/kT}$ is responsible for $G(E)$ falling off at high energies.

From a quantum physics perspective, what is wrong with the Maxwell–Boltzmann energy distribution described above?

One problem is that it is based on the idea of a *phase cell*, a bin of very closely specified position and velocity. This is a classical concept, alien to quantum physics. You know that the uncertainty principle ($\Delta x\,\Delta p_x \geq \hbar/2$) forbids the simultaneous specification of both position and velocity. Although the product of quantum uncertainties is extremely small, the uncertainty principle undermines the concept of a phase cell, and certainly prevents us from specifying very small phase cells. If we are interested in finding a quantum-mechanical derivation of the Maxwell–Boltzmann energy distribution, it is essential to abandon the classical notion of a phase cell. Fortunately, quantum mechanics immediately suggests a sensible alternative. In quantum mechanics, the state of a particle is naturally specified by a quantum state. The switch from *phase cell* to *quantum state* is one of the main differences between the classical and quantum derivations of the Maxwell–Boltzmann distribution. Thus, a quantum-mechanical derivation will be based on the following interpretation of Equation 1.5:

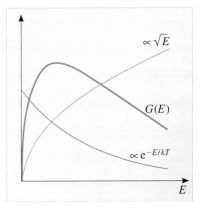

**Figure 1.6** The Maxwell–Boltzmann distribution $G(E)$ is the product of two energy-dependent factors, one proportional to $e^{-E/kT}$ and the other proportional to $\sqrt{E}$.

$$\begin{pmatrix} \text{average number of molecules} \\ \text{with energies in the range} \\ \text{from } E \text{ to } E + \Delta E \end{pmatrix} \propto \begin{pmatrix} \text{number of quantum states} \\ \text{within the energy range} \\ \text{from } E \text{ to } E + \Delta E \end{pmatrix} \times \begin{pmatrix} \text{average number of molecules} \\ \text{in each quantum state} \end{pmatrix}$$

where, of course, the key change is the replacement of *phase cell* by *quantum state* in both the brackets on the right-hand side. In Section 2.2 we will look at the quantity in the first brackets on the right-hand side, counting up the number of quantum states in a small range of translational energies.

There is another fundamental problem with the classical derivation of the Maxwell–Boltzmann distribution. The Boltzmann factor in the second bracket of Equation 1.5 is derived by an argument that assumes that the individual molecules can be distinguished from one another and labelled, A, B, C, etc. This argument is outlined in Section 2.3. In quantum physics identical particles cannot be distinguished from one another and cannot be labelled. In Section 3 we shall see what the implications of this are for the second bracket on the right-hand side of the above equation, the average number of molecules in each quantum state.

Given the fact that the Maxwell–Boltzmann energy distribution works well enough, and that classical physics is capable of deriving it, you might think we are being rather finicky, insisting on a quantum-mechanical derivation. Couldn't we just accept that classical physics emerges, somehow, as a limiting case of quantum physics, and so accept that the Maxwell–Boltzmann energy distribution is true, at least to a good approximation? Perhaps, but we have other aims in mind. By deriving the Maxwell–Boltzmann distribution directly from quantum-mechanical ideas, we will be able to reveal criteria for its validity. Later in this chapter, especially when we discuss gases of electrons and photons, you will see that there are situations in which these criteria are not met, and a quantum-mechanical approach is absolutely essential.

So let's get started.

## 2.2 Quantization of translational energy and the density of states

You have seen in *QPI* that the total energy of a confined particle is quantized, i.e. it can have certain discrete energies only. The most familiar example is the electron in a hydrogen atom, confined by the Coulomb attraction of the positively-charged nucleus. The same idea applies to a molecule confined in a container. The molecule can occupy discrete translational energy levels determined by the size and shape of the container. For an ideal gas, the particles do not interact with one another and so the quantum states and energy levels of identical molecules can be found by solving the Schrödinger equation for a single molecule in the container.

A rigid container, for example a glass bottle, can be represented by a *three-dimensional infinite square well*. It is convenient to take the potential energy inside the well to be zero, so that the total energy $E$ of the molecule is equal to its translational kinetic energy. At the container boundaries the potential energy rises abruptly to infinity, representing the fact that the particles cannot penetrate into the walls. You have seen (in *QPI*) that a particle confined to a three-dimensional cubic region (a hollow cubical box) can occupy discrete translational quantum states with allowed energies given by

$$E = \frac{h^2}{8mL^2}(n_1^2 + n_2^2 + n_3^2)$$

(1.6)

where $L$ is the length of each side of the box, $h$ is Planck's constant and $m$ is the mass of the particle. The quantum numbers $n_1$, $n_2$ and $n_3$ can be any positive integers, 1, 2, 3, etc. Each ordered set of three quantum numbers $(n_1, n_2, n_3)$ defines a translational quantum state of the particle. The energies of the seventeen lowest energy states are shown in Figure 1.7.

Note that the typical spacing of the energy levels is $h^2/(8mL^2)$ and most of the energy levels are degenerate, i.e. there is more than one quantum state with the same energy. In fact the degeneracy, (the number of quantum states with the same energy), tends to increase with increasing energy. In general, the states get more and more closely packed with increasing $E$.

This close packing suggests that for very large energies we can expect the quantized energy levels to be so numerous and so close together that they may be treated as a gapless continuum of allowed energies, of just the type we are familiar with in classical physics. When the energy levels (and the corresponding quantum states) are treated in this way, we say we are making the **classical continuum approximation**. For this continuum approximation to hold good, the typical spacing between energy levels, $h^2/(8mL^2)$, must be very small compared with the width of the

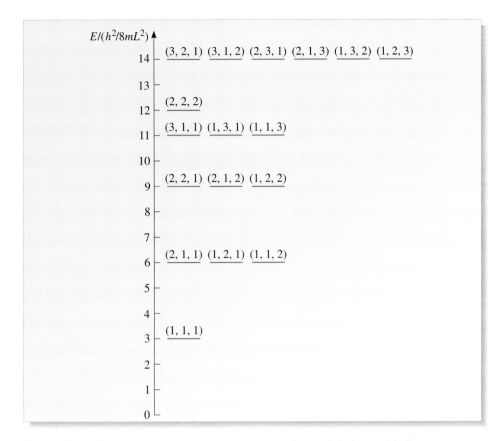

**Figure 1.7**   The seventeen lowest energy states of a particle in a cubic box.

Maxwell–Boltzmann distribution, the typical thermal energy per particle of order $kT$ (see Figure 1.2). Thus, the criterion for a classical continuum approximation is given by

$$\frac{h^2}{8mL^2} << kT. \tag{1.7}$$

The symbol << means 'is very much less than'.

**Question 1.1**   A sample of nitrogen gas is contained in a cubic box of side 10 cm. Determine the typical spacing of translational energy levels. Would you expect the continuum approximation to be valid at room temperature? (Take the mass of a nitrogen molecule to be $4.65 \times 10^{-26}$ kg.)

**Question 1.2**   By reading Equation 1.7, state qualitatively the conditions of temperature, container size and particle mass where you might expect the classical continuum approximation to fail.  ■

We can express the criterion of Equation 1.7 in another way, in terms of the typical de Broglie wavelength $\lambda_{dB} = h/mv$, where $mv$ is the magnitude of the momentum of a typical molecule. Making $v$ the subject of this equation yields

$$v = \frac{h}{m\lambda_{dB}}.$$

Now we can take the rms speed, given by Equation 1.4, as the speed of a typical molecule,

$$v = v_{\mathrm{rms}} = \sqrt{\frac{3kT}{m}}.$$

Equating these two expressions for $v$ gives

$$\frac{h}{m\lambda_{\mathrm{dB}}} = \sqrt{\frac{3kT}{m}}.$$

If we make $kT$ the subject of this equation, we obtain

$$kT = \frac{h^2}{3m\lambda_{\mathrm{dB}}^2}.$$

You can move factors from one side of the 'very much less than' symbol ($<<$) to the other side in the same way as you move symbols from one side of an equal sign ($=$) to the other, provided the factors you are moving represent *positive* quantities as they do here.

Substituting this expression for $kT$ into the right-hand side of Equation 1.7, and then cancelling $m$ and $h^2$, we find $1/(8L^2) << 1/(3\lambda_{\mathrm{dB}}^2)$. Ignoring the factors 8 and 3, we can write the criterion simply as $\lambda_{\mathrm{dB}}^2 << L^2$, or more simply still as

$$\lambda_{\mathrm{dB}} << L. \tag{1.8}$$

Thus, the validity of the classical continuum approximation for the translational energy states and energy levels requires the de Broglie wavelengths of the molecules to be very short compared to the size of the container.

Assuming the continuum criterion to be satisfied, we can now consider an energy range $\Delta E$ that is narrow compared to $kT$, but wide enough to contain a very large number of states, as illustrated in Figure 1.8. The number of translational energy states in this range will then be proportional to the range $\Delta E$ and can be written as $D(E)\,\Delta E$, where $D(E)$ is a function of $E$, that is:

number of quantum states with energies between $E$ and $E + \Delta E = D(E)\,\Delta E$.

The function $D(E)$ is called the **density of states** function; it describes the number of states per unit energy interval at energy $E$, and thus describes the way in which the states of allowed translational energy are distributed.

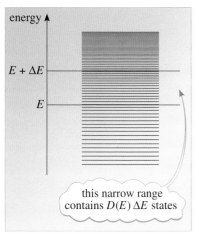

Starting from Equation 1.6 and using the classical continuum approximation, it is possible to derive an expression for $D(E)$. The result of the derivation is

$$D(E) = B\sqrt{E} \tag{1.9}$$

where $B$ is a constant:

$$B = \frac{2\pi L^3 (2m)^{3/2}}{h^3}.$$

**Figure 1.8** Defining the density of states function. Each horizontal line indicates the energy of a particular quantum state or a set of degenerate quantum states, i.e. a set of states all having the same energy.

(If you would like to see the details of the derivation you can find them in the Appendix to Chapter 1. You will not be asked questions about the details of the derivation.)

Note that the symbols $L^3$ and $m^{3/2}$ appear in the numerator of the expression for the constant $B$ in the density of states function, Equation 1.9. This shows that the density of states $D(E)$ is largest for particles of large mass in containers of large volume $V = L^3$.

It is important to realize that the validity of Equation 1.9 rests on the classical assumption of a continuum of energy levels, i.e. Equation 1.7 or 1.8. Equation 1.9 then gives a very good approximation to the density of states. In practice, container sizes are usually large enough for the discreteness of the energy levels to be completely ignored, as Question 1.1 demonstrates, and so the classical continuum approximation is almost always valid for molecular gases and Equation 1.9 can be used.

To summarize so far, we have:

1   The translational energy of a particle in a container is quantized with allowed energies $E$ given by Equation 1.6.

2   For large energies $E$, the translational energy levels are packed very close together and may be regarded as a continuum. The criterion for the classical continuum approximation is given by Equation 1.7 or 1.8 ($\lambda_{dB} \ll L$), and is always satisfied for gases in containers of macroscopic size and at normal temperatures.

3   Within the classical continuum approximation, we can define a density of states function $D(E) = B\sqrt{E}$, where the constant $B$ depends only on the size of the container and the mass of the particle (see Equation 1.9). The number of translational energy states with energies between $E$ and $E + \Delta E$ is given by $D(E)\,\Delta E$.

You have seen that the Maxwell–Boltzmann energy distribution, Equation 1.1, is a product of two energy-dependent factors. We can now write it as

$$G(E) = (B\sqrt{E}) \times (ANe^{-E/kT})$$

where $A$ and $B$ are constants. We have seen that the density of states function, $D(E) = B\sqrt{E}$, specifies the energy distribution of the allowed states provided the classical continuum approximation is valid. In the next section you will investigate the quantum-mechanical basis of the Boltzmann factor, $e^{-E/kT}$, which tells us how the molecules are distributed amongst these states.

## 2.3  Origins of the Boltzmann factor

You met Boltzmann's law and the Boltzmann factor in *CPM* in the context of classical statistical mechanics. We now quote Boltzmann's law in the context of quantized energy levels. This essentially involves replacing the term *phase cell* by *quantum state*.

### Boltzmann's law

Consider any system in thermal equilibrium at temperature $T$. Then the probability $p$ of finding a given particle in a particular quantum state of energy $E$ is

$$p = Ae^{-E/kT}. \tag{1.10}$$

$A$ is a normalization constant determined by the requirement that the sum of the probabilities $p$ over all states is equal to unity.

The exponential factor $e^{-E/kT}$ is called the **Boltzmann factor**. At any fixed temperature $T$ the Boltzmann factor is a function of energy $E$ (Figure 1.9).

This quantum form of Boltzmann's law is very general. The *quantum states* referred to may be those of particles in solid, liquid or gaseous systems in thermal equilibrium, although we shall only study the case of *translational quantum states* in a gas.

Note that the probability $p$ in Boltzmann's law depends only on the energy $E$ of the quantum state and the temperature $T$. In particular, it *doesn't* depend on whether or not the quantum state in question is already occupied by another particle. Thus if $N$ is the total number of particles in a system, the average number of particles occupying a particular quantum state of energy $E$ is given by $Np$. In thermal equilibrium, the actual number of particles in a particular quantum state fluctuates about this average value, but these fluctuations are relatively small when $N$ is very large. We shall denote this average occupancy, $Np$, by $F(E)$. Thus, an alternative statement of Boltzmann's law is

$$\left(\begin{array}{c}\text{average number of particles} \\ \text{in a quantum state of energy } E\end{array}\right) = F(E) = NA\mathrm{e}^{-E/kT}. \tag{1.11}$$

We shall refer to the function $F(E)$ as the **Boltzmann occupation factor** since it describes the average number of particles in each state, or the *occupancy* of each state. It is proportional to the Boltzmann factor since $N$ and $A$ are constants for a given gas at a fixed temperature, i.e $F(E) \propto \mathrm{e}^{-E/kT}$.

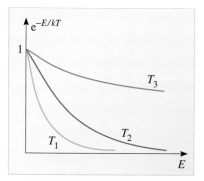

**Figure 1.9** The Boltzmann factor $\mathrm{e}^{-E/kT}$ shown for three different temperatures, $T_3 > T_2 > T_1$.

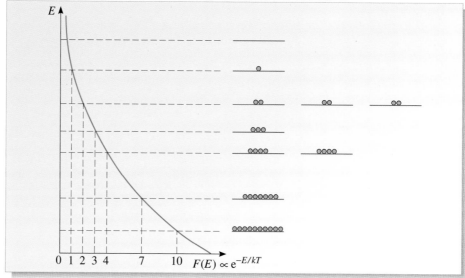

**Figure 1.10** A hypothetical system consisting of 35 particles in thermal equilibrium at some temperature $T$. The energy axis is plotted vertically and the horizontal lines on the right of the diagram show the energies of ten quantum states. Some of the states are degenerate (have the same energy). The small circles on the lines indicate the average occupancy of each quantum state in thermal equilibrium. Thus there is an average of 10 particles in the lowest energy state, etc. These occupancies of each state are described by the Boltzmann occupation factor $F(E)$ for this system which is plotted along the horizontal axis with the energy axis vertical so that the energies $E$ correspond with the vertical plot of the 10 energy levels.

We emphasize that Boltzmann's law refers to individual *quantum states*, not individual energies. Equation 1.11 gives the average number of particles in each quantum state of energy $E$; it does *not* give the average number of particles with energy $E$. This is because there is generally more than one quantum state for each allowed energy $E$, i.e. the energy levels are usually degenerate. This is illustrated in Figure 1.10 for a hypothetical system of 35 particles in thermal equilibrium. Note that there is the same average number of particles in each of the degenerate states of the same energy. (In the continuum approximation, all degenerate states are included in the density of states function $D(E)$.)

In classical statistical mechanics, Boltzmann's law can be derived by counting *configurations* of molecules distributed amongst the possible phase cells, subject to the classical definition of a configuration and the two principles listed below.

### Classical definition of a configuration

A configuration of molecules in a gas is a particular arrangement of all the molecules among the phase cells. It is assumed that the molecules can be labelled so that it matters which molecule is in which phase cell, and it is assumed that there is no restriction on the number of molecules that can occupy a given phase cell.

1  **The conservation of energy**  The total translational energy of the molecules in a gas has a constant fixed value. The only allowed configurations are those with this fixed energy.

2  **Equal probabilities of allowed configurations**  Each allowed configuration is equally likely.

In *CPM* it was shown that when a given molecule occupies a phase cell of low energy there is more energy left to share out amongst the other molecules and therefore there are more configurations. This is the physical basis for the exponential behaviour of Boltzmann's law in classical statistical mechanics. How does this change in quantum mechanics?

You have seen that in quantum mechanics we work with the fundamental concept of quantum states rather than the classical notion of phase cells. Thus the classical definition of a configuration can be replaced by the following quantum-mechanical definition in which the term *phase cell* is replaced by *quantum state* everywhere:

### Quantum-mechanical definition of a configuration

A configuration of molecules in a gas is a particular arrangement of all the molecules among the translational quantum states. It is assumed that the molecules can be labelled so that it matters which molecule is in which quantum state, and it is assumed that there is no restriction on the number of molecules that can occupy a given quantum state.

If you read through the justification of Boltzmann's law given in *CPM*, replacing the term *phase cells* everywhere by the term *translational quantum states*, then you have an argument for Boltzmann's law that is consistent with energy quantization. This argument will not be fully quantum-mechanical because it still assumes that identical gas molecules are distinguishable, but before we come to that you should try Questions 1.3 and 1.4. These questions, and the discussion following them, are designed to remind you of the argument leading to Boltzmann's law. The questions, however, are phrased in the language of quantum mechanics and you should use the above definition of a configuration. In the interests of simplicity the questions refer to rather unrealistic hypothetical gases consisting of just a few molecules and an artificially restricted number of quantum states.

Question 1.3 tests your understanding of what is meant by a configuration and Question 1.4 illustrates the effect of energy conservation.

**Question 1.3**   (a) Consider a hypothetical gas consisting of just three particles, labelled A, B and C. Assume that they are distributed between only two translational quantum states of energies $E = 0\,\mathrm{J}$ and $E = \varepsilon$. Draw eight diagrams showing the eight different possible configurations for this gas. Does the gas always have the same total energy? (b) How many different configurations are there for a gas of 20 particles distributed between the two states? (By specifying an energy of $E = 0\,\mathrm{J}$, we are assuming that the arbitrary zero of energy has been chosen to be at this level.)

**Question 1.4**   A hypothetical gas consists of four particles, labelled A, B, C and D. Assume that each particle has six allowed translational quantum states, of energies $\varepsilon$, $2\varepsilon$, $3\varepsilon$, $4\varepsilon$, $5\varepsilon$, and $6\varepsilon$. Suppose the total energy of the gas is fixed at $E_T = 9\varepsilon$. (a) How many configurations, with the fixed value of $E_T$, correspond to particle A being in the state of energy $5\varepsilon$? (b) How many configurations, with the fixed value of $E_T$, correspond to particle A being in the state of energy $2\varepsilon$? (c) Is particle A more likely to have energy $5\varepsilon$ or $2\varepsilon$?   ■

The conclusions of Question 1.4 bear repetition. When the total energy of the gas is fixed at $9\varepsilon$, there are fifteen different configurations in which particle A is in a state of energy $2\varepsilon$ and only three configurations in which it is in a state of energy $5\varepsilon$. All these configurations are supposed to be equally likely. It follows from simple probability arguments alone that particle A is five times more likely to have energy $2\varepsilon$ than it is to have energy $5\varepsilon$.

Extending these calculations, we can work out the number of configurations that correspond to particle A being in each of the states in turn. We can also determine the corresponding probability $p$ of particle A being in each of these states. The complete set of results is shown in Table 1.1.

Notice that the number of configurations, and hence the probabilities, steadily decreases as the energy of particle A rises. This is not surprising. The total energy of the gas is constant, so if particle A moves to a state of higher energy, less energy is left to share amongst particles B, C and D. With less energy available, fewer configurations can be produced. This, in turn, makes the higher-energy states of A more unlikely.

**Table 1.1**   The results of Question 1.4, extended to cover all the allowed energies of particle A, under the restriction $E_T = 9\varepsilon$. Note that the total number of configurations is 56 and so the probabilities are $21/56 = 0.375$, etc.

| Energy of particle A | Number of configurations | Probability |
| --- | --- | --- |
| $\varepsilon$ | 21 | 0.375 |
| $2\varepsilon$ | 15 | 0.268 |
| $3\varepsilon$ | 10 | 0.179 |
| $4\varepsilon$ | 6 | 0.107 |
| $5\varepsilon$ | 3 | 0.054 |
| $6\varepsilon$ | 1 | 0.018 |

It is worth remembering the analogy developed in *CPM* between counting configurations and counting scores in a special game of dice. The situation described in Question 1.4 is reminiscent of a game in which four dice labelled A, B, C and D are rolled. The dice and their possible scores correspond to the molecules and their allowed energies. When the rule is adopted that only throws with a total score of 9 are accepted, the results of Table 1.1 are echoed precisely. Under these circumstances, die A is five times more likely to score 2 spots than 5 spots.

The fall-off of the probabilities $p$ (in the third column of Table 1.1) with increasing energy of particle A is nearly exponential. When similar calculations are done with much larger numbers of particles and energy levels, the fall-off of $p$ is almost exactly exponential. This, in essence, is Boltzmann's law.

Let's now summarize what we have done in Section 2.

- We have reviewed the Maxwell–Boltzmann energy distribution for a molecular gas, pointing out that the distribution consists essentially of two energy-dependent factors, the density of states function and the Boltzmann factor.

- By considering the quantized translational energy states of a particle in a container, we have found the density of states function to be $D(E) = B\sqrt{E}$. This is valid for an ideal gas in a rigid container, provided $kT$ is very large compared with the spacing between energy levels. This continuum condition (Equation 1.7 or 1.8) is normally well satisfied for molecular gases.

- We have expressed Boltzmann's law in terms of translational quantum states and shown that it is based on the law of energy conservation and the idea of equal probabilities of allowed configurations.

Having spent some considerable time on the justification of Boltzmann's law, we now point out an important restriction on its validity. Embedded in our discussion of Boltzmann's law is an assumption concerning the fundamental nature of particles. We have assumed that identical particles in a gas can, in principle at least, be distinguished from one another; we have labelled them A, B, C, etc. This assumption appeared explicitly in the above definition of a configuration where we stated: 'It is assumed that the molecules can be labelled so that it matters which molecule is in which quantum state, and there is no restriction on the number of molecules that can occupy a given quantum state.'

This assumption will be examined critically in Section 3 and shown to be false in quantum physics. This will have profound implications for molecular gases at very low temperatures and for the photon gases and free-electron gases to be studied in Sections 4 and 5.

# 3    Identical particles in quantum mechanics

Boltzmann himself pointed out that his work on the statistical mechanics of gases in thermal equilibrium assumes that identical gas particles can, in principle at least, be distinguished from one another. This was a prescient observation, since the development of quantum mechanics in the 1920s, after Boltzmann's death, showed that identical particles are *fundamentally indistinguishable* from one another, and cannot, even in principle, be labelled. You will see that this development placed limitations on the validity of Boltzmann's law.

In this section, we explore the effect of adopting rules for counting configurations that take account of the indistinguishability of identical particles. We shall find it necessary to replace the Boltzmann factor with the quantum-mechanical factors that describe how indistinguishable particles in thermal equilibrium are distributed amongst the available quantum states.

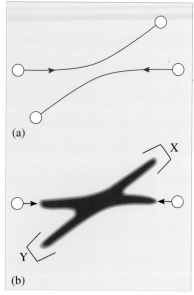

**Figure 1.11** (a) A collision between two identical particles in classical physics. (b) A collision between two identical particles in quantum physics. The shading indicates that the trajectories are not well defined because of the uncertainty principle. It is impossible to say which of the two particles enters the detector at X and which enters the detector at Y.

## 3.1 Indistinguishability of identical particles

In the industrial world of mass manufacturing, objects are produced that appear to be identical. Supermarkets stock apparently identical cans of food, cars produced by robots seem to be replicas of one another. Yet, on closer inspection minor distinctions can be seen. A can of beans may have a scratch on it, each car has its own number plate, and so on. In the everyday world, objects that seem to be identical turn out to be different after all.

Another sense in which objects can be distinguished, according to classical physics, is illustrated in Figure 1.11a. It shows a collision between two particles. According to Newtonian mechanics, the particles trace out continuous paths that can in principle be accurately predicted. It follows that there is never any doubt about which particle is which — each is characterized by its own trajectory.

A very different situation prevails in quantum physics. Consider two hydrogen atoms in their ground states in empty space. These atoms are not just similar. They are *identical*. They have the same mass, size, ionization energy and so on. This is not an approximation. In quantum physics, identical particles are **indistinguishable**, they are *exactly* alike. There is no such thing as a hydrogen atom with a scratch on it, or one that is slightly oversize. All hydrogen atoms are the same. Moreover, the idea of a particle tracing out a definite path is undermined by the uncertainty principle. Figure 1.11b depicts a collision between two identical particles in quantum mechanics. After colliding, the particles enter detectors at X and Y, but we cannot say *which* particle has entered *which* detector. The particles might have been deflected through a small angle, or through nearly 180 degrees. In quantum mechanics, there is no way of distinguishing between these two possibilities. Thus, when particles collide it is impossible to characterize each one by its trajectory.

We conclude that the concept of identical particles goes much further in quantum physics than in classical physics. In quantum physics, identical particles are completely indistinguishable from one another. If two identical particles are free to move around within the same region of space, as in a gas, it is impossible to distinguish between them by means of their physical properties or their trajectories, and so it is not possible to label them.

## 3.2 Counting configurations of indistinguishable particles

You may remember from the discussion of Boltzmann's law in the previous section that it does matter *which* particle is in *which* state. For example, the thre configurations X, Y, and Z in Figure 1.12a would be counted as being different However, you can see that if you remove the labelling of the particles and accept tha they are indistinguishable from one another, there is nothing to distinguish thos three configurations. They are in fact just a single configuration of indistinguishabl particles. We can specify this single configuration by saying: 'there are two particle in the lower state and one particle in the upper state', as illustrated in Figure 1.12b.

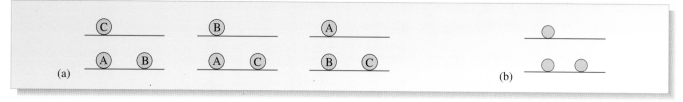

**Figure 1.12** (a) According to classical statistical theory these three configurations are all different. (b) A single configuration of indistinguishable particles: there are two particles in the lower state and one in the upper state.

Thus, the indistinguishability of identical particles requires us to revise our definition of a configuration in quantum mechanics:

> The **configuration of a system of identical particles** in quantum mechanics is defined by giving the numbers of particles in each quantum state.

As before, we assume that the law of conservation of energy restricts the number of possible configurations, and that all configurations are equally likely. However, before we can use our new definition of a configuration, we must take account of a surprising consequence of indistinguishability.

You know from *Quantum physics: an introduction* (*QPI*) that in quantum mechanics an individual particle is described by a wavefunction that represents the quantum state of the particle. In fact, it is possible to describe an entire system of identical particles, a gas of hydrogen atoms for example, by a *single* many-particle wavefunction. This many-particle wavefunction is constructed from the individual particle wavefunctions by methods that are outside the scope of this book. However, the important point is that when we do this, and put in the condition that the particles are indistinguishable from one another, we find that *all* many-particle wavefunctions can be classified into *two distinct types*.

One type of many-particle wavefunction always vanishes when two or more of the constituent particles have the same single-particle wavefunction, i.e. when two or more particles occupy the same state. It is found that systems of electrons, for example, are always described by many-particle wavefunctions of this type. What this means is that it is not possible for two or more electrons to occupy the same quantum state. You will of course recognize this as a statement of the Pauli exclusion principle. Identical particles, such as electrons, that obey the Pauli exclusion principle are collectively called **fermions**. Examples of fermions are given in the first column of Table 1.2.

The other type of many-particle wavefunction is not subject to this restriction. Systems of particles described by this second type of wavefunction can have any number of constituent particles sharing the same quantum state — the exclusion principle does not apply. It is found that systems of photons, for example, are always described by many-particle wavefunctions of this type. Particles that do not obey the exclusion principle, such as photons, are called **bosons**. Examples of bosons are shown in the second column of Table 1.2.

*Every* particle in physics is *either* a boson *or* a fermion. Thus the quantum-mechanical definition of a configuration needs to be used with the following rule:

> In any system of identical fermions one particle only can occupy any particular quantum state — this is the Pauli exclusion principle.
>
> In any system of identical bosons any number of particles can occupy any quantum state.

● Consider the configuration of identical particles in Figure 1.12b. Is this a possible configuration of (a) identical fermions, (b) identical bosons?

○ The configuration shown in Figure 1.12b cannot be a configuration of identical fermions because it shows two particles in the lower state, but it is a possible configuration of identical bosons. ■

**Table 1.2** Examples of fermions and bosons.

| Fermion | Boson |
|---|---|
| electron | photon |
| proton | $^{23}_{11}$Na atom |
| neutron | $^{85}_{37}$Rb atom |
| $^{3}_{2}$He atom | $^{4}_{2}$He atom |
| $^{13}_{8}$O atom | $^{14}_{8}$O atom |

The superscripts on the chemical symbols specify the total number of protons and neutrons in the nucleus, thereby distinguishing between different isotopes. The subscripts specify the atomic number.

Let's now look at an example illustrating how we count configurations of identical particles.

### Example 1.1

Consider a hypothetical gas consisting of just two identical particles. Assume that each particle has only three quantum states, of energies $0\,\text{J}$, $\varepsilon$ and $2\varepsilon$ (see Figure 1.13) and suppose the total energy of the gas is fixed at $E_T = 2\varepsilon$. List the configurations for each of the following cases: (a) The particles are distinguishable, i.e. they can be treated classically and labelled as in Section 2. (b) The particles are identical bosons. (c) The particles are identical fermions.

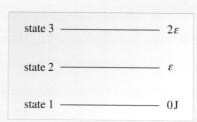

**Figure 1.13** Three quantum states of energies $0\,\text{J}$, $\varepsilon$ and $2\varepsilon$.

### Solution

(a) If the particles are distinguishable, they can be labelled A and B. Three configurations then have total energy $E_T = 2\varepsilon$, as shown in Figure 1.14a.

(b) If the particles are identical bosons, they cannot be labelled and there are only two distinct configurations with $E_T = 2\varepsilon$. One configuration has a particle in each of the lowest and highest energy states, and the other configuration has both particles in the middle state. This is shown in Figure 1.14b.

(c) If the particles are identical fermions, they cannot be labelled and the exclusion principle applies to them. Only one configuration has total energy $E_T = 2\varepsilon$, as shown in Figure 1.14c.

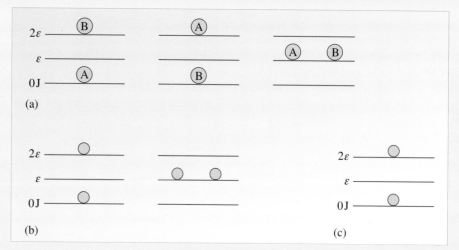

**Figure 1.14** Configurations for (a) two distinguishable particles, (b) two identical bosons, (c) two identical fermions.

It should not surprise you that Example 1.1 has different answers for the three different cases. The rules specifying configurations depend on whether the particles are distinguishable or not and the exclusion principle is an extra constraint for fermions. The main point to stress is that the different ways of specifying configurations produce different distributions of particles among quantum states.

It is instructive to use the answers of Example 1.1 to work out the *average number of particles in the middle state* (state 2 of energy $\varepsilon$) for each of the three cases: distinguishable particles, identical bosons and identical fermions.

(a) For *distinguishable particles* (Figure 1.14a) there are three configurations with the required value of the total energy. Two of these configurations have no particles in the state 2 and one has two particles in this state. All these configurations are equally likely, and so each has a probability of 1/3. It follows that the average number of particles in the middle state is

$$\langle N_2 \rangle = 0 \times \frac{1}{3} + 0 \times \frac{1}{3} + 2 \times \frac{1}{3} = \frac{2}{3}.$$

(b) For *identical bosons* (Figure 1.14b) there are only two distinct configurations with total energy $E_T = 2\varepsilon$. One has no particles in state 2 while the other has two particles in this state. The two configurations are equally likely, so each has a probability of 1/2. The average number of bosons in the middle state is therefore

$$\langle N_2 \rangle = 0 \times \frac{1}{2} + 2 \times \frac{1}{2} = 1.$$

(c) For *identical fermions* (Figure 1.14c) there is only one configuration with the required energy. This configuration has no particles in the middle state, so in this case

$$\langle N_2 \rangle = 0.$$

An important lesson can be learnt from the above example. The probability of finding two particles in *different* states is greater for distinguishable, i.e. labelled, particles than for identical bosons. This is a general rule. It comes about because interchanging two labelled particles produces a new configuration if the particles are in different states, but not if they are in the same state. Interchanging two identical bosons, on the other hand, produces no new configurations whatever. Thus, labelled particles have a higher proportion of their configurations with particles in *different* states. This rule can be interpreted by turning it around: the probability of finding two identical bosons in the same state is greater than the corresponding probability for labelled particles. In other words, bosons have a tendency to congregate together in the same state. This important point can be expressed as a slogan: *bosons are sociable*. Fermions, of course, display exactly the opposite behaviour. The exclusion principle prevents two identical fermions from sharing the same quantum state. In this respect, *fermions are unsociable*.

**Question 1.5** A hypothetical gas contains three particles. Assume that each particle has three quantum states of energies $0\,\text{J}$, $\varepsilon$ and $2\varepsilon$. Suppose the total energy of the gas is fixed at $E_T = 3\varepsilon$. What is the probability of finding all three particles in the same quantum state if the particles are (a) distinguishable, (b) identical bosons, (c) identical fermions? ∎

You have seen that *every* particle in physics is either a boson or a fermion. In the following sections of this chapter we concentrate on photons and electrons, but Table 1.2 gives several other examples and it is easy to extend this list to composite particles by using the following rule:

**Composite particles**

A composite particle, such as an atom or a molecule, is a fermion if it contains an *odd* number of fermions, and it is a boson if it contains an *even* number of fermions.

For example, a $^3_2$He atom consists of a nucleus containing two protons and one neutron, with two orbiting electrons. According to Table 1.2, each of these five constituent particles is a fermion, so a $^3_2$He atom is itself a fermion. A $^4_2$He atom contains an extra neutron in the nucleus. It therefore contains six fermions and is a boson.

**Question 1.6** Decide whether each of the following particles is a fermion or a boson: (a) a hydrogen atom $^1_1$H, (b) a deuterium atom $^2_1$H (the nucleus contains one proton and one neutron), (c) a nitrogen atom $^{14}_7$N (the nucleus contains 7 protons and 7 neutrons). (d) a diatomic nitrogen ($^{14}_7$N) molecule, (e) a rubidium atom $^{87}_{37}$Rb (the nucleus contains 37 protons and 50 neutrons). ■

## 3.3  Replacing the Boltzmann factor

The Boltzmann occupation factor, $F(E) = NAe^{-E/kT}$, was obtained in Section 2 by counting configurations of distinguishable particles. Its exponential form is illustrated in Figure 1.15a. (It was also shown in Figure 1.10 but with the energy axis plotted vertically.) You have seen in this section that the definition of a configuration has been redefined for indistinguishable particles. Example 1.1 and Question 1.5 show that this new definition has a profound effect on the way the particles are distributed amongst the available quantum states, as summed up by the slogans *bosons are sociable* and *fermions are unsociable*. This means that the probability of a given atom occupying a particular quantum state depends on whether or not there is another particle in that state, which clearly contradicts Boltzmann's law. Clearly then, before we can study the energy distributions in more general types of gases, such as photon gases and electron gases, we shall need to know the factors that replace the Boltzmann occupation factor for identical bosons and for identical fermions.

The factor that replaces the Boltzmann occupation factor and gives a precise expression of the sociability of bosons is called the *Bose occupation factor*, while the corresponding factor for fermions is the *Fermi occupation factor*. These factors can be derived by counting configurations of indistinguishable particles in much the same way as the Boltzmann factor was obtained in Section 2 by counting configurations of distinguishable particles. We shall not attempt to give the details of these derivations. Instead we simply quote the Bose and Fermi occupation factors and discuss their properties.

### The Bose occupation factor

For a system of identical bosons in thermal equilibrium, the average number of particles in a quantum state of energy $E$ is given by:

$$F_B(E) = \frac{1}{e^{(E - \mu_B)/kT} - 1}. \tag{1.12}$$

The energy $\mu_B$ is a characteristic energy for the system and is in general a function of $N$, the total number of bosons in the system, and the temperature $T$. For a photon gas, the only boson system we shall consider in any detail, $\mu_B$ is equal to zero. We then have the Bose occupation factor for photons:

$$F_B(E) = \frac{1}{e^{E/kT} - 1}. \tag{1.13}$$

Figure 1.15b shows a graph of the Bose occupation factor for photons as a function of energy $E$. It shows that the number of photons in a quantum state of energy $E$ falls off with increasing $E$. Qualitatively this behaviour is similar to that of the Boltzmann occupation factor (compare Figures 1.15a and b). However, there are fundamental differences. The Bose occupation factor is very much steeper than the Boltzmann occupation factor and is larger for small energies. This expresses the fact that bosons have a stronger tendency than labelled particles to congregate together in the lowest energy states. This is another indication of the sociability of bosons.

### The Fermi occupation factor

For a system of identical fermions in thermal equilibrium, the average number of particles in a single quantum state of energy $E$ is given by the Fermi occupation factor

$$F_F(E) = \frac{1}{e^{(E-\mu_F)/kT} + 1}.$$

where $\mu_F$ is a characteristic energy for the system. In general, the constant $\mu_F$ depends on the number $N$ of fermions in the system and the temperature $T$. In this book we shall consider the details of only one kind of system of identical fermions: the case of the free electrons in a metal. In this case $\mu_F$ increases as $N$ increases but is almost independent of temperature. The value of $\mu_F$ at $T = 0\,\text{K}$ is a constant called the *Fermi energy* $E_F$. Thus in the case of the free electrons in a metal, the Fermi occupation factor is given, to a very good approximation, by

$$F_F(E) = \frac{1}{e^{(E-E_F)/kT} + 1}. \tag{1.14}$$

A graph of the Fermi occupation factor, Equation 1.14, is shown in Figure 1.15c.

The characteristic energies $\mu_B$ and $\mu_F$ are generally called *chemical potentials*.

You will *not* be required to do any manipulations with Equations 1.12, 1.13 and 1.14. Our discussions will be based on the graphs of those functions.

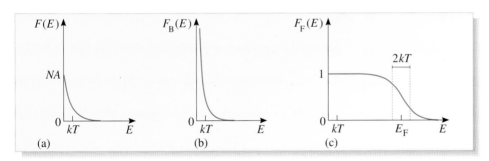

**Figure 1.15** The average number of particles in a quantum state of energy $E$ at temperature $T$: (a) the Boltzmann occupation factor $F(E)$, (b) the Bose occupation factor $F_B(E)$ and (c) the Fermi occupation factor $F_F(E)$.

The most important feature of the Fermi occupation factor is most easily revealed at $T = 0\,\text{K}$, as shown in Figure 1.16. You can see that $F_F(E) = 1$ for all energies $E$ less than the Fermi energy $E_F$. This means that each of these states is occupied by just *one* particle. Looking at higher energy values, you can see that $F_F(E) = 0$ for all energies $E$ greater than the Fermi energy $E_F$, showing that all of these states are unoccupied. You can now see the significance of the **Fermi energy** $E_F$. At $T = 0\,\text{K}$, the Fermi energy marks the boundary between the filled states and the empty states. Thus the Fermi occupation factor is the embodiment of the exclusion principle. In any gas of fermions at $T = 0\,\text{K}$, the states fill up, one particle to each state, until at $E = E_F$ all the particles are accommodated, and the higher energy states remain empty. You are of course familiar with this kind of behaviour from the way in which electrons fill shells and subshells in atoms.

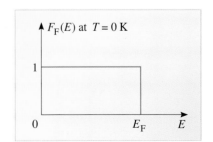

**Figure 1.16** The Fermi occupation factor at $T = 0\,\text{K}$.

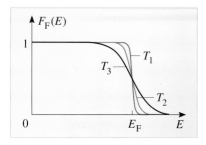

**Figure 1.17** The Fermi factor shown at three temperatures: $T_3 > T_2 > T_1$. The graphs fall from $F_F(E) = 1$ to $F_F(E) = 0$ over a region of order a few $kT$, where $T$ is the temperature.

In fact, the absolute zero of temperature, $T = 0\,\text{K}$, cannot be reached in practice. When the Fermi occupation factor is plotted at non-zero temperature, as in Figure 1.15c and Figure 1.17, there is a blurring of the boundary at $E_F$ as particles just below the Fermi energy are thermally excited into empty states just above. This blurring of the boundary occurs over a region of order a few $kT$, and so increases with temperature as illustrated in Figure 1.17. The value of $kT$ is usually very much smaller than the Fermi energy itself, as indicated in all the cases shown Figure 1.17. We shall have much more to say about the effect of temperature and the significance of the characteristic Fermi energy $E_F$ in Section 5 where we study electron gases in metals.

We shall refer to the three factors shown in Figure 1.15 as **occupation factors** since they give the *average number of particles occupying a quantum state of energy E* in each of the three types of gas in thermal equilibrium. Thus, the Boltzmann occupation factor $F(E)$ is the classical occupation factor for gases of distinguishable particles, while the Bose occupation factor and the Fermi occupation factor are the occupation factors for quantum gases of bosons and fermions respectively.

Of course, the occupation factors alone do not give the energy distributions of gas particles. For this we also need the density of states. The Maxwell–Boltzmann energy distribution, you'll remember, is proportional to the product of the occupation factor (i.e. the Boltzmann factor) and the density of states function. You will see in Sections 4 and 5 that the energy distributions that apply to gases of bosons and fermions have the same generic structure:

energy distribution = (density of states) × (occupation factor).

Before going further, however, we should first clear up a mystery.

## 3.4 When does indistinguishability matter?

If every atom and molecule is either a boson or a fermion, you might expect that this would influence the properties of ordinary gases. In fact it doesn't, except at temperatures very near absolute zero — we shall have more to say about that in Sections 4 and 5. At normal temperatures, no differences can be detected and atomic and molecular gases can be treated as if they consisted of distinguishable (i.e. labelled) particles, as in Section 2, and it is irrelevant whether they consist of bosons or fermions. Why is this? When are the effects of indistinguishability small enough to be ignored? We need a criterion.

Consider a situation in which the quantum states of a gas are *sparsely occupied* (Figure 1.18). Under these circumstances, fermions are scarcely affected by the exclusion principle and bosons have little chance to display their social behaviour. Thus, the effects discussed in this section are expected to be very small when the states are sparsely occupied; that is, when the total number of particles $N$ in a system is very much less than the total number of states of energies up to a few $kT$.

We know that for a gas of molecules, each of mass $m$ in a rigid container of volume $V$, the density of states $D(E)$ is given by Equation 1.9. Knowing $D(E)$, the number of states with energy between $0\,\text{J}$ and $kT$ can be evaluated by means of a definite integral. (We do not include states above $kT$ because these are much less likely to be occupied than those below.) When this is done, we obtain the following criterion for neglecting the effects of indistinguishability,

$$N << \frac{V}{3\pi^2}\left(\frac{8m\pi^2 kT}{h^2}\right)^{3/2}. \tag{1.15}$$

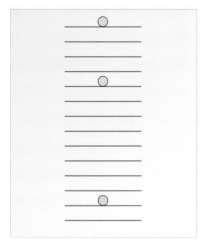

**Figure 1.18** Sparsely occupied quantum states. Most states are unoccupied and the probability of any one quantum state being occupied at any instant is very much less than 1.

Reading the right-hand side, you can see that factors $V$, $m^{3/2}$ and $T^{3/2}$ are all in the numerator, showing that large containers, heavy particles and high temperatures all make the right-hand side large and so help to reduce the importance of indistinguishability.

Equation 1.15, however, is not very memorable. Fortunately we can recast it into a simpler form by introducing the typical de Broglie wavelength $\lambda_{dB} = h/(mv)$, and taking $v$ to be the rms speed $v = v_{rms} = \sqrt{3kT/m}$ of Equation 1.4. We can use these results to express $T$ in terms of $\lambda_{dB}$ on the right-hand side of Equation 1.15. We can also express $N$ on the left-hand side in terms of the typical distance $d$ between molecules. The molecules in a gas are of course at random positions at any instant, but if you imagine the molecules fixed in a cubic lattice with distance $d$ between neighbours, then each molecule would occupy a volume $d^3$, so we can say $N = V/d^3$. Using this relationship to eliminate $N$ from Equation 1.15, we find, after a little algebra, that the criterion for neglecting indistinguishability is equivalent to

$$\lambda_{dB} << d. \tag{1.16}$$

This form of the criterion is illuminating. It shows that the effects of indistinguishability become irrelevant when the de Broglie wavelength of a typical gas particle is very much less than the typical spacing between particles.

Equation 1.16 refines a criterion obtained in the previous section where it was found that the classical assumption of a continuum of translational energy levels is valid when $\lambda_{dB} << L$, where $L$ is the size of the container confining the gas. We have now found that for a molecular gas, indistinguishability is a more persistent quantum-mechanical feature than energy quantization — it becomes negligible only when $\lambda_{dB} << d$.

We can easily show directly that the criterion of Equation 1.16 (or equivalently Equation 1.15) is satisfied for molecular gases under normal conditions. For example, taking the mass of a nitrogen molecule to be $4.7 \times 10^{-26}$ kg, the typical speed of a nitrogen molecule in air at 300 K is found from $v = v_{rms} = \sqrt{3kT/m}$ to be about 500 m s$^{-1}$. Hence the typical de Broglie wavelength is

$$\lambda_{dB} = h/(mv) \approx 3 \times 10^{-11} \text{ m}. \qquad \text{(nitrogen gas)}$$

The number density $n = N/V = 1/d^3$ of nitrogen molecules in air is about $2 \times 10^{25}$ m$^{-3}$, and so

$$d \approx (2 \times 10^{25} \text{ m}^{-3})^{-1/3} = 4 \times 10^{-9} \text{ m}. \qquad \text{(nitrogen gas)}$$

Hence the criterion of Equation 1.16 is easily satisfied. This confirms that nitrogen gas in air can be treated as if it consisted of labelled particles obeying the Maxwell–Boltzmann distribution.

**Question 1.7**   Consider a sample of 20 000 rubidium atoms ($^{87}_{37}$Rb) confined to a volume of $1.0 \times 10^{-14}$ m$^3$ at a temperature of $1.0 \times 10^{-7}$ K. Estimate the typical distance $d$ between atoms and the typical de Broglie wavelength $\lambda_{dB}$. Take the mass of a rubidium atom to be $1.45 \times 10^{-25}$ kg. Show that the effects of indistinguishability may *not* be ignored in this case by confirming that Equation 1.16 is *not* satisfied for this sample.  ∎

The answer to Question 1.7 shows that the effects of indistinguishability cannot be neglected for the very cold rubidium sample considered, since $\lambda_{dB}$ and $d$ are of comparable size (of order 1 µm). You have also seen, from Question 1.6, that $^{87}_{37}$Rb atoms are bosons. The sample conditions specified in Question 1.7 are in fact those at which a Bose–Einstein condensate of atoms was first produced in 1995. We shall have more to say about this in Section 4.4.

In the next section we study gases of photons. You have seen in Table 1.2 that photons are bosons. Photons are of course very different from atoms or molecules. One crucial difference is that photon gases in thermal equilibrium *never* behave classically. The indistinguishability of photons, as expressed by the Bose occupation factor, always plays a determining role.

# 4 The photon gas — blackbody radiation explained

Hot matter emits light. For example, a burning candle illuminates a room and the bars of an electric fire glow red. People also glow in the dark. The radiation we emit lies in the infrared region of the electromagnetic spectrum, but it can be detected by infrared surveillance equipment, as the Army well knows. Even an iceberg emits radiation, also in the infrared. In fact, these examples illustrate a universal phenomenon:

Blackbody radiation was discussed in Chapter 1 of *QPI*.

> All matter emits some electromagnetic radiation and the amount and nature of the radiation depend on the temperature and composition of the matter.

Not surprisingly, a cold object emits a small number of low-energy photons, while a hot object emits a large number of high-energy photons. The number of photons is related to the intensity of the radiation, and the energy $E$ of a photon is related to the frequency $f$ of the radiation by $E = hf$. This means that the intensity and frequency of the radiation tend to increase with the temperature of the emitting object. People (at 310 K) are not hot enough to emit much visible light, but we do emit infrared radiation. A poker taken from a furnace (at say 1000 K) emits enough visible light to be called 'red hot'.

A typical light source, such as an electric light filament, is at a much higher temperature than its surroundings. It emits a copious supply of photons which are absorbed by surrounding matter. The net effect, then, is to transfer energy away from the hot source and deposit it in the cooler surroundings. Such a situation may be described as a steady-state flow of photons from the hot source to the surroundings; the photons are not in thermal equilibrium. From here on, we shall restrict ourselves to a simpler case in which the *electromagnetic radiation is in thermal equilibrium with the matter around it*. Let's see what this means.

Figure 1.19a shows a large hollow block of matter with a cavity at its centre. The block of matter is assumed to be kept at a constant uniform temperature. The matter walls surrounding the cavity emit photons, so the cavity is bound to contain some radiation. Does this radiation build up indefinitely? Of course not.

Matter not only emits electromagnetic radiation; it absorbs electromagnetic radiation too.

Some of the photons are absorbed by the walls of the cavity and a competition develops between the opposing processes of emission and absorption. At some point a balance must be struck. This happens when, at each wavelength, the average number of photons being emitted per second is equal to the average number being absorbed per second. When no net change takes place, the cavity contains radiation that is in thermal equilibrium with the matter that surrounds it. Such radiation is called *blackbody radiation* or *thermal radiation*. It is also called *cavity radiation*.

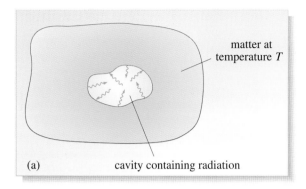

**Figure 1.19** (a) The radiation inside a cavity, surrounded by matter at a constant uniform temperature. (b) The space inside a furnace is a good approximation to such a cavity.

Thus

> **Blackbody radiation** (also called **thermal radiation** or **cavity radiation**) is radiation that is in thermal equilibrium with matter at a fixed temperature. By convention, thermal radiation is said to be at temperature $T$ if it is in thermal equilibrium with matter at temperature $T$.

In the above discussion, the cavity is only an artificial device which prevents photons from leaving the system and allows thermal equilibrium to be reached. In some cases, this can be achieved without a cavity. For example, in the core of a star, photons are continually being emitted and absorbed by the stellar matter and so they are in thermal equilibrium with it. In fact the radiation emitted by the Sun has the same wavelength distribution, i.e. the same spectrum, as thermal radiation at 5800 K, the temperature of the Sun's surface. Nevertheless, we shall continue to refer to the radiation inside a cavity because this is such a clear-cut way of achieving thermal equilibrium.

In this section we shall investigate the properties of thermal radiation by treating it as a *gas of photons*, in a state of *thermal equilibrium*. The gas model is a good one because photons do not interact with one another; they interact only with the walls of the cavity where they are continually emitted and absorbed.

**Question 1.8** Which of the following situations correspond to thermal radiation? (a) The radiation inside an uninhabited cave. (b) The radiation emitted by a glow-worm. (c) The radiation emitted by an electric fire. (d) The radiation inside a hot oven. ■

## 4.1 Energy distribution of a photon gas

Just as white light is a mixture of different colours, so the thermal radiation inside a cavity is a mixture of different wavelengths or frequencies. Since the frequency $f$ of radiation is related to the photon energy by $E = hf$, it follows that the photons have a spread of energies. This energy spectrum depends on temperature. A furnace glows dull red at 1000 K, but shines with bright orange or yellow light at higher temperatures. In general, high energy photons become more abundant as the temperature rises, but what are the details?

The *quantum states* of a photon in a cavity are sometimes called the *modes* of the electromagnetic radiation in the cavity.

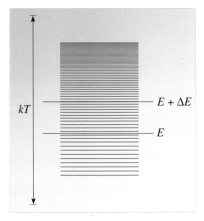

**Figure 1.20**   The energy range $\Delta E$ is narrow compared to $kT$ but wide compared to the spacing between photon energy levels.

According to quantum mechanics, the energy of *any* confined particle is quantized. This applies to a photon in a cavity just as it does to an atom in a container. For cavities of macroscopic size, the energy levels are very closely packed and a density of states function can be defined in the continuum approximation. However, photons are relativistic particles and the density of states function for photons in a cavity is different from the density of states for gas molecules. Also, photons are bosons, and so we need to consider the influence of the Bose occupation factor. Clearly then, the energy distribution law for photons in thermal equilibrium is expected to be different from the Maxwell–Boltzmann energy distribution that applies to the molecules in a gas.

As usual, we shall consider thermal radiation inside a cavity, surrounded by matter at temperature $T$. We shall focus on a range of energies from $E$ to $E + \Delta E$ (Figure 1.20). This range is assumed to be narrow compared to $kT$, but is still wide enough to contain many quantum states (thus satisfying the classical continuum approximation). The average number of photons in the given range will be denoted by $G_p(E)\,\Delta E$. This defines what we mean by the energy distribution $G_p(E)$ for the photon gas. Like the Maxwell–Boltzmann energy distribution of a gas of molecules $G(E)$, the distribution $G_p(E)$ has units J$^{-1}$, and represents the number of photons per unit energy range at energy $E$.

The form of the function $G_p(E)$ can be found by direct experiment. We shall not say much about the experimental details. A light detector is needed and also some filters to restrict attention to narrow energy ranges. By counting photons with energies between $E$ and $E + \Delta E$ it is possible to measure $G_p(E)\,\Delta E$ and hence find $G_p(E)$. Figure 1.21 shows some typical results for a cavity of volume 1 m$^3$ at 200 K, 300 K and 400 K.

**Question 1.9**   A one litre cavity contains thermal radiation at 300 K. Use Figure 1.21 to estimate the number of photons with energies between $1.00 \times 10^{-20}$ J and $1.01 \times 10^{-20}$ J. Repeat your calculation for radiation at 400 K. (1 litre $= 10^{-3}$ m$^3$) ■

An obvious feature of Figure 1.21 is the way in which the scale of the curves grows with temperature. Thermal radiation at 400 K contains more photons, at each energy, than thermal radiation at 200 K. This is because the atoms in the walls surrounding the cavity emit photons more profusely at higher temperatures. Note also how the peak moves towards higher energies as the temperature increases, showing that the typical photon energies are larger at higher temperature.

The function $G_p(E)$ is sometimes called the *spectrum of blackbody radiation*. Blackbody spectra are often plotted as a function of wavelength rather than energy, as in Figure 1.7 of *QPI*.

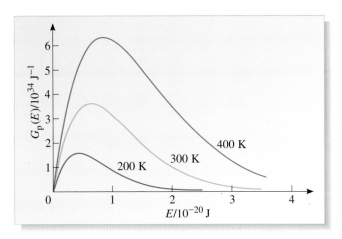

**Figure 1.21**   The photon energy distribution for photons in thermal radiation at three different temperatures in a cavity of volume 1 m$^3$.

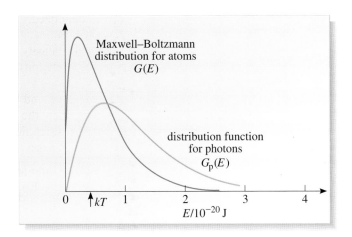

**Figure 1.22**
A comparison between the energy distribution functions for a gas of photons (green curve) and that for a gas of molecules (blue curve). Both gases are at 300 K and have the same average number of particles.

Figure 1.22 compares the energy distribution function for photons with the Maxwell–Boltzmann energy distribution for molecules. The green curve is the energy distribution function for photons at 300 K. The blue curve is the Maxwell–Boltzmann energy distribution function for an equivalent number of molecules, also at 300 K. The two curves have entirely different shapes. Clearly, the Maxwell–Boltzmann energy distribution law does *not* apply to photons! This is no surprise.

## 4.2 Planck's radiation law

The energy distribution law that applies to photons in thermal radiation was first formulated by Max Planck at the turn of the century. Known as *Planck's radiation law*, it has the same generic form as the Maxwell–Boltzmann energy distribution, i.e. it is a product of a *density of states function* and an *occupation factor*. However, the occupation factor in Planck's radiation law is the Bose occupation factor $F_B(E)$, Equation 1.13, and in place of the density of states function $B\sqrt{E}$ that applies to gas molecules, we have the **density of states function for photons**, which we shall denote by $D_p(E)$. Hence, the energy distribution law for photons has the form

$$G_p(E) = D_p(E) \times F_B(E). \tag{1.17}$$

The density of states for a photon gas is in fact given by

$$D_p(E) = CE^2 \tag{1.18}$$

where $C$ is a constant. For thermal radiation in a cavity of volume $V$, the constant $C$ has the value

$$C = 8\pi V/h^3 c^3 = (3.206 \times 10^{75}\,\text{J}^{-3}\,\text{m}^{-3})V \tag{1.19}$$

where the numerical value is given to four significant figures.

We shall not attempt to justify this result beyond saying that it can be found by a method similar to that in the Appendix to this chapter. The result $D_p(E) \propto E^2$ is obtained, rather than $D(E) \propto \sqrt{E}$ for molecules, because of the different way in which energy $E$ is related to the magnitude $p$ of the momentum, i.e. $E = cp$ for photons and $E = p^2/2m$ for ordinary non-relativistic particles.

Thus, using Equations 1.13, 1.17 and 1.18, we have the **energy distribution law for photons**, known as **Planck's radiation law**,

$$G_p(E) = CE^2 \times \frac{1}{e^{E/kT} - 1}. \tag{1.20}$$

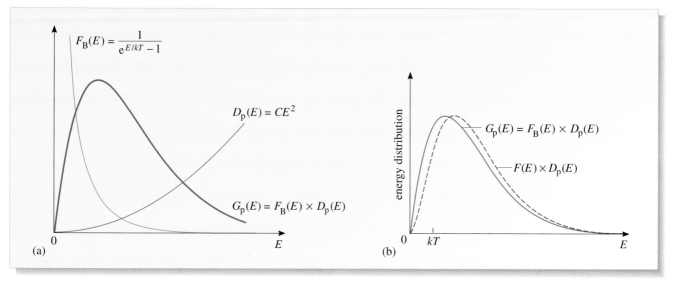

Figure 1.23 plots the functions. For the curve (a):

$$F_B(E) = \frac{1}{e^{E/kT} - 1}$$

$$D_p(E) = CE^2$$

$$G_p(E) = F_B(E) \times D_p(E)$$

For curve (b), the axis is labelled "energy distribution" with markers at $kT$:

$$G_p(E) = F_B(E) \times D_p(E)$$

$$F(E) \times D_p(E)$$

**Figure 1.23** (a) The energy distribution law $G_p(E)$ for photons is the product of the density of states $D_p(E)$ and the Bose occupation factor $F_B(E)$. (b) The full curve is the photon energy distribution, $G_p(E) = D_p(E) \times F_B(E)$. The dashed curve shows the reduction in photon density in states of low energy obtained by wrongly using the Boltzmann occupation factor $F(E)$ in place of the Bose occupation factor. At $E/kT = 0.01$ (very close to the origin), for example, the actual photon density is nearly 100 times greater than it would be if the Boltzmann occupation factor applied.

Figure 1.23 plots each factor, $D_p(E)$ and $F_B(E)$, and their product $G_p(E)$. This graph of $G_p(E)$, when suitably scaled for cavity volume $V$ and temperature $T$, is in excellent agreement with the experimental measurements such as those shown in Figure 1.21.

The graphs in Figure 1.23a show how the effect of the sociability of photons, expressed by the very large values of the Bose occupation factor near $E = 0$, is partly masked by the density of states function which goes to zero at $E = 0$. Bosonic behaviour is still an important feature of the photon energy distribution however, as can be seen in Figure 1.23b where the effect is shown of *wrongly* using the Boltzmann occupation factor in place of the Bose occupation factor in Equation 1.20;

**Figure 1.24** The energy distribution law for photons goes back to the work of Max Planck (left) at the turn of the century. In 1900, Planck proposed an equation that was equivalent to Equation 1.20. Planck's work was a major landmark because it introduced the constant $h$ into physics; it was here that quantum physics was born. In 1924, the Indian physicist S. N. Bose (right) developed the modern theory of thermal radiation by treating it as a gas of particles. The word *photon* was not coined until 1926.

photons do have a stronger tendency to congregate together in the low-energy states than labelled particles.

Planck's radiation law, Equation 1.20, summarizes all the data on Figure 1.21 and much more besides. It applies to all thermal radiation, irrespective of its temperature, the shape of the cavity, or the composition of the surrounding walls. The only requirement is that the radiation should be in thermal equilibrium with the matter around it. We can in fact use Planck's radiation law as a good approximation to describe the photon energy distribution in the radiation escaping from the open door of a furnace and the radiation emitted by a hot body such as the Sun.

**Question 1.10**   A greenhouse is a construction that allows sunlight to enter during the day, but prevents thermal radiation from escaping at night. How should the transparency of the glass depend on wavelength? Use Figure 1.21 to justify your answer.  ◼

## 4.3  Consequences of Planck's radiation law

Planck's radiation law is a very powerful statement. We can use it to answer a number of questions about thermal radiation in a cavity. How many photons does it contain? What is the total energy of the radiation? What pressure does the radiation exert? The answers are obtained as definite integrals involving the energy distribution function (i.e. Planck's radiation law). We shall simply quote the results of the integrations, just as we did in Section 2.

### How many photons does thermal radiation contain?

As usual, we consider the case of a cavity of volume $V$ that contains thermal radiation at temperature $T$. By definition, this cavity contains $G_p(E)\,\Delta E$ photons with energies between $E$ and $E + \Delta E$; this number of photons is equal to the area of the very narrow strip in Figure 1.25.

Thus the total number of photons $N$ is equal to the total area under the graph of $G_p(E)$. This area can be evaluated by integrating the function $G_p(E)$, Equation 1.20, over all energies $E$,

i.e.  $$N = \int_0^\infty G_p(E)\,dE.$$

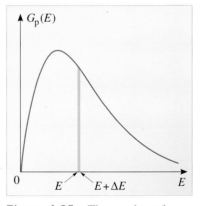

**Figure 1.25**   The number of photons with energies between $E$ and $E + \Delta E$ is equal to the area of the narrow strip.

The result (to 2 significant figures) is

$$N = 2.4C(kT)^3. \tag{1.21}$$

Using the value of $C$ given in Equation 1.19, this becomes (to two significant figures)

$$N = (2.0 \times 10^7\,\text{m}^{-3}\,\text{K}^{-3})VT^3. \tag{1.21a}$$

Strictly speaking, this is an average value. The walls surrounding the cavity absorb and emit individual photons in an erratic, unpredictable way, so the actual number of photons fluctuates. In general, though, the number of photons is so large that the fluctuations are completely negligible in comparison to $N$.

Equation 1.21a shows that the number of photons is proportional to the volume of the cavity and the cube of the temperature. Here we see one of the big differences between a gas of photons and a gas of molecules. The total number of molecules in a gas does not change when the temperature of the container increases, assuming there are no leaks — they just move faster. The number of photons, by contrast, is not conserved. When the temperature increases, the cavity walls emit and absorb photons at a faster rate, and so the equilibrium number of photons inside the cavity increases.

**Question 1.11** How many thermal photons are there in each cubic metre of your room when the temperature of the walls is (a) 300 K and (b) 380 K? (We are of course excluding photons from non-equilibrium light sources such as electric lights or daylight from a window.) ■

Equation 1.21a can be applied on the grandest possible scale. Following the Big Bang, the early Universe was bathed in electromagnetic radiation. This radiation was in thermal equilibrium with the electrons, protons and other material particles that were present. It was thermal radiation at a very high temperature. Gradually, the matter and the radiation cooled, keeping in step with one another, until stable hydrogen atoms were formed when the temperature fell to between 3000 K and 5000 K. This is thought to have happened when the Universe was about 300 000 years old. At this point, the Universe became almost transparent and so the link between the radiation and the matter was almost completely broken. In the absence of any significant absorption or emission, the ancient radiation has survived to the present day. It is called the **cosmic background radiation**. As the Universe has expanded, the background radiation has expanded with it; and because the radiation has had little chance to interact with matter, the number of photons has remained fixed. With $N$ constant, Equation 1.21a shows that $VT^3$ is constant, and so the temperature of the radiation has fallen in proportion to $1/V^{1/3}$. Thus, if the background radiation had an initial temperature of 3000 K and the Universe has since expanded by a factor of 1000 in every direction, the current temperature of the radiation should be

$$T = \frac{1}{(1000^3)^{1/3}} \times 3000\,\text{K} = 3\,\text{K}.$$

Thermal radiation with this temperature was detected in 1964 by Penzias and Wilson, who received the 1978 Nobel Prize in Physics. This discovery was of great importance to cosmology. Not only does the background radiation fill the entire Universe, but it also contains a vast number of photons. It is estimated that there are about $10^9$ photons of background radiation for every single proton or neutron!

### What is the energy of thermal radiation?

To calculate the total energy $U$ of thermal radiation in a cavity we again make use of Planck's radiation law. The energy of photons in a small energy interval $\Delta E$, i.e. with energies between $E$ and $E + \Delta E$, is equal to the number of photons in this range, $G_p(E)\,\Delta E$, times the energy $E$ of each photon. Thus the contribution to the total energy $U$ from photons in this small energy interval is $G_p(E)\,\Delta E \times E$. The total energy $U$ of thermal radiation is found by summing similar contributions over all energy intervals. Using Equations 1.18 and 1.19 this sum can be carried out by means of a definite integral to give (to two significant figures)

$$U = \frac{\pi^4}{15} C(kT)^4 = (7.5 \times 10^{-16}\,\text{J m}^{-3}\,\text{K}^{-4})VT^4. \tag{1.22}$$

Again, this is an average value but, as before, the fluctuations are small enough to be neglected. Equation 1.22 shows that the average total energy of thermal radiation is a very sensitive function of the temperature ($U \propto T^4$). This fourth-power law was discovered by C. Stefan in 1879 and is contained in the *Stefan–Boltzmann law*. It shows, for example, that if the temperature is doubled, the total energy increases by a factor of 16.

**Question 1.12**   The average energy of a photon, $\langle E \rangle$, is equal to the total energy $U$ of the photon gas divided by the total photon number $N$. Use Equation 1.22 and Equation 1.21 to show that the average energy of a photon is $\langle E \rangle = 2.7kT$. At approximately what temperature does this average energy correspond to a photon of visible light?   ■

## Radiation pressure

A photon gas exerts a pressure due to the momentum transferred to the walls of the cavity when photons are absorbed and emitted. This pressure can be calculated by a model similar to that used to derive the pressure of a gas of molecules in classical statistical mechanics (see *CPM*). The model is based on *Joule's classification* in which the particles in a cubic box of side length $L$ are notionally divided into three classes moving parallel to the three Cartesian axes and colliding elastically with the walls of the container or cavity.

In fact photons do not collide elastically with the cavity walls. When a photon strikes a wall it is absorbed, delivering its momentum to the wall, as well as its energy. However, in thermal equilibrium at a constant temperature, the average number of photons in the cavity remains constant, and the average rate of absorption of photons at each wavelength is equal to the average rate of emission. Thus, for each photon that is absorbed, an identical one is emitted and the wall receives a recoil momentum. There is a similar situation with molecules in a container. The molecules do not in fact simply collide with the container walls elastically. On striking the container wall a molecule usually sticks for a while before being ejected back into the gas. However, you have seen that the assumption of elastic collisions works well with molecules, and so we adopt it here for photons.

An important difference between gas molecules and photons is the relationship between the magnitude of the momentum $p$ and energy $E$. For photons, $p$ $E/c$. Another difference is that photons always travel with the speed of light $c$. Now refer to Figure 1.26 which shows one photon belonging to the class of photons imagined to be travelling parallel to the $x$-axis. During each elastic collision with the right-hand wall, the momentum of a photon of energy $E$ changes from

$$p_{x,1} = E/c$$

to   $$p_{x,2} = -E/c$$

and so the change of the photon's momentum is $\Delta p_x = p_{x,2} - p_{x,1} = -2E/c$. The photon, moving to and fro between the walls, takes a time $\Delta t = 2L/c$ between collisions with the right-hand wall, and so the *average* rate of change of the photon's momentum due to collisions with the right-hand wall is

$$\Delta p_x/\Delta t = (-2E/c)/(2L/c) = -E/L.$$

By Newton's second law, this average rate of change of momentum is equal to the average force exerted by the wall on the photon. By Newton's third law, there is an average force exerted *on* the right-hand wall by the photon, of the same magnitude but in the opposite direction. Thus the average force on the wall due to this one photon is

$$F_x = E/L.$$

The average pressure on the wall due to this photon is obtained by dividing this average force by the area $L^2$ of the wall

$$\text{average pressure due to one photon of energy } E = E/L^3 = E/V \qquad (1.23)$$

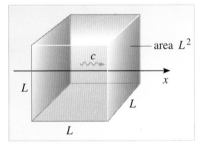

**Figure 1.26**   A photon moving parallel to the $x$-axis between two cavity walls.

where $V = L^3$ is the volume of the container. All we need to do now is to sum over all $N/3$ photons moving parallel to the $x$-axis. These photons have a range of energies (described by Planck's radiation law). Taking this into account, the total pressure on the wall is found simply by multiplying Equation 1.23 by $N/3$ and replacing the photon energy $E$ by the average photon energy $\langle E \rangle$. Thus the total pressure is $P = N\langle E \rangle/3V$. Recognizing that $N\langle E \rangle = U$, the total energy of the photon gas, we have the pressure exerted by a photon gas, or radiation pressure,

$$P = U/3V. \tag{1.24}$$

This shows that the radiation pressure is one-third of the energy density. (It is interesting to note that this result is similar to the expression for the pressure of a gas of molecules which was found in *CPM* to be $P = 2U/3V$. A factor 2 difference is not surprising in view of the differences between photons and molecules.)

Combining the result $P = U/3V$ with Equation 1.22, gives the radiation pressure as a function of temperature

$$P = (2.5 \times 10^{-16}\,\mathrm{J\,m^{-3}\,K^{-4}})T^4. \tag{1.25}$$

This shows that the pressure of thermal radiation is independent of the volume of the cavity, but depends sensitively on the temperature. Equation 1.25 is in fact the *equation of state* of a photon gas. Note how different this is from the equation of state of an ideal gas of molecules ($PV = nRT$). The main reason for this difference is the fact that the number of photons in the cavity is not conserved; it changes when the temperature or the volume changes (Equation 1.21a).

The pressure of the Earth's atmosphere at sea-level is about $10^5$ Pa.

Under normal circumstances, the pressure of thermal radiation is very small. As you sleep at night, you are bombarded by thermal photons (mostly invisible infrared photons) that exert a pressure of about $2 \times 10^{-6}$ Pa. Even at the surface of the Sun, the pressure of radiation is only about 0.3 Pa, but deep in the Sun's core there are very large photon densities and the pressure of radiation is about $10^{10}$ Pa, i.e. equivalent to about $10^5$ Earth atmospheres. In large stars, radiation pressure plays an important role in preventing the star from collapsing under the enormous pull of gravity.

**Question 1.13** It is sometimes said that the typical spacing between photons in thermal radiation is comparable to the wavelength of a typical photon. Knowing that $N = 2.4C(kT)^3$ (Equation 1.21) and the average energy per photon is $\langle E \rangle = 2.7kT$ (see Question 1.12) show that this statement is a reasonable rule of thumb.

**Question 1.14** The total number of accessible quantum states for a photon gas at temperature $T$ can be roughly estimated by summing all the states up to an energy of $kT$. This sum can be evaluated from the density of states $D_p(E)$ given by Equations 1.18 and 1.19, by means of a definite integral, to give $(2.8 \times 10^6\,\mathrm{m^{-3}\,K^{-3}})VT^3$. By comparing this number with the right-hand side of Equation 1.21a, say whether or not you would expect the indistinguishability of photons to be important in a photon gas. ∎

## 4.4 Other gases of bosons

Most atomic and molecular gases, in their commonest forms, are boson gases. The common molecules of nitrogen, oxygen and carbon dioxide, for example, are all bosons, and so are the vapours of the most common isotopes of the alkali metals lithium, sodium, potassium, rubidium and caesium. In Section 3.4 you saw that a gas of molecules behaves classically when the typical de Broglie wavelength is very much smaller than the typical spacing between gas molecules ($\lambda_{dB} << d$), which is always the case except at very near the absolute zero of temperature (see Question 1.7).

It was Einstein who first predicted that at a very low but finite temperature, where the de Broglie wavelengths of gas atoms are as large as the spacing between atoms ($\lambda_{dB} \approx d$), almost all of the particles in a gas of bosons would congregate in the lowest energy state, losing their individual identity and moving together as a single quantum system of macroscopic dimensions. We now call this phase of matter a **Bose–Einstein condensate**. Einstein showed that the congregation of the bosons into the lowest energy state is expected to occur quite suddenly as the temperature is lowered through a characteristic critical temperature near 0 K. This phenomenon is an example of the sociability of bosons. The more bosons there are in the lowest energy state the more likely it is for the remaining bosons to move into that state. As the temperature is lowered, the build-up of bosons in the lowest energy state is such that the remaining bosons find that state irresistible; they all avalanche into the lowest energy state and a Bose–Einstein condensate is formed.

Einstein's prediction was considered to be a mathematical artefact until the 1930s when Fritz London (1900–1954) studied superfluid liquid helium. Atoms of the common helium isotope $^{4}_{2}$He are bosons. When helium is cooled to about 2.17 K it begins to exhibit very strange properties — it offers almost no resistive drag to moving objects and it can pass through minute capillaries too small for normal liquids to pass through. These properties can be interpreted in terms of a component of the liquid having zero viscosity and an infinite thermal conductivity! It was realized that Bose–Einstein condensation played a role in this, but so did the large interactions between helium atoms in the liquid, and it is difficult to disentangle the two effects.

For many years scientists attempted to create a Bose–Einstein condensate in a less complicated system, preferably a tenuous gas where the interactions between atoms, being very much smaller than in the liquid state, would not obscure the pure Bose–Einstein behaviour. The problem was that to achieve $\lambda_{dB} \approx d$ in a tenuous gas, temperatures of less than $10^{-6}$ K were needed, and these ultra-low temperatures had to be achieved without the gas liquefying.

The technical breakthrough came with the development of methods for slowing atoms down using the forces exerted on them by laser beams. These laser-cooling techniques are now routinely used to cool vapours of alkali atoms to temperatures as low as $10^{-5}$ K. The laser-cooled vapours are extremely tenuous, having densities about $10^{-10}$ times that of air, and so the atoms never come close enough to condense into a liquid. The cold gas can then be further cooled to temperatures of $10^{-7}$ K or even lower by a sophisticated version of 'cooling by evaporation'. (In its simplest form, evaporative cooling is the main process by which a hot cup of tea cools.) During the evaporative cooling stage the atoms are contained by an arrangement of magnetic fields known as a magnetic trap. This avoids collisions of gas atoms with the walls of any material container at a higher temperature, which would of course heat the gas.

In 1995 a group of American physicists used these techniques to produce a Bose–Einstein condensate of rubidium vapour. The condensate appeared quite suddenly, as expected, when the evaporative cooling took the sample below a critical temperature of $1.7 \times 10^{-7}$ K. The sudden change from a classical gas to a pure macroscopic quantum state can be observed as a change in the shape of the sample, (see Figure 1.27). Nowadays, Bose–Einstein condensates of quite large size, with dimensions in the millimetre range, can be produced. Among the many fascinating quantum

**Figure 1.27** The formation of a Bose–Einstein condensate. The density of the sample is shown by the intensity of light scattered from it. This is shown by false colour as a function of position. Blue corresponds to low sample density and red to high density. At the left the temperature is just above the critical temperature and the particle distribution is almost classical. As the temperature is lowered through the critical temperature $T_c$, the density of the condensate increases dramatically.

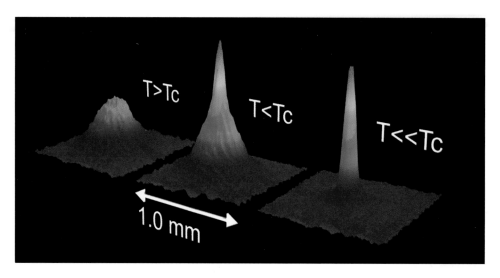

phenomena that have since been observed on the macroscopic scale, is the interference of two Bose–Einstein condensates (see Figure 1.1). The fringes are produced by the interference of wavefunctions representing the two quantum gases, the constituent atoms themselves having lost their individual identities.

Bose–Einstein condensation has, at the time of preparation of this book, also been achieved in sodium vapour, lithium vapour and in a gas of atomic hydrogen. These achievements have provided us with novel forms of quantum gases, and the overlapping sub-fields of physics involved in this study have made exciting rapid progress, which continues vigorously in laboratories across the world.

You may wonder why there is no Bose–Einstein condensation of photons in the theory of a photon gas that we have described in Sections 4.1 and 4.2. As you have seen (Equations 1.22 and 1.25 for example) the properties of a photon gas vary smoothly with temperature. There is no sign of an abrupt change at low temperatures. This is because of a crucial difference between atoms and photons. The number of atoms in a gas is conserved. When the temperature of an atomic gas changes, there is no change in the number of atoms, assuming there are no leaks from the container. Photons, on the other hand, are created and absorbed by matter. You have seen that there are fewer photons at all energies at lower temperatures (Figure 1.21). Equation 1.21a shows that $N \propto VT^3$. Thus if you cool a photon gas by reducing the cavity temperature the number of photons falls rapidly. For example, in a cavity of volume $1\,\mathrm{m}^3$ at a temperature of $1\,\mathrm{K}$, the average number of thermal photons in the cavity is about $2 \times 10^7$, while at a temperature of $10^{-2}\,\mathrm{K}$, the average number is only about 20, and at $10^{-3}\,\mathrm{K}$ it is 0.02.

A photon gas is of course a particular kind of photon system. It is in thermal equilibrium. Congregations of huge numbers of photons in the same quantum state, very similar in principle to Bose–Einstein condensation, can be achieved in non-equilibrium situations. An example is the laser. The photons produced by stimulated emission inside a laser all have the same wavelength, polarization and direction of propagation. You may have wondered why it is that stimulated emission has this special property. The answer is simple. Photons are sociable. Photons are preferentially emitted into any quantum state that is already occupied by photons. In a laser, the emitted photons build up in the particular quantum state corresponding to the wavelength, polarization and direction of propagation supported by the laser medium and mirrors. All the photons emitted by the laser are in this same quantum state. It is this that gives laser light its special properties of coherence and intensity.

# 5  The electron gas in metals

A metallic object, such as a silver spoon or a crow bar, may seem an unlikely arena for a gas model, but any metal object contains electrons that have detached themselves from their atoms and move freely through the whole region occupied by the metal. These electrons are called **free electrons**. In 1900, the German physicist Paul Drude (1863–1906) developed a model in which he treated the free electrons in a metal as a classical ideal gas. We begin this section by describing this model and some of its successes and failures. We then go on to describe a more successful, quantum-mechanical, model of the free electrons introduced in 1927 by Wolfgang Pauli. In Pauli's quantum model, the free electrons in a metal are treated as a gas of fermions. Our main aim in this section is to obtain Pauli's energy distribution law for the electron gas and use it to describe some properties of metals.

The *free electrons* in a metal are sometimes called *conduction electrons* because they are responsible for the metal's ability to conduct electricity.

## 5.1  Drude's classical model of the free electrons in metals

The essential idea of **Drude's free-electron model** is that each atom in a metal loses one or more of its electrons and becomes a positively charged ion. These ions form a fairly rigid lattice. The free electrons lost by the ions are free to wander throughout the body of the metal but are confined within the metal by the attractive forces of the positively charged ions.

The free electrons are assumed to collide occasionally with the lattice ions, but other than that the details of the interactions between electrons and ions are neglected. The Coulomb repulsion between electrons is also neglected. Left undisturbed, the electron gas settles down to a state of thermal equilibrium, characterized by a particular temperature and number density. The temperature of the electron gas is the same as the temperature of the metal. The number density of free electrons, i.e. the number of free electrons per unit volume, depends on the choice of metal, but is not difficult to calculate.

Paul Drude (1863–1906).

The calculation uses the concept of **valency**. In Drude's model, the valency of a metal is simply the number of free electrons released per atom. For example, the valency of aluminium is three, so each aluminium atom releases three of its 27 electrons into the electron gas. To obtain a general formula for the number density of free electrons, we shall consider a metal of valency $z$, density $\rho$ and relative atomic mass $M_r$. One mole of this metal contains $N_m$ ions and $zN_m$ free electrons, where $N_m$ is the Avogadro constant. These electrons roam throughout the molar volume, $V_m$, and so their number density is

$$n = \frac{zN_m}{V_m}.$$

As usual, the molar volume can be found by noting that the molar mass is $M_r \times 10^{-3}\,\text{kg mol}^{-1}$. For a metal of density $\rho$, the molar volume is therefore

$$V_m = \frac{M_r \times 10^{-3}\,\text{kg mol}^{-1}}{\rho}$$

and the number density of free electrons is,

$$n = \frac{z\rho N_m}{M_r \times 10^{-3}\,\text{kg mol}^{-1}}. \tag{1.26}$$

Note that the units of $N_m$ are $\text{mol}^{-1}$ while those of $V_m$ are $\text{m}^3\,\text{mol}^{-1}$. Since $z$ is unitless it follows that the number density $n$ has units $\text{m}^{-3}$.

**Question 1.15**   (a) Sodium has a valency of one, its relative atomic mass is 23 and its density is $970\,\mathrm{kg\,m^{-3}}$. What is the number density of free electrons in sodium? (b) Compare your answer with the number density of molecules in air, under normal conditions. (Take normal air pressure to be $10^5\,\mathrm{Pa}$ and the temperature to be $300\,\mathrm{K}$.)   ■

**Table 1.3**   The valency $z$, density $\rho$ and relative atomic mass $A_\mathrm{r}$, of selected metals. The last two columns give the number density of the free electrons and the typical spacing $d$ between free electrons.

| Metal | $z$ | $\rho/10^3\,\mathrm{kg\,m^{-3}}$ | $M_\mathrm{r}$ | $n/10^{28}\,\mathrm{m^{-3}}$ | $d/10^{-10}\,\mathrm{m}$ |
|---|---|---|---|---|---|
| sodium | 1 | 0.97 | 23.0 | 2.54 | 3.4 |
| copper | 1 | 8.93 | 63.6 | 8.45 | 2.3 |
| silver | 1 | 10.50 | 107.9 | 5.86 | 2.6 |
| magnesium | 2 | 1.74 | 24.3 | 8.62 | 2.3 |
| zinc | 2 | 7.14 | 65.3 | 13.2 | 2.0 |
| iron | 2 | 7.87 | 55.9 | 17.0 | 1.8 |
| aluminium | 3 | 2.70 | 27.0 | 18.1 | 1.8 |
| tin | 4 | 7.30 | 118.7 | 14.8 | 1.9 |
| lead | 4 | 11.34 | 207.2 | 13.2 | 2.0 |
| antimony | 5 | 6.70 | 121.8 | 16.6 | 1.8 |

Table 1.3 shows the number densities $n$ of free electrons in a variety of metals, calculated from Equation 1.26. The number density you have just obtained for sodium is quite small, by the standards of iron or aluminium.

Once the number density of the free electrons is known, it can be used to estimate other quantities. For example, the typical spacing, $d$, between the electrons is found by using $Nd^3 = V$, where $N$ is the total number of free electrons and $V$ is the volume of the metal. It follows that

$$d = \left(\frac{V}{N}\right)^{1/3} = \left(\frac{N}{V}\right)^{-1/3} = n^{-1/3}.$$

The last column of Table 1.3 uses this equation to calculate $d$ in our selection of metals. Evidently, the free electrons are very close to one another, at least as close to one another as the metal ions are — about one atomic diameter.

The close spacing of the free electrons is rather embarrassing for Drude's model. To see why, consider two electrons at the typical spacing of $d = 0.2\,\mathrm{nm}$. The potential energy between these two electrons is

$$\frac{1}{4\pi\varepsilon_0}\frac{e^2}{d} \approx 7\,\mathrm{eV}. \tag{1.27}$$

This energy can be put into perspective by comparing it with a typical kinetic energy. According to a classical ideal gas model, the average kinetic energy of an electron is $3kT/2 \approx 0.04\,\mathrm{eV}$ at room temperature. On this basis, the potential energy of $7\,\mathrm{eV}$ looks very large and it seems reasonable to infer that the free electrons interact powerfully with one another. Thus, Drude's model, with its neglect of interactions between electrons, could be criticized as being too simple.

It is not at all easy to answer this criticism. It is not apparent that Drude was able to do so and we can assume that he advanced his model for pragmatic reasons. In this respect, however, Drude's optimism paid off. We now know that the interactions play a smaller role than the above estimates would suggest — the classical gas model turns out to be surprisingly useful. A reason for this will be given later, but for the moment, we shall simply accept Drude's model and explore its consequences.

**Question 1.16**  A particle in a classical gas has an average speed of $\langle v \rangle = (8kT/\pi m)^{1/2}$. Estimate the average speed of a free electron in a metal at room temperature. Is this speed large enough for relativistic effects to be significant?  ∎

Drude's model can be tested against experiment. We shall now describe two of its early successes.

### The conduction of heat and electricity

If you warm one end of a poker by placing it in a fire, the other end soon becomes unbearably hot. If you connect a piece of resistance wire across the terminals of a battery, a steady current flows. In general, metals are very good conductors of heat and electricity, much better than non-metals such as wood or asbestos.

Drude's model explains these facts in terms of the mobile free electrons. When a poker is heated, free electrons from a hot region collide with electrons and ions in cooler regions and communicate their excess energy to them. In this way, heat is transferred from the hot end of the poker to the cold end. The process is so effective that the thermal conductivity of a metal is almost entirely due to its free electrons. These electrons are also responsible for electrical conduction. When a metal wire is connected to a battery, the negatively charged free electrons are attracted towards the positive terminal. The electron gas flows through the metal, but the speed of flow is limited by collisions with the lattice of fixed ions. An equilibrium is reached where the accelerating influence of the battery is balanced by the retarding effects of collisions. A steady current is then established.

Although all metals are good conductors of heat and electricity, some are better than others. One of the main predictions of Drude's model was that metals like copper, that conduct heat very well, should also be very good conductors of electricity. Metals like lead, that conduct heat less well, should be inferior electrical conductors. This prediction agrees very well with experimental measurements.

A second piece of evidence for Drude's electron gas was obtained from the experiment of Tolman and Stewart in 1915. This experiment, which was described in *Static fields and potentials* (*SFP*), demonstrated that the electric currents in wires of *different* metals were carried by electrically charged particles with the *same* charge to mass ratio which, to within the accuracy of their measurements, was found to be equal to the ratio $e/m_e$ for the electron.

To set alongside these two successes of Drude's model, we shall now mention one failure.

A serious objection to Drude's model is that it greatly overestimates the heat capacities of metals, even at room temperature. If the free electrons really behaved like a classical gas, they would make a significant contribution to the heat capacity of a metal. For example, a mole of lead (valency 4) contains $4N_m$ free electrons. According to classical physics, each electron has an average energy of $3kT/2$, so the total energy of a mole of Drude's electron gas is $E = 4N_m \times 3kT/2 = 6N_m kT$, and the corresponding molar heat capacity is $C = \Delta E/\Delta T = 6N_m k \approx 50\,\text{J K}^{-1}$.

In fact, lead obeys the Dulong–Petit law (see *CPM*). At room temperature its molar heat capacity is only $25\,\mathrm{J\,K^{-1}}$ and this experimental value can be explained in terms of the thermal vibrations of the lattice ions, making no allowance for the free electrons. It follows that the heat capacity of the electron gas must be small, very much smaller than Drude's model predicts.

You might wonder whether this defect in Drude's model can be remedied by including the details of the interactions between electrons, within a classical framework. It cannot. In classical physics, interacting electrons would have an even larger heat capacity than Drude's non-interacting gas.

There are other failures of Drude's model, some of which we shall describe in Chapter 2 where we examine the details of how the free electrons conduct electricity in metals.

The main fault of Drude's model lies not in its use of a gas model, but in the analysis based on classical physics. A quantum-mechanical theory is needed. In 1927, Wolfgang Pauli devised a model of the electron gas as a quantum gas of fermions. Here, we shall give an outline of Pauli's model.

## 5.2 Pauli's quantum model of an electron gas

In **Pauli's quantum free-electron model**, the free electrons are still treated as a gas confined to the volume of the metal by the positively charged lattice ions, but this gas is analysed by the methods of quantum physics, recognizing that electrons are fermions.

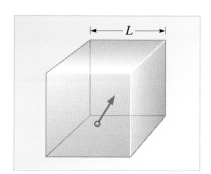

**Figure 1.28** A conduction electron is treated as a particle trapped inside a cubic box of volume $V = L^3$.

Each free electron in a block of metal is treated as a quantum-mechanical particle in a box. Since the electrons are free to roam throughout the metal, the box has the same size and shape as the block of metal. For simplicity, we shall take this to be a cube, with sides of length $L$ (Figure 1.28), and assume that the potential energy of an electron is zero everywhere inside the cube. Then the translational energy levels of an electron trapped inside this box are given by Equation 1.6

$$E = \frac{h^2}{8m_e L^2}(n_1^2 + n_2^2 + n_3^2)$$

where $m_e$ here is the electron's mass.

As usual, the quantum states become more tightly packed at higher energies and the **density of states function for electrons** takes the form

$$D_e(E) = B'\sqrt{E} \tag{1.28}$$

where 

$$B' = \frac{4\pi V(2m_e)^{2/3}}{h^3} = (1.06 \times 10^{56}\,\mathrm{J^{-3/2}\,m^{-3}})V. \tag{1.29}$$

$D_e(E)$ has the familiar $\sqrt{E}$ variation as we found in Section 2 for gas molecules, but there is an extra factor of 2 in the constant $B'$ to account for the two possible electron spin states associated with each translational energy state, and of course $m_e$ here is the electron mass and $V$ is the volume of the metal.

If Pauli's model went no further than this, it would offer few surprises. It would be the standard model of an ideal gas with quantized energy levels and it would lead to the Maxwell–Boltzmann distribution. At this point, however, Pauli introduced the vital new ingredient. He insisted that the free electrons obey the exclusion principle. Thus no more than two free electrons (with opposite spins) in a block of metal are allowed to occupy the same translational quantum state.

**Question 1.17**   Consider a cubic block of copper, with sides of length $L$. What is the maximum number of free electrons that can have an energy of $11h^2/8mL^2$ according to Pauli's model? (Refer to Figure 1.7 of Section 2.)   ■

You saw in Section 3 that the exclusion principle has a large influence on the properties of a Fermi gas, since it means that the electrons fill up the available states, two electrons per translational energy state, in much the same way as electrons fill shells and subshells in atoms. As a result, the electrons pile up into higher and higher energy states and most of the electrons are in states of energy very much greater than $kT$. To illustrate this, consider a 1 cm cube of copper at room temperature. The number of free electrons $N$ can be found from Table 1.3 to be $N = nV = 8.45 \times 10^{28}\,\text{m}^{-3} \times 10^{-6}\,\text{m} = 8.45 \times 10^{22}$. The total number of quantum states up to energy $kT$, (found by using the density of states $D_e(E)$ in a definite integral) has the value $2 \times 10^{19}$. You can see that this number of states can accommodate only about 0.02% of the free electrons. The rest have to pile up into states of higher energy, a long way above $kT$. If we ask how far up the energy scale we have to go to accommodate all the free electrons, we obtain the amazing answer of about 7 eV. This is about $300kT$ at room temperature. Thus we conclude that the free electrons have a huge spread of energies, greatly exceeding the range of a few $kT$ that would apply if the electrons were a classical gas obeying the Maxwell–Boltzmann distribution.

**Question 1.18**   The calculations above show that the translational energy states are not sparsely occupied and so the principle of indistinguishability and the exclusion principle apply. Use Equation 1.16 to confirm this by calculating the de Broglie wavelength of a typical electron of energy $kT$ and comparing it with the typical distance $d \approx 0.2$ nm, as given in Table 1.3.   ■

### Energy distribution function for an electron gas

The distribution of electron energies can actually be measured by high-energy photoemission experiments. This is simply the photoelectric effect carried out with high-energy photons. The general experimental arrangement is sketched in Figure 1.29. A sample of metal is illuminated with high-frequency monochromatic radiation, either ultraviolet or X-rays. Some of the photons are completely absorbed. The energy of each absorbed photon is acquired by an electron which then escapes from the metal. The escaping electrons are detected and their distribution of kinetic energies recorded.

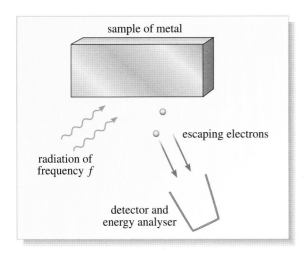

**Figure 1.29**
The experimental arrangement for high-energy photoemission. The radiation is either ultraviolet ($hf \approx 50$ eV) or X-rays ($hf \approx 1000$ eV). The photons rarely penetrate more than 2 nm below the surface. It is therefore important to ensure that the surface layers are free from impurities and the experiments are always conducted in a vacuum.

Each escaping electron has a total energy $E + hf - \phi$, where $E$ is the energy it had in the metal, i.e. the energy we want to measure, $hf$ is the photon energy, and $\phi$ is the energy needed to escape from the metal surface. ($\phi$ is the *work function* of the metal.) Hence the electron energy distribution in the metal is given by the measured kinetic energy distribution of the escaping electrons minus the constant energy $(hf - \phi)$.

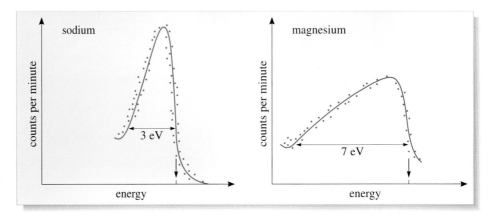

**Figure 1.30** High-energy photoemission results for sodium and magnesium. (The fact that the counts per minute do not go to zero at low energies is not relevant to the present discussion.)

Figure 1.30 shows experimentally determined energy distributions for electrons in sodium and magnesium at room temperature. We make two observations about these distributions:

- They decrease abruptly near the energies marked by arrows.
- They cover a *very* wide range of energies. In the case of sodium, the width is about 3 eV, while for magnesium it is 7 eV.

The second point is remarkable, but consistent with our estimates above of the occupied quantum states in copper. For comparison, molecules in a gas have a spread of energies of a few $kT$ (about 0.06 eV at room temperature). To put it another way, the Maxwell–Boltzmann distribution function reaches a width of 3 eV at a temperature of 15 000 K, yet free electrons have this spread of energies at room temperature! This is a complete failure of the Maxwell–Boltzmann distribution and provides impressive evidence against Drude's classical model. With Pauli's quantum model in mind, the classical failure is easy to understand. It arises because Boltzmann's distribution law takes no account of the exclusion principle and allows particles to congregate in low-energy states. This is precisely what the exclusion principle forbids for electrons.

### Pauli's distribution

The energy distribution function for the electron gas was first obtained by Pauli in 1927. It has the same generic form as the Maxwell–Boltzmann energy distribution and Planck's radiation law, i.e. it is the product of a *density of states function* and an *occupation factor*. Electrons are fermions and so the occupation factor is the *Fermi occupation factor* $F_F(E)$ given by Equation 1.14 in Section 3, the embodiment of the exclusion principle. The density of states function for electrons is given by Equation 1.28. Thus **Pauli's distribution**, also known as the **energy distribution law for free electrons in metals**, is given by

$$G_e(E) = B'\sqrt{E}\,\frac{1}{e^{(E-E_F)/kT} + 1} \tag{1.30}$$

where $B'$ is given by Equation 1.29 as $B' = (1.06 \times 10^{56}\,\mathrm{J^{-3/2}\,m^{-3}})V$.

The two factors in Pauli's distribution are plotted separately in Figure 1.31, together with their product $G_e(E)$, for the free electrons copper at 1000 K. The determining feature is the Fermi occupation factor. (The Fermi energy of copper is 7.0 eV and $kT$ at 1000 K is 0.086 eV.) You have already seen a graph of the Fermi occupation factor at $T = 0$ K in Figure 1.15 of Section 3. There you saw that the Fermi energy $E_F$ marked the boundary between filled and unfilled states. Here, in Figure 1.31, we have the Fermi occupation factor plotted at $T = 1000$ K. Note that the fall-off is not perfectly sharp at $E = E_F$. At this finite temperature, the Fermi occupation factor falls off near $E = E_F$ over a range of about $2kT$. This indicates that some of the electrons from states just below $E_F$ have been thermally excited to states just above $E_F$.

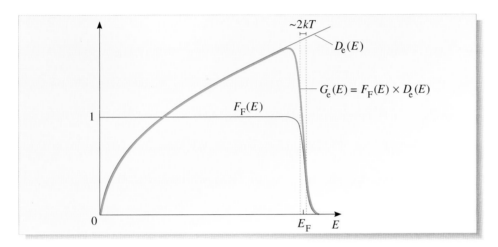

**Figure 1.31** The energy distribution function for an electron gas is the product of the Fermi occupation factor and the density of states function. The green graph shows the Fermi occupation factor $F_F(E)$ for copper at a temperature of 1000 K. The red graph shows the density of states function $D_e(E)$ for the free electrons in copper. The blue graph is the product $G_e(E) = F_F(E) D_e(E)$. This is the energy distribution for free electrons in copper at 1000 K.

The graph of the energy distribution, $G_e(E)$, follows closely the density of states $D_e(E)$ over the region where $F_F(E) \approx 1$, but then falls to zero as the Fermi occupation factor falls to zero. At a non-zero temperature the Fermi energy $E_F$ is equal to the value of $E$ where $F_e(E)$ has fallen to 1/2.

It is useful to compare the Pauli distribution at $T = 1000$ K with that at $T = 0$ K (Figure 1.32). At $T = 0$ K all the electrons are piled up in the states of lowest possible energy consistent with the Pauli exclusion principle. This is indicated by the sharp cut-off at the Fermi energy $E_F$ in Figure 1.32a. At the higher temperature, Figure 1.32b, a small proportion of the electrons, those of energy just a few $kT$ below the Fermi energy, are thermally excited to states just above the Fermi energy, and so the energy distribution cuts off less dramatically. This means that most of the electrons in filled states below the Fermi energy are unaffected by the increase in temperature because there are no empty states immediately above them. Only those electrons within a few $kT$ of the Fermi energy can be excited into empty states.

The Fermi energy is easily calculated by referring to the case where $T = 0$ K. Then all states up to the Fermi energy are filled and all those above are empty. Hence the number of states with energies below the Fermi energy must be equal to the total number $N$ of free electrons. Knowing the function $D_e(E)$, Equation 1.28, the total number of states up to the Fermi energy can be evaluated by means of a definite

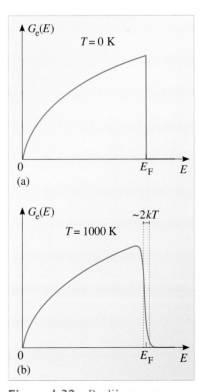

**Figure 1.32** Pauli's energy distribution $G_e(E)$ at (a) $T = 0$ K and (b) $T = 1000$ K.

integral to give $2B'E_F^{3/2}$. With the expression for $B'$ in Equation 1.29, this gives

$$\frac{8\pi V}{3h^3}(2m_e E_F)^{3/2} = N$$

and further rearrangement leads to

$$E_F = \frac{h^2}{8m_e}\left(\frac{3n}{\pi}\right)^{2/3} \tag{1.31}$$

where $n = N/V$ is the number density of the free electrons. This shows that, apart from various constants, the Fermi energy depends only on the number density of the free electrons. Using the number densities $n$ given in Table 1.3, we can work out the Fermi energies of some common metals (Table 1.4). Notice that $E_F$ varies from metal to metal, but is generally between 5 eV and 12 eV. This is hugely greater than the room-temperature value of the typical thermal energy $kT$ of an electron which is about 0.025 eV.

**Question 1.19** Use data from Table 1.3 to estimate the Fermi energies of sodium and magnesium. Are your answers consistent with the experimental results of Figure 1.30.

**Question 1.20** Figure 1.32 shows the electron energy distributions in a metal at $T = 0$ K and at $T = 1000$ K. Explain by reference to these graphs why the electron gas has only a very small heat capacity. ■

## 5.3 Consequences of Pauli's distribution

In parallel with our treatment of the photon gas, we shall now use Pauli's distribution to answer some questions about the electron gas.

### What is the energy of the electron gas?

There are $G_e(E)\,\Delta E$ electrons with energies in the range from $E$ to $E + \Delta E$. These electrons contribute an amount of translational energy $EG_e(E)\,\Delta E$. Hence the total translational energy of the electron gas, $U$, is the sum of all such contributions. As usual, this sum is evaluated by means of a definite integral. To keep matters as simple as possible, we shall consider the case of $T = 0$ K. Then the energy distribution function $G_e(E)$ cuts off sharply at the Fermi energy, see Figure 1.32a. The definite integral can then be easily evaluated to give

$$U = \tfrac{2}{5} B' E_F^{5/2}.$$

Using Equations 1.29 and 1.31, we obtain, after some algebra,

$$U = \tfrac{3}{5} N E_F \quad \text{(at } T = 0 \text{ K)} \tag{1.32}$$

where $N = nV$, the total number of electrons in the gas.

Thus, even at absolute zero, the electron gas has a large energy. This is a direct result of the Pauli exclusion principle which, as you have seen, forces many of the electrons into high translational energy states.

**Question 1.21** Use data from Table 1.4 to estimate the average kinetic energy of a free electron in zinc at $T = 0$ K. Is your answer likely to be a good approximation to the average kinetic energy of an electron at room temperature? ■

**Table 1.4** Fermi energies of selected metals, calculated from Equation 1.31 using data from Table 1.3,

| Metal | $E_F$/eV |
|---|---|
| copper | 7.0 |
| silver | 5.5 |
| zinc | 9.5 |
| iron | 11.1 |
| aluminium | 11.7 |
| tin | 10.2 |
| lead | 9.5 |
| antimony | 10.9 |

The answer to Question 1.21 goes some way towards explaining the success of Drude's free-electron gas model. This model was embarrassed by a calculation of the potential energy between two electrons of about 7 eV. This was about a hundred times greater than the total translational energy of a Maxwell–Boltzmann gas at room temperature, which we found in Section 2 to be $U = 3nkT/2$, so it was difficult to justify the decision to neglect the interactions altogether. You can now see that in Pauli's model the kinetic and potential energies are comparable. It is still a rather drastic step to neglect the interactions, but it is no longer an unreasonable one.

### What is the pressure of the electron gas?

It is easy to calculate the pressure of the electron gas. Because the electrons move with non-relativistic speeds, the relationship between translational energy and momentum, $E = p^2/2m$, is the same as for molecules in a ordinary gas, and so the pressure of the electron gas is related to the total translational energy in the same way as for a molecular gas,

$$P = \frac{2U}{3V}.$$

Using Equation 1.32 for the total translational energy $U$ at absolute zero, the pressure of the electron gas at absolute zero is

$$P = \frac{2}{3V}\left(\frac{3NE_F}{5}\right) = \frac{2}{5}nE_F \quad (\text{at } T = 0 \text{ K}) \tag{1.33}$$

where $n = N/V$, the number density of electrons. Again, for the reasons explained in the answer to Question 1.20, this result gives a good approximation to the pressure at room temperature. This is a very large pressure. It helps to explain why metals are so hard.

**Question 1.22** Estimate the electron gas pressure in copper at room temperature. Give your answer in Pa and in atmospheres. (Assume 1 atmosphere = $10^5$ Pa.) ■

The quantum theory of electrons in solids will be developed further in Chapter 2.

## 5.4 Other gases of fermions

We have looked extensively at the electron gas and how the Pauli exclusion principle, expressed by the Fermi occupation factor, is the main determinant of its properties. Other collections of fermions can be considered. A good example is a neutron star, which can be considered as a Fermi gas of neutrons.

Most stars consist mainly of protons (hydrogen nuclei) and electrons. Neutron stars are formed at the termination of a massive star's life. When the core of a massive star runs out of nuclear fuel, the core begins to compress under gravity, and when the core finally collapses the entire star collapses. The star's surface falls down until it hits the now incredibly dense core. If it rebounds it can blow apart in a supernova. The core itself, consisting mainly of protons and electrons, tries to resist gravity with electron pressure. This pressure is due to the exclusion principle and is essentially the same in origin as the electron pressure you calculated in Question 1.22. However, because the density of the core is so high, the gravitational attraction is just too strong. The star collapses as the electrons combine with the protons to form neutrons. Neutrons are also fermions whose quantum-mechanical pressure is actually smaller than that of electrons for the same number density $n$ due to the larger neutron mass (see Equations 1.33 and 1.31). Hence the collapse is catastrophic and continues until the number density $n$ of neutrons is high enough for the neutron pressure to finally resist gravity. The star, which can be of the order of 10 km in diameter at this stage, is then almost completely made up of a dense collection of neutrons. (One spoonful of neutron star substance would weigh as much as all the cars on Earth put together!)

Finally, we should mention what may be the most important fermion gas of all. In Section 4.3 we remarked that the Universe as a whole contains a photon gas (the cosmic background radiation) with about $10^9$ photons for every proton or neutron. It is also believed to contain a fermion gas of neutrinos at roughly the same temperature as the cosmic background photons. For every electron in the Universe, there are about $10^9$ neutrinos, with an energy distribution predicted by Equation 1.30 with $E_F$ put equal to zero.

*Neutrinos will be discussed in Chapter 4 of this book.*

# 6    Closing items

## 6.1    Chapter summary

1    The translational energy of a molecule of mass $m$ confined to a cubical container of side length $L$ is quantized. The allowed energies of the translational quantum states are given by

$$E = \frac{h^2}{8mL^2}(n_1^2 + n_2^2 + n_3^2) \tag{1.6}$$

where the quantum numbers $n_1$, $n_2$ and $n_3$ can be any positive integers. Each ordered set of three quantum numbers defines a translational quantum state. Most of the allowed energies are degenerate and the quantum states become more closely packed with increasing $E$.

2    The classical continuum approximation applies when the typical spacing between energy levels is small compared with $kT$, i.e.

$$\frac{h^2}{8mL^2} \ll kT \tag{1.7}$$

or    $\lambda_{dB} \ll L.$ \hfill (1.8)

3    The energy distribution of the translational quantum states is described by the density of states function

$$D(E) = B\sqrt{E} \tag{1.9}$$

where

$$B = \frac{2\pi V(2m)^{3/2}}{h^3}.$$

The number of quantum states with energies in the range $E$ to $E + \Delta E$ is $D(E)\,\Delta E$.

4    The distribution of distinguishable particles amongst the allowed quantum states is given by Boltzmann's law which tells us that the average number of particles occupying a single quantum state of energy $E$ is given by

$$F(E) = NAe^{-E/kT} \tag{1.11}$$

where $N$ is the total number of particles, $A$ is a constant, and the factor $e^{-E/kT}$ is called the Boltzmann factor. We call $F(E)$ the Boltzmann occupation factor.

5    The Maxwell–Boltzmann energy distribution, $G(E)$, is the product of the density of states and the Boltzmann occupation factor:

$$G(E) = B\sqrt{E} \times NAe^{-E/kT}.$$

Graphs of the Maxwell–Boltzmann energy distribution function, $G(E)$, the Boltzmann occupation factor $F(E) = NAe^{-E/kT}$ and the density of states $D(E) = B\sqrt{E}$ can be found in Figure 1.6.

6    A configuration of a gas of distinguishable particles is a particular arrangement of the particles among their allowed quantum states. In thermal equilibrium all configurations are equally likely.

7    Boltzmann's law can be traced back to three fundamental assumptions:

(i)   the law of conservation of energy;

(ii)  the idea that all configurations are equally likely;

(iii) the idea that the particles are distinguishable from one another.

8    In quantum mechanics, identical particles cannot be distinguished from one another. The indistinguishability of identical particles invalidates Boltzmann's law and requires a revision of the definition of a configuration.

A configuration of identical particles in quantum mechanics is defined by giving the numbers of particles in each quantum state.

Every particle in physics is either a boson or a fermion. Identical fermions obey the exclusion principle, and so only one fermion can occupy any quantum state. Bosons do not obey the exclusion principle and so any number of bosons can occupy any quantum state. A composite particle is a fermion if it contains an odd number of fermions, or a boson if it contains an even number of fermions.

9    The average number of identical bosons in a single quantum state of energy $E$ is given by the Bose occupation factor. The Bose occupation factor for photons is

$$F_B(E) = \frac{1}{e^{E/kT} - 1}. \tag{1.13}$$

The Bose occupation factor expresses the fact that bosons have a tendency to congregate together in the low-energy quantum states (bosons are sociable).

10   The average number of identical fermions in a single quantum state of energy $E$ is given by the Fermi occupation factor,

$$F_F(E) = \frac{1}{e^{(E-E_F)/kT} + 1} \tag{1.14}$$

where $E_F$ is the Fermi energy.

The Fermi occupation factor is the embodiment of the exclusion principle. At $T = 0\,\text{K}$, $F_F(E) = 1$ for $E$ less than $E_F$ and $F_F(E) = 0$ for $E$ greater than $E_F$. This shows that, at $T = 0\,\text{K}$, states of energy up to the Fermi energy are all occupied and those above are all empty.

11   At higher temperatures the fall-off of $F_F(E)$ at the Fermi energy $E_F$ is less abrupt and occurs over an energy range of a few $kT$, showing that a few electrons just below $E_F$ have been excited to empty states just above $E_F$.

12   The effects of indistinguishability and the exclusion principle can be neglected when the quantum states are sparsely occupied. The criterion for this is

$$N \ll \frac{V}{3\pi^2}\left(\frac{8m\pi^2 kT}{h^2}\right)^{3/2} \quad \text{or} \quad \lambda_{dB} \ll d. \tag{1.15 or 1.16}$$

Here $\lambda_{dB}$ is the typical de Broglie wavelength and $d$ is the typical distance between molecules. This criterion is satisfied for gases of molecules under normal conditions, but is not satisfied for gases of photons or electrons.

13 Thermal radiation (also called blackbody radiation or cavity radiation) is radiation that is in thermal equilibrium with matter at a fixed temperature $T$. A clean-cut example of thermal radiation is the radiation inside a cavity, for example an oven.

14 Photons are bosons and so the photons of thermal radiation can be treated as a boson gas. The energy distribution law for photons, called Planck's radiation law, has the form

$$G_p(E) = D_p(E) \times F_B(E) \tag{1.17}$$

where $D_p(E)$ is the density of states for photons (Equation 1.18) and $F_B(E)$ is the Bose occupation factor (Equation 1.13). Thus Planck's radiation law is

$$G_p(E) = CE^2 \times \frac{1}{e^{E/kT} - 1}. \tag{1.20}$$

Here $C = 8\pi V/h^3 c^3 = (3.206 \times 10^{75}\,\text{J}^{-3}\,\text{m}^{-3})V$ and $V$ is the volume of the cavity.

Graphs of the energy distribution function $G_p(E)$, the Bose occupation factor $F_B(E)$ and the density of states $D_p(E)$ are shown in Figure 1.23a.

15 The number of photons of thermal radiation in a volume $V$ and at a temperature $T$ is (to two significant figures)

$$\begin{aligned} N &= 2.4C(kT)^3 \\ &= (2.0 \times 10^7\,\text{m}^{-3}\,\text{K}^{-3})VT^3. \end{aligned} \tag{1.21 and 1.21a}$$

The energy of thermal radiation is

$$U = \frac{\pi^4}{15}C(kT)^4 = (7.5 \times 10^{-16}\,\text{J}\,\text{m}^{-3}\,\text{K}^{-4})VT^4. \tag{1.22}$$

The pressure of thermal radiation is $P = U/3V$.

16 Other Bose gases include the recently discovered Bose–Einstein condensates of rubidium and sodium atoms. These can be produced at extremely low densities and at temperatures between about $10^{-6}$ K and $10^{-7}$ K.

17 Some of the electrons in a metal are free to wander around within the entire volume occupied by the metal and can be regarded as a gas of electrons. These free electrons are responsible for the electrical conductivity of metals.

Drude's classical theory of the electron gas had some success in describing the electrical and thermal conductivities of metals but was unable to explain the fact that the electron gas makes a very small contribution to the heat capacities of metals, or the fact that the electrons have an extremely large range of energies (several eV) very much greater than $kT$.

18 Pauli's quantum theory of the electron gas recognizes that electrons are fermions and therefore obey the exclusion principle. The energy distribution function for the electron gas, called Pauli's distribution, has the form

$$G_e(E) = D_e(E) \times F_F(E).$$

Here $D_e(E)$ is the density of states for free electrons in the metal (Equation 1.28) and $F_F(E)$ is the Fermi occupation factor (Equation 1.14).

Thus Pauli's distribution is

$$G_e(E) = B'\sqrt{E} \times \frac{1}{e^{(E-E_F)/kT} + 1} \qquad (1.30)$$

where $\quad B' = \dfrac{4\pi V(2m_e)^{3/2}}{h^3}.$ $\qquad\qquad\qquad$ (1.29)

Graphs of the energy distribution function $G_e(E)$, the Fermi occupation factor $F_F(E)$ and the density of states $D_e(E)$ are shown in Figure 1.31.

19  The Fermi energy $E_F$ is found by equating the total number of electrons $N$ to the total number of electron states up to $E_F$, at $T = 0\,$K. The result is

$$E_F = \frac{h^2}{8m_e}\left(\frac{3n}{\pi}\right)^{2/3} \qquad (1.31)$$

where $n$ is the number density $n = N/V$. Note that $E_F$ is typically between $5\,$eV and $12\,$eV (see Table 1.4).

The total translational energy of the electron gas at $T = 0\,$K is

$$U = \tfrac{3}{5}NE_F$$

and the pressure of the electron gas at $T = 0\,$K is

$$P = \tfrac{2}{5}nE_F.$$

20  The electron energy distribution changes very little with temperature since only those electrons with energies within a few $kT$ of $E_F$ can be excited into empty states above $E_F$. For this reason the above values of $E_F$, $U$ and $P$, which are calculated for $T = 0\,$K, remain good approximations at higher temperatures. The free electrons contribute only slightly to the heat capacities of metals for the same reason.

21  Another example of a Fermi gas is a neutron star where the neutron pressure prevents the star from collapsing.

## 6.2 Achievements

Now that you have completed this chapter, you should be able to:

A1  Explain the meanings of all the newly defined (emboldened) terms introduced in the chapter.

A2  Determine the configurations of simple systems of distinguishable particles, identical bosons and identical fermions

A3  Know a few examples of bosons and fermions, and determine whether composite particles are bosons or fermions.

A4  Discuss the properties of the occupation factors for classical gases, Bose gases and Fermi gases (i.e. the Boltzmann occupation factor, the Bose occupation factor and the Fermi occupation factor).

A5  Use criteria to determine whether it is appropriate to use a classical or a quantum description of a gas.

A6  Recognize the photon energy distribution law (Planck's radiation law) for a gas of photons in thermal equilibrium, and, given the cavity volume and temperature, calculate the number of photons in the gas, the total energy of the gas and the pressure exerted by it.

A7 Describe Drude's model of the free-electron gas in metals and state its successes and shortcomings. Given the valency of the metal and its volume, calculate the number density of free electrons in metals.

A8 Recognize the energy distribution law for free electrons in a metal (Pauli's distribution); explain the significance of the Fermi energy; calculate the energy and pressure of the electron gas.

A9 Describe briefly examples of boson and fermion gases other than photon and electron gases.

## 6.3 End-of-chapter questions

**Question 1.23** Consider a hypothetical gas of three identical particles, each of which has only four quantum states, of energies $\varepsilon$, $2\varepsilon$, $3\varepsilon$ and $4\varepsilon$. List all the different configurations of this gas that have a total energy of $8\varepsilon$, assuming that the particles are (a) identical bosons, (b) identical fermions, and (c) distinguishable particles.

**Question 1.24** The translational kinetic energy of a gas of $10^{23}$ hydrogen molecules confined in a cubic box of volume $0.2\,\text{m}^3$ at a temperature of $300\,\text{K}$ can reasonably be studied on the basis of classical theory alone (i.e. without recourse to quantum theory). Give a quantitative explanation of why this is so. (Take the mass of the hydrogen molecule to be twice the mass of the proton.)

**Question 1.25** Assuming the temperature in part of the Sun's core is $3 \times 10^6\,\text{K}$, estimate the number of photons in one cubic metre, the photon energy in one cubic metre and the photon pressure.

**Question 1.26** Decide whether each of the following particles is a fermion or a boson: (a) the nucleus of the helium ($^3_2\text{He}$) atom, (b) the neutral helium ($^3_2\text{He}$) atom, (c) a singly-ionized helium ($^3_2\text{He}^+$) atom (i.e. a helium atom that has lost one electron).

**Question 1.27** Knowing that the Fermi energy at $T = 0\,\text{K}$ is given by Equation 1.31, evaluate the Fermi energy of silver (where the number density of free electrons is $5.86 \times 10^{28}\,\text{m}^{-3}$). Find also the total translational energy and the pressure of the electron gas in a silver cube of side length $1.0\,\text{cm}$. ■

# Appendix to Chapter 1: Derivation of the density of states function for gas molecules

Refer to Figure A1 where the three integer quantum numbers $n_1$, $n_2$ and $n_3$ in Equation 1.6 are marked off along three mutually orthogonal axes. The 'space' defined by these axes is a set of cubic lattice points of coordinates $(n_1, n_2, n_3)$. Each elementary cube of the lattice has unit volume and is associated with a translational quantum state defined by the quantum numbers $(n_1, n_2, n_3)$ with energy given by Equation 1.6. (Note that the space defined by the three axes is a mathematical space of quantum numbers, not real space, and so distances, areas and volumes in this space are dimensionless numbers.)

Now imagine a thin spherical shell of radius $R$ (a dimensionless number) and thickness $\Delta R$, centred at the origin. Since $n_1$, $n_2$ and $n_3$ are positive integers, we are only interested in the one-eighth part of the shell shown in Figure A1. We shall assume that $R$ is very large compared with the separation of lattice points. This allows us to regard the lattice points as being so close together that they constitute a continuum. (This is the classical continuum approximation.) Thus we can write $R = \sqrt{n_1^2 + n_2^2 + n_3^2}$ .

Equation 1.6 tells us that the energy $E \propto n_1^2 + n_2^2 + n_3^2$, so all states with the same $R$ have the same energy. This gives us a nice picture of the degeneracy of the translational states: all states on the surface of a sphere centred on the origin have the same translational energy.

The volume (dimensionless) of the thin shell of radius $R$ is approximately equal to its surface area times its thickness, i.e.

$$\text{volume of shell} = \frac{1}{8} \times 4\pi R^2 \Delta R.$$

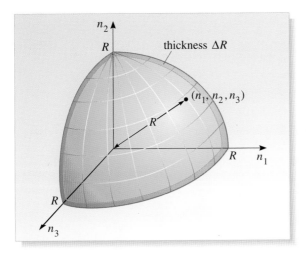

**Figure A1**   The quantum numbers of the translational states, $n_1$, $n_2$, and $n_3$ (see Equation 1.6), are plotted along three mutually orthogonal axes, so that each point with coordinates $(n_1, n_2, n_3)$ represents a quantum state.

Since each quantum state is associated with a cube of unit volume, the above volume is equal to the number of states in the thin shell between radii $R$ and $R + \Delta R$. We can denote this number of states by $D(R)\,\Delta R$, i.e.

$$D(R)\,\Delta R = \frac{1}{8} \times 4\pi R^2\,\Delta R.$$

You have seen that the radius $R$ is an abstract quantity representing 'distance' from the origin in the quantum-number space of Figure A1. It will be much more useful to refer instead to the energy $E$ of states represented by lattice points at a distance $R$ from the origin. We can easily do this. We have

$$R = \sqrt{n_1^2 + n_2^2 + n_3^2}\;.$$

Now using Equation 1.6 we can relate $R$ to the corresponding energy $E$,

$$R = \sqrt{\frac{8mL^2E}{h^2}}\;.$$

We shall also need

$$\frac{\mathrm{d}R}{\mathrm{d}E} = \sqrt{\frac{8mL^2}{h^2}}\,\frac{E^{-1/2}}{2}$$

so that we may write $\Delta R = \dfrac{\mathrm{d}R}{\mathrm{d}E}\,\Delta E.$

We can now replace $D(R)\,\Delta R$, the number of states in a shell between radii $R$ and $R + \Delta R$, by $D(E)\,\Delta E$, the number of states with energies between $E$ and $E + \Delta E$. Thus, using the above expressions for $R$ and $\Delta R$, we can rewrite the equation $D(R)\,\Delta R = (1/8) \times 4\pi R^2\,\Delta R$ as

$$D(E)\,\Delta E = \frac{1}{8} \times 4\pi R^2\,\Delta R$$

$$= \frac{1}{8} \times 4\pi\left(\frac{8mL^2E}{h^2}\right)\sqrt{\frac{8mL^2}{h^2}}\,\frac{E^{-1/2}}{2}\,\Delta E.$$

Tidying this up and cancelling the $\Delta E$, we obtain

$$D(E) = B\sqrt{E} \qquad\qquad\text{(Eqn 1.9)}$$

where $\qquad B = \left(\dfrac{2\pi V(2m)^{3/2}}{h^3}\right) \quad$ and $\quad V = L^3$.

The function of energy, $D(E)$, is the density of states function.

# Chapter 2   Solid-state physics

## 1   Beyond tin disease

Captain R. F. Scott's tragic expedition to the South Pole in 1911–12 is a well-known epic story. One of the mishaps to befall the expedition was the mystery of the leaking fuel cans. The expedition had left supply dumps at strategic points for their journey back across the Antarctic wastes from the Pole. On arrival at the dumps, they found many of their precious fuel cans to be only partly full. Fuel had somehow leaked out. In fact, the cans were afflicted with 'tin disease'. The cans were soldered with tin, a well-established manufacturing process at the time. What Scott did not know is that at temperatures below about −40 °C, the familiar so-called 'white' form of tin undergoes a change in crystal structure, becoming a brittle form of tin known as 'grey' tin. Scott's fuel had leaked out through the soldered joints as the grey tin had crumbled. Ironically, we now know that the arrangement of tin atoms in the crystals of grey tin is the same as the arrangement of carbon atoms in diamond — the hardest substance known.

Most of the material on Earth is solid. Yet solids were very little studied at the start of the twentieth century and even less well understood in a fundamental sense, because without quantum physics no depth of comprehension was possible. Even today there are still many mysteries, and research on solids provides challenge and excitement for a high proportion of the physicists working at the start of the twenty-first century. In this chapter we explore some of the achievements in the last 100 years.

Of course, many of the macroscopic properties of solids were known before 1900. It had long been established, for example, that glass is a good thermal and electrical insulator and will break under a modest stress. By contrast, copper is a good thermal and electrical conductor which will bend rather than break under a similar loading and can be hammered into shape in a way impossible for brittle materials like glass. However, the physical reasons for such differences between materials were not known. Scientists wanted to know *why* solids had these properties, but such knowledge was long in coming.

As you saw in *Classical physics of matter* (*CPM*) the atomic nature of matter was suspected for many years and finally confirmed in the late nineteenth century via the statistical mechanics of gases. So solids must be made of atoms, but what holds the atoms together and how are they arranged within the solid? This started to become known only when studies of the diffraction of X-rays (from 1912) and electrons (from 1927) by crystalline solids revealed the structures.

You may have been struck by the picture in the upper part of the front cover of this book. This amazing image was obtained using a scanning tunnelling microscope, invented in 1986. The image shows a ring, 14 nm in diameter, consisting of 48 iron atoms placed on a copper surface. The concentric rings on the surface inside show the density distribution of electrons trapped inside the ring. We are now into the age where physicists are developing systems for the routine engineering of solids literally on an atomic scale. We've come a long way from the days when tin disease was a problem, but there are still plenty of mysteries left to keep researchers occupied. (Front cover photo courtesy of D. M. Eigler, IBM Research Division.)

We begin in Section 2 with a short survey of how atoms bond together to form molecules and, on a larger scale, solids. We also investigate how the mechanical properties of solids depend on their structure. This builds on your knowledge of the electronic structure of atoms and the stability of filled electron shells (*Quantum physics: an introduction*). The rest of the chapter looks in some detail at the electrical properties of solids, especially the technologically-important semiconductors. Section 3 reviews and extends the classical and quantum free-electron models of electrical conduction in metals that you studied in Chapter 1 of this book. Section 4 introduces the *band theory of solids* which modifies Pauli's quantum free-electron model by taking account of the ordered structure of atoms in crystalline solids. In Section 5 the band theory is applied to a study of semiconductors in a variety of devices such as solar cells and diode lasers. Finally, Section 6 describes the basic physical ideas behind superconductivity and looks at some of its applications.

# 2 The bonding and structure of solids

This section introduces the principles of atomic bonding in solids and explains how this determines their mechanical properties, a subject of immense interest to mechanical engineers. We only shall give a brief survey of the main physical principles, touching on a few selected examples, as preparation for the study of the electrical properties of solids in the subsequent sections.

At a sufficiently high temperature almost all matter can exist in the form of an almost ideal gas. At lower temperatures, the gas departs from ideal gas behaviour, and typically atoms join to form molecules and then the molecules condense into a solid, perhaps via a liquid phase. As a first step towards discussing bonding and structure in solids it is useful to examine the bonding mechanisms by which atoms come together to form molecules.

## 2.1 Bonding in molecules

The bonding of atoms into molecules has been discussed ever since John Dalton (1766–1844) proposed an atomic basis for chemistry, but it was only after the publication of Bohr's atomic theory (1913) that progress really began towards understanding molecules. All bonds involve electrical interactions between electrons in the atoms which are joined, and have to be described by using quantum mechanics. The electrons involved in bonding are usually in the highest energy levels occupied in the ground state of the atoms; these are the **valence electrons**.

A big step was taken in 1916 when Gilbert Lewis (1875–1946) suggested that one way in which individual atoms might bond together to form molecules was by sharing electrons. The shared electrons are said to form a **covalent bond** between the atoms. Such bonds provide the basis for molecular formation in many materials. The shared electrons are generally unpaired in their individual atoms but form pairs with opposite spins in the bonded molecule.

We can describe the covalent bond in terms of electric forces. The wavefunction describing a bonding pair of electrons is strongly localized between the bonded atoms, and less localized elsewhere in the molecule. This concentration of negative charge tends to pull the positive nuclei together, thereby reducing the repulsion between nuclei and allowing the atoms to bond. The covalent bonding of two hydrogen atoms in the diatomic hydrogen molecule $H_2$ is the easiest case to consider. This problem was first solved in a simplified approximation by Heitler and London in 1927.

Figure 2.1a shows the total energy of two hydrogen atoms, each in its ground state with a single (unpaired) 1s electron, as the atoms are gradually brought towards each other. There are two cases to consider. If the two atoms have their 1s electrons in the same spin state, the total energy rises as the two atoms approach, and they do not bond. This antibonding case is shown in the upper curve in Figure 2.1a and is a result of the Pauli exclusion principle. The quantum state of an electron includes a spin state as well as a spatial wavefunction. Because, in this case, the two electrons have the same spin state, the exclusion principle forbids them from also having the same spatial wavefunction, and so they avoid one another, staying mostly outside the region between the two nuclei (Figure 2.1b) which then repel one another so that no bond is formed.

If, on the other hand, the atoms approach one another with opposite spin states the reverse is true. The most likely place for the electrons is between the two nuclei, linking the positive nuclei by electric attraction. The energy for this bonding case is shown in the lower curve of Figure 2.1a. As the atoms approach from a large distance, the energy decreases at first, corresponding to attraction between the charged parts of the atoms. The energy falls to a minimum at the equilibrium distance $r_0$, with the negative charge concentrated in the region between the two nuclei (Figure 2.1c). This is the stable configuration of the molecule. If the atoms were to approach closer than $r_0$, the bonding electrons would be squeezed out of the central region and the nuclei would repel one another strongly.

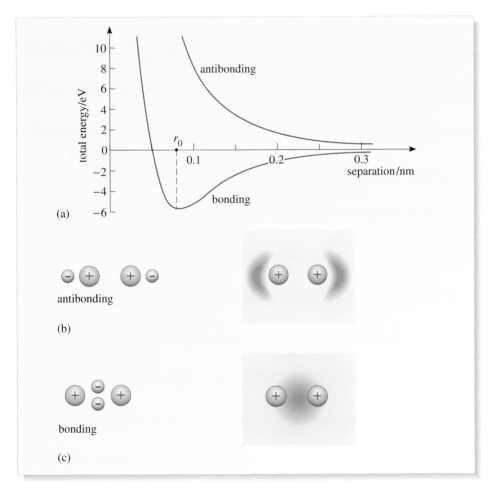

**Figure 2.1**  (a) Graph of the total energy as a function of the distance between two hydrogen atoms. The energy is taken to be zero when the two atoms are at a very large distance from one another where they do not interact. The two separate curves correspond to the two cases of bonding (opposite spin states) and antibonding (same spin state). The electron clouds, i.e. probability distributions, are indicated schematically for (b) the antibonding case and (c) the bonding case.

**Question 2.1** The binding energy of a hydrogen molecule is the energy required to separate the two atoms to a very large distance from one another where they no longer interact. By referring to Figure 2.1, estimate the binding energy of the hydrogen molecule. ■

Similar covalent bonding is often found for other atoms that have unpaired electrons. Consider the elements in Group VII of the Periodic Table (see Figure 3.2). This group includes chlorine Cl, which has the ground-state electronic structure: $1s^2 2s^2 2p^6 3s^2 3p^5$. The 3p subshell can accommodate 6 electrons but in chlorine it contains only 5, four paired electrons and one unpaired. This allows Chlorine to form diatomic molecules via covalent bonding to make the chlorine molecule, $Cl_2$. Here the pairing is between the single unpaired electrons in the outer 3p subshells of the two atoms. By sharing these two electrons, each atom is, in a sense, surrounded by a 3p subshell with its full compliment of 6 electrons (three pairs) and the covalent bond is formed.

Once the pairing has formed in such a case, no unpaired electrons are left to make more bonds and no further atoms can be added to make molecules larger than the diatomic ones. This is an example of **bond saturation**.

In Group IV of the Periodic Table we find oxygen. This has the electronic structure $1s^2 2s^2 2p^4$. *Hund's rule* tells us that two of the four electrons in the 2p subshell are unpaired and so two oxygen atoms can form a double covalent bond involving a doubled pairing of electrons to make the oxygen molecule $O_2$. Its neighbour nitrogen in Group V has electronic structure $1s^2 2s^2 2p^3$. Here all 3 of the 2p electrons are unpaired (Hund's rule again), so nitrogen performs a triple pairing to make $N_2$, i.e. it forms a triple covalent bond in which each atom is surrounded by a full 2p subshell by sharing its three 2p electrons with its partner. These double and triple bonds are illustrated in Figure 2.2. Double and triple bonds are not saturated because they can be unpicked and arranged across more atoms allowing more and more complex molecules to be formed, $O_3$, $NO_2$, $NO$, $N_2O$, etc., giving plenty of variety to keep the chemists happy!

Another important property of covalent bonding is **bond directionality**. The *valence electrons* that form the bonds are not free to move away from the bonded atoms but are strongly attached to them. This is apparent when one atom is bonded to two or three others, and the resulting molecule has a definite geometrical structure with definite angles between the bonds. Consider the molecule ammonia $NH_3$, with three hydrogen atoms covalently bonded onto a nitrogen atom. The outer 2p subshell of nitrogen has only 3 electrons, all unpaired. To make $NH_3$ we must add three hydrogen atoms with electron spin states opposite to those in the nitrogen 2p subshell, to make three single covalent bonds. Since these three bonds are identical, the three hydrogen atoms must be arranged symmetrically in space. In fact the ammonia molecule has the shape shown in Figure 2.3a. The bonds are separated by equal angles, which happen to be slightly greater than 90° in this case.

Another important example is the water molecule $H_2O$. The oxygen atom has four electrons in the 2p subshell, two of which are unpaired. These two unpaired electrons can form directed bonds with hydrogen atoms, with a separation angle of about 105°. This is shown in Figure 2.3b.

The atom next to nitrogen in the first row of the Periodic Table is the Group IV element carbon, electronic structure $1s^2 2s^2 2p^2$. This is the most versatile element of all in forming molecules and we cannot explore its many tricks here. However, it is important that one detail of carbon's behaviour is known. In forming the molecule $CH_4$ (the natural gas methane, which we use in cooking), carbon uses *all* four of its electrons in the outer $n = 2$ shell, not just the two in the 2p subshell, but also those in

Hund's rule was given in *QPI*. It requires that the spins of the electrons in unfilled shells are arranged so that, within the constraints of the Pauli exclusion principle, as many as possible are unpaired, i.e. have opposite spin states.

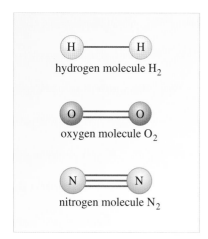

**Figure 2.2** Examples of single, double and triple covalent bonds. Each rod linking two atoms represents a pair of bonding electrons.

the lower 2s subshell. In effect, one of the 2s electrons in carbon is raised into the 2p subshell, giving *four* unpaired electrons. The wavefunctions for the single 2s and the three 2p electrons are mixed together to form new kinds of electron wavefunctions known as sp$^3$ hybrid wavefunctions, or simply **hybrid wavefunctions,** in which all four electrons are unpaired and behave similarly. By joining an electron in each of these four states in carbon to an electron of opposite spin from a hydrogen atom, we get four pairs of electrons to make the bonds. The four identical bonds are arranged symmetrically in space with the hydrogen atoms at the corners of a tetrahedron (i.e. a triangular pyramid) as shown in Figure 2.3c. This symmetry is also found when carbon is in its solid form (diamond) and for other elements in the Group IV column of the Periodic Table, such as silicon and germanium.

In symmetric diatomic molecules like $O_2$, $N_2$ and $Cl_2$, the bonding electrons are shared equally between the two atoms. But now consider an asymmetric molecule like sodium chloride, NaCl. The ground-state electronic structures of the atoms are: Na ($1s^2 2s^2 2p^6 3s^1$); Cl ($1s^2 2s^2 2p^6 3s^2 3p^5$). In a covalent bond, the Na atom would share its single 3s valence electron and the Cl atom would share its single unpaired 3p electron, giving both atoms a complete outer shell. However, the chlorine atom has a much greater affinity for electrons than sodium; chlorine is said to be more **electronegative** than sodium. As a result, what actually happens is that the chlorine atom takes nearly all of the electron pair and effectively ends up as a negatively charged chlorine ion Cl$^-$, with a full 3p subshell and a net electric charge of $-e$. This leaves a sodium ion Na$^+$ with a net electric charge $+e$. Thus the molecule is best described as Na$^+$Cl$^-$, with the two atoms joined by electrostatic attraction. This kind of bond, which involves the transfer of electrons, is known as an **ionic bond**.

The above picture of an ionic bond suggests an electron transfer which in reality is never quite as complete as the symbols Na$^+$ and Cl$^-$ indicate. There is never a complete electron transfer and the bonding is always partly covalent in nature. We see the partial nature of electron transfer very clearly in the case of water, $H_2O$. This is predominantly a covalently-bonded molecule, but oxygen is more electronegative than hydrogen, and so it takes rather more than a half-share of the bonding electrons while the hydrogens must make do with less than half. This preferential accumulation of negative electric charge at the oxygen end of the molecule leaves a small positive charge at the hydrogen ends — a fact of momentous importance, as you will see later in this section.

This **mixed bonding**, partly covalent and partly ionic, is found in practice for all cases except the bonding of molecules made up of identical atoms such as $H_2$, $O_2$, etc, which are covalently bonded. However, we often ignore all this complexity and class the bonds as covalent or ionic according to the dominant part of the mixture.

## 2.2 Bonding in solids

A further stage of bonding occurs when atoms and molecules condense to the solid form. In this subsection we draw the broad picture of bonding within solids, showing how the type of bonding determines some of the mechanical properties of solids.

Whatever the bonding mechanism, the maximum binding energy, and hence the most stable structure, is one where the atoms or molecules are arranged with total regularity into a **crystalline solid**, ideally a single crystal where the regularity extends throughout the entire material. The regular internal arrangement of atoms may cause the external shape of the crystal to have a particular geometric form, with the surface covered with the beautiful facets for which crystals are famed.

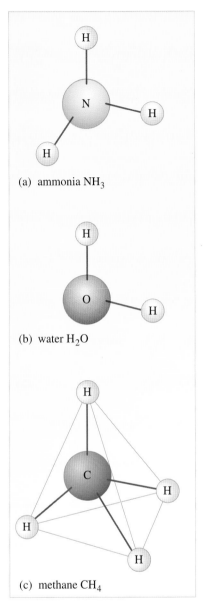

(a) ammonia $NH_3$

(b) water $H_2O$

(c) methane $CH_4$

**Figure 2.3** (a) The ammonia molecule $NH_3$. (b) The water molecule $H_2O$. (c) The methane molecule $CH_4$. Water is shown in a direct two-dimensional view, but the others are represented as three-dimensional structures seen in perspective view. (The thin lines are not bonds but construction lines drawn to show the three-dimensional structure.)

The single crystal form is in fact relatively rare. In natural or in laboratory conditions single crystal growth can be achieved only by relatively slow formation of the solid into a regular ordered structure extending normally from a single starting point. In this way, ice crystals tens of miles long sometimes form on the surface of lakes. Rates of growth that are too fast will lead to irregularities in the atomic arrangements. In an extreme case there is total loss of long range regularity and the result is an **amorphous solid**, like glass, but this is unusual as the drive towards local order is so strong. If solid formation starts in many places independently, many small individual crystals will form, joined at their boundaries into a **polycrystalline solid**. Most solids are polycrystals, and therefore not in their lowest energy state. A lower energy could in principle be achieved by a reordering of the atoms throughout the polycrystal to form a single crystal, but the rate at which the atoms are able to exchange positions is, in practice, so slow that no significant progress towards that equilibrium occurs.

To give even a modest description of all the possible crystal structures would occupy all the space available to this chapter. However, many of the common structures can be understood by imagining the atoms to consist of small hard spheres of the same size stacked in simple arrays. The simple cubic array of Figure 2.4a is perhaps the most obvious, but in fact crystals with a simple cubic structure are quite rare, partly because it makes an inefficient use of space. Putting additional atoms at the centres of each cube gives the more space-efficient *body-centred cubic* structure shown in Figure 2.4b, where 68% of the space is occupied when the atoms are modelled by spheres of the same size and in contact with one another. The most efficient use of space however is obtained in the two so-called *close-packed structures*, the *face-centred cubic* and the *hexagonal close-packed* structures shown in Figure 2.5. Both are based on hexagonal layers (Figure 2.5a). This is obvious in the hexagonal close-packed case (Figure 2.5b) but can also be seen by looking at a diagonal slice of the face-centred cubic case (Figure 2.5c). Your grocer is likely to stack oranges in hexagonal layers thereby building up a hexagonal close-packed or face-centred cubic structure depending on how the successive layers are placed. In each case the fruit occupies 74% of the space available, the maximum possible for identical spheres.

For most materials we cannot predict the equilibrium crystal structure reliably. Our theoretical models of solids do not discriminate well enough between rival crystal structures, because the subtleties of the atomic interactions over long ranges are not modelled with sufficient precision. For this reason an experimental approach is required.

Much of what we know about crystal structures has come from experimental studies of the diffraction of electrons, neutrons and especially X-rays. This work began in 1912 when Max von Laue realized that X-rays have wavelengths that are comparable with the size of atoms. Consequently if crystals existed as arrays of atoms, as had been discussed for many years, the regular planes of atoms in a crystal could act like a three-dimensional optical diffraction grating.

Generally, an X-ray beam will pass though a crystal undeflected, but strong reflections can occur at certain angles as a result of constructive interference between waves scattered by parallel planes of atoms separated by a spacing $d$ (Figure 2.6). The condition for a strong reflection to occur is expressed by **Bragg's law**,

$$n\lambda = 2d \sin \theta \tag{2.1}$$

where $n$ is an integer, $\lambda$ is the X-ray wavelength and $\theta$ is the angle between the beam direction and the atomic planes in the crystal. Diffraction studies based on this equation have revealed a great deal about the structure of a variety of materials.

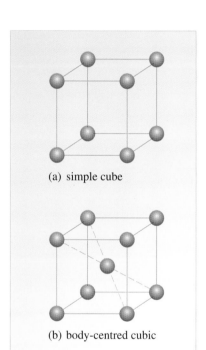

(a) simple cube

(b) body-centred cubic

**Figure 2.4** (a) The simple cubic structure and, (b) the body-centred cubic structure. Each sphere in this figure represents an atom but the size of the spheres has been chosen to give an easily visualized picture of the structure rather than a realistic indication of atomic size. Examples of metals forming body-centred cubic structures are: iron (Fe), tungsten (W) and the alkali metals (Li, Na, K, Rb and Cs).

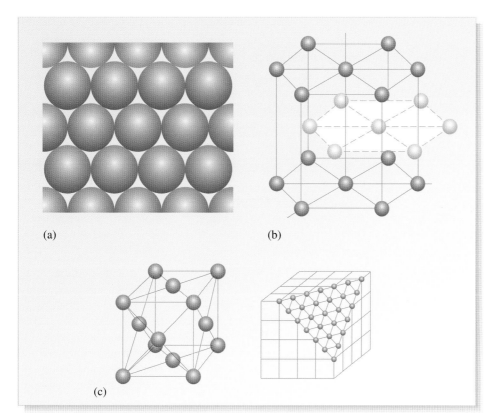

(a)

(b)

(c)

**Figure 2.5** The two close-packed structures, both based on hexagonal layers. (a) A hexagonal layer of spheres. (b) The *hexagonal close-packed* structure. Examples of metals having this structure are: magnesium (Mg), zinc (Zn) and nickel (Ni). (c) The face-centred cubic structure. Shown also is a diagonal slice across a large sample of a face-centred cubic structure showing a hexagonal layer. Examples of metals having this structure are: copper (Cu), silver (Ag) and gold (Au). As in Figure 2.4, the sizes of the spheres in (b) and (c) have been chosen to give an easily visualized picture.

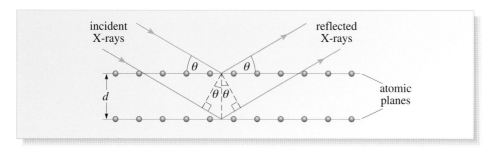

**Figure 2.6** A Bragg reflection occurs when the path difference, $2d \sin \theta$, between beams reflected from adjacent planes of atoms is equal to an integer number of wavelengths, $n\lambda$.

**Question 2.2** A monochromatic beam of X-rays of wavelength 0.5520 nm is used to study a crystal with a set of atomic planes spaced 0.4250 nm apart. Use Bragg's law to calculate the angles between the X-ray beam and the crystal planes at which you would expect to see reflections. ∎

## Ionic solids

You have seen that the ionic bond in sodium chloride $Na^+Cl^-$ is formed by the electrostatic attraction between the positively charged $Na^+$ ion and the negatively charged $Cl^-$ ion. At normal temperatures, bonding between different sodium chloride molecules occurs via electrostatic attraction between their positive and negative ends. Each molecule can become surrounded by other molecules to form a regular bound crystal structure containing huge numbers of $Na^+Cl^-$ units. This is the origin of an **ionic solid** such as common salt, NaCl. Ionic bonds, unlike covalent bonds, are not directional and do not saturate, and so the ions can pack together as closely as possible, the only limitation being one of space. A sodium chloride crystal consists of $Cl^-$ ions arranged in the face-centred cubic lattice, with the smaller $Na^+$ ions fitting in the gaps, as shown in Figure 2.7a. You can also think of this structure as consisting of two similar interpenetrating face-centred cubic lattices, as illustrated in Figure 2.7b.

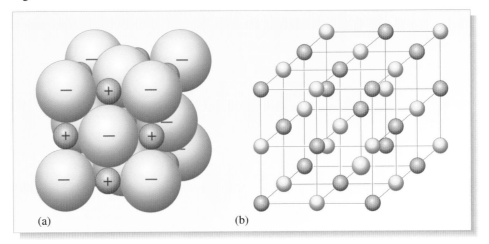

(a)                                (b)

**Figure 2.7**   (a) The sodium chloride (common salt) crystal structure consists of a face-centred cubic lattice of $Cl^-$ ions (blue) with the smaller $Na^+$ ions (red) fitting in the gaps. The relative sizes of the spheres are chosen to be representative of the actual sizes of the ions. They are effectively in contact in the solid. (b) An alternative view of the sodium chloride structure as two interpenetrating face-centred cubic lattices, one of $Cl^-$ ions and one of $Na^+$ ions. In this case, the sizes of the spheres are chosen to give an easy visualization of the lattice structure.

The ionic bonding is weak or strong depending on the separation distance $d$ of any adjacent pair of ions. This length is equal to the sum of the effective radii of the two ions. An alkali metal ion, such as $Na^+$, is very small, so the $d$ value is small for sodium chloride, giving strong bonding.

**Question 2.3**   (a) What is the electrostatic potential energy $E_{pot}$ for a positive ion with a charge $e$ and a negative ion with a charge $-e$, if the two ions are separated by a fixed distance $d = 0.282$ nm (the distance between an $Na^+$ ion and a $Cl^-$ ion in sodium chloride)? Give your answer in J and in eV. (b) Show that, in general, the product $d \times E_{pot}$ is a constant, independent of the value of $d$.   ■

We can take the magnitude of the electrostatic potential energy $E_{pot}$ that you calculated in Question 2.3, as an indication of the ionic bond strength. How does this compare with observation? We can regard the melting temperature of a solid $T_m$ as an easily measurable indication of bond strength, since the stronger the bond the higher the temperature needed to disrupt it. Thus, in view of the result in part (b) of

Question 2.3, we would expect to find $d \times T_m$ to be a constant, independent of spacing $d$. In fact this relationship is illustrated fairly well for typical ionic solids formed by sodium and the Group VII elements, as listed in Table 2.1. Thus we conclude that for a given type of ionic compound, the bond strength is inversely proportional to the separation $d$ of adjacent ions.

All the ionic solids in Table 2.1 have the interpenetrating face-centred cubic structure and are directly comparable. We could easily extend the type of calculation of Question 2.3a to find the total electrostatic potential energy of all the ion pairs in the crystal. The magnitude of this would give the total *binding energy* of the crystal, i.e. the minimum energy required to separate the constituent ions to a large distance from one another. Other ionic compounds may have a different crystal structure, but because the bonding is via the well understood electrostatic force between ions, it is a simple matter to calculate the binding energies. Thus our understanding of ionic solids and ability to predict their behaviour are relatively great.

**Table 2.1** The ionic spacing ($d$) and melting temperature ($T_m$) of some ionic compounds with the sodium chloride structure.

| Ionic solid | $d$ = Spacing/nm | $T_m$/K | $d \times T_m$/nm K |
|---|---|---|---|
| NaF | 0.232 | 1260 | 292 |
| NaCl | 0.282 | 1075 | 303 |
| NaBr | 0.299 | 1015 | 303 |
| NaI | 0.324 | 935 | 303 |

## Covalent solids

For covalent bonding, the initial electron pairing within the molecules is usually re-arranged during condensation into the solid crystalline form. This is especially likely if the bonding is unsaturated, i.e. if there are double or triple bonds in the molecule. In effect, covalently bonded molecules become larger by the addition of more and more atoms, all similarly bonded together via pairing of electrons, until a macroscopic **covalent solid** forms. However, the number of neighbours is limited by the directionality and saturation of the covalent bonds, and so covalent solids tend to have a low-density open structure unlike close-packed ionic solids. A good example is diamond, a crystalline form of carbon. You have seen that each carbon atom can make 4 covalent bonds, and this can give a tetrahedral structure (as in methane, Figure 2.3c). In diamond, each carbon atom is bonded with four others in a tetrahedral structure, as illustrated in Figure 2.8a. Actually, this structure is equivalent to two interpenetrating face-centred cubic structures, as is shown in Figure 2.8b. The other Group IV elements germanium and silicon also crystallize in the diamond structure, and so does grey tin.

Table 2.2 shows the products $d \times T_m$ for some covalent solids with diamond-type structures. You can see immediately that in contrast to the case of ionic solids (Table 2.1) these values vary considerably. We find approximately that $T_m \propto 1/d^3$ rather than $T_m \propto 1/d$. This indicates that the covalent bond strength depends much more strongly on distance; it gets very rapidly weaker as $d$ increases. In fact, the strength differences of the four covalent solids listed in Table 2.2 are vast, from diamond, the strongest of all materials, to grey tin, a very weak metal.

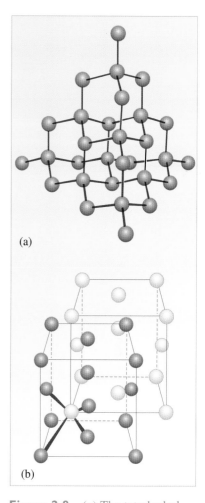

(a)

(b)

**Figure 2.8** (a) The tetrahedral diamond structure. (b) This is equivalent to two interpenetrating face-centred cubic lattices. (The sizes of the spheres are chosen to give an easy visualization of the lattice structure.)

**Table 2.2** The atomic spacing ($d$) and melting temperature ($T_m$) of some covalent materials with the diamond structure. (Sn is the chemical symbol for tin.)

| Material | $d$ = Spacing/nm | $T_m$/K | $d \times T_m$/nm K |
|---|---|---|---|
| C (diamond) | 0.154 | 4100 | 631 |
| Si | 0.235 | 1685 | 396 |
| Ge | 0.245 | 1230 | 301 |
| Sn (grey) | 0.281 | 505 | 142 |

The sodium chloride and diamond-type structures we have considered in the above paragraphs represent the extremes of ionic and covalent behaviour. Many solids have mixed bonding, i.e. partly covalent and partly ionic. There is also a very wide range of crystal structures. While we have no room here to discuss the many common crystal structures, it is worth noting that most of them, like sodium chloride and diamond, can be understood in terms of interpenetrating forms of the simple structures shown in Figures 2.4 and 2.5. In general though, it is difficult to calculate the binding energies of mixed-bond crystals and so their properties are much less well understood than those of ionic solids.

### Metallic solids

You have seen that in covalent materials the individual atoms share their valence electrons, and in ionic materials electrons are simply transferred to the strongly electronegative atoms. In metals by contrast, the valence electrons escape completely from their parent atoms when the solid forms, and are confined only by the surface of the metal itself. You have already studied this electron gas in Chapter 1 of this book, and we shall develop that model further in the next section.

The bonding in a **metallic solid** can be partly understood using Equation 1.6, which shows that the kinetic energy of a particle confined in a cubical box of length $L$ is inversely proportional to $L^2$. When bound to an atom, a valence electron is essentially confined to a box of atomic size, but when it is dissociated from the atom and is free to roam throughout the metal, it is effectively in a very much larger box determined by the size of the metal sample. Thus, according to Equation 1.6, the average kinetic energy of the valence electrons is decreased when they become free electrons. While this lowering of the energy is to some extent offset by the requirements of the Pauli exclusion principle, there is an overall lowering of the total energy, which binds the metal together. There are also contributions from covalent and van der Waals bonding (described below). There is no directionality in metallic bonding and so most metals crystallize in the body-centred cubic structure or in one of the two close-packed structures, as indicated in the captions of Figures 2.4 and 2.5.

Because metallic bonding via the free electrons is distributed throughout the metal rather than localized between individual atoms, the nature of individual lattice ions is of relatively little significance. As a result, a metal can accommodate much larger quantities of impurity atoms (of both metallic and non-metallic elements) than any other kind of crystalline solid. This allows a metal to be modified in many ways, a trait exploited by traditional metal craftsmen (such as blacksmiths) and in the more recent scientific techniques devised by metallurgists.

The majority of the elements solidify as metals. We tend to think of metals as strong and relatively hard materials, for that is the form in which we usually encounter them. In fact, there is a huge range of metals with very different properties. Mercury is liquid at room temperature, but tungsten melts at about 3400 °C and tungsten light bulb filaments can be heated white hot without softening.

## Molecular solids

When none of the mechanisms for ionic, covalent or metallic solidification is effective, molecules can still condense into solids thanks to a much weaker type of bonding originally proposed in 1875 by Johannes van der Waals (1837–1923) to explain departures from the ideal gas law. It was not until 1930 that this was applied to the formation of solids in a detailed theory by Fritz London (1900–1954). London showed that when two atoms or molecules are sufficiently close to one another, the electric charges in one (i.e. the nuclei and electrons) perturb those in the other, in such a way that a very weak short-range attractive force exists between them. These forces are called **van der Waals forces** (or London forces) and they give rise to a very weak type of bonding between all atoms, molecules and ions. However, they are only of prime importance when the other forms of bonding are absent, and make only a very small contribution to the bonding in mainly ionic or covalent materials.

The simplest van der Waals materials are the solids formed at low temperatures by the inert gases, such as neon and argon. Despite their lack of practical significance as materials, the inert-gas solids have been extensively studied as prime examples of van der Waals bonding.

Many molecular gases, such as ammonia and chlorine, also solidify by van der Waals bonding of molecules into a molecular solid. Thus diatomic chlorine molecules $Cl_2$, in which the two chlorine atoms are covalently bonded, form a solid with each molecule weakly bound via the van der Waals interaction with its neighbours. Solid chlorine melts at 172 K and boils at 238 K, releasing $Cl_2$ molecules.

Graphite crystals (another form of carbon) are formed in a similar way. Strong covalent bonds rigidly join carbon atoms into a hexagonal array of plane sheets, with only weak van der Waals bonds holding the sheets together (Figure 2.9), making graphite crystals easy to shear parallel to the layers. (This was once thought to explain the very important lubricating properties of graphite, that allow its use in pencils and lubricating greases, but these properties are now known to be an effect of oxygen and water impurities between the hexagonal layers.)

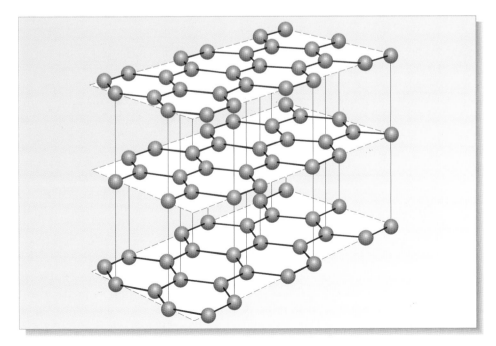

**Figure 2.9** The graphite structure. The carbon atoms in each hexagonal layer are held by strong covalent bonds and the layers are weakly bound by van der Waals bonds.

The physical behaviour of solid *polymers* (such as plastics) and biological materials (such as skin and muscle) is largely determined by the van der Waals bonding between the molecules. The large number of bonds between their large molecules ensures that the material is solid to quite large temperatures, while the weakness of the individual links between the molecules allows these solids to be flexible and elastic and permits them to be transformed in various ways at relatively low temperatures. Thus we can mould plastics and cook meat!

The **molecular solids** we have considered so far are weakly bonded by van der Waals forces. A stronger bonding effect between molecules, but still much weaker than ionic or covalent bonding, can occur when hydrogen forms a covalently bonded molecule with one or more electronegative atoms, such as oxygen. In these molecules the bonding electrons are pulled away slightly from the hydrogen. This results in a small net positive charge at the hydrogen end of the molecule and a negative charge at the other end. This results in so-called **hydrogen bonding** which allows the molecule to link its positive hydrogen end to a negative end of a nearby molecule. Large numbers of molecules can be linked together in this way to form a hydrogen-bonded molecular solid.

> The strongly electronegative elements are those near the top right-hand region of the Periodic Table (but excluding the noble gases).

Hydrogen bonding is strong in $H_2O$. Water is such a familiar liquid that we take its properties for granted. However, thanks to hydrogen bonding, it is a quite remarkable material. If water molecules were linked by van der Waals bonding alone, water would be expected to boil at about 200 K ($-73\,°C$). The linkage between molecules due to hydrogen bonding raises the boiling point and allows water to be such a conspicuous feature of our planet. Many of the peculiar properties of ice, the solid form of $H_2O$, are also a result of this kind of bonding. Hydrogen bonding is vital also in creating the solubility of ionic compounds in water, and in linking the DNA strands which carry our genetic codes. Without hydrogen bonding there would be no organic life on Earth, and there would be neither physics nor anybody to study it!

## 2.3 Imperfections and mechanical strength

Real crystals are not perfect. They generally contain various **defects** in their regular structure, such as chemical impurities, polycrystal boundaries, vacancies (i.e. missing atoms) and *dislocations* (i.e. missing rows of atoms). These defects can have a marked effect on the physical properties of the material.

An important mechanical property of a material is its strength, which we can measure by stressing it harder and harder until it breaks. There are of course different types of 'strength' depending on how the material is stressed: by tension, compression or shear. Generally it is found that the strength of a sample is much less than the theoretical value that can be calculated for a pure ideal crystal. The strength of a real sample is very dependent on its shape, the nature of its surface and its impurity content.

Strength and hardness go together, and are usually connected with a material being brittle. These properties may be contrasted with weakness (low strength), softness and ductility. Crudely speaking, a brittle material breaks before it bends. Hence the difference between the effect of a blow on a glass bottle and a metal can. Metals are usually **ductile**, i.e. they can be hammered or bent into a new shape and will flow a bit if the stress applied is great enough.

Although the study of materials began in the Stone Age, until about 80 years ago the behaviour of a sample under stress was in essence a mystery. Selection, preparation and processing of metals and other materials were a mixture of custom, cookery and witchcraft! Two ideas changed all that: firstly, the explanation of why brittle substances such as glass can fracture quite suddenly and without warning when stressed, and secondly, the discovery of *dislocations*.

The explanation of **brittle fracture** came when A. A. Griffith (in 1920) considered the energy changes inside materials during fracture. This led him to point out that, in contrast to the perfect crystal samples considered in theoretical studies, where the stress is shared all through the volume, real samples contain invisible tiny cracks, either within the bulk or on the surface, which act as a source of weakness. Most of the stress is concentrated right at the end of a crack, tending to open it more widely. If the crack is large enough, quite small stress can cause the crack to suddenly and catastrophically extend right across the sample, causing brittle failure. This is the trick exploited by glass-cutters, when they put a fine scratch along a glass sheet and then break it along the crack that develops under a slight load. The method works only with a new glass sheet, free from accidental surface damage, and requires a single continuous scratch on the surface, not the deep multiple groove attempted by amateur glass-cutters! Any sharp corners or holes in load-bearing components cause *stress concentration* and today this is carefully taken into account by mechanical engineers. It was not always so, and many engineering failures were due to neglect of the effects of stress concentration. Examples are the collapse in storms of Scotland's first Tay Bridge and (much more recently) of a huge North Sea Platform, with great loss of life in each case.

The second crucial idea in understanding strength is dislocation, which was suggested (in 1934) as the cause of ductility for metals. **Dislocations** are incomplete sheets of atoms ending at a line, as illustrated in Figure 2.10a. When a material contains dislocations, the ductile flow accompanying a deformation can pass gradually through the material, with the extra stress concentrated at the dislocation 'unstitching' the bonding one plane at a time. By contrast, to go directly from state A to state D in Figure 2.10b would require all the bonds in the whole section of the solid to be broken at once. (This is a bit like the difference in the strength needed to open a zip fastener by moving the slide in the usual way versus that for tearing the whole zip apart all at once by brute force.)

There is a fundamental difference between dislocations in covalently bonded solids and in other cases. The non-directional bonding in metallic and ionic crystals allows a dislocation to spread over wide regions of the lattice, leaving relatively little lattice distortion on each plane. A modest applied stress can cause a significant local change in the stress pattern, and prompts the dislocation to move. In contrast, the directionality of the bonds in covalent crystals causes the dislocations to be very narrow, with considerable local distortion. Any further applied stress has relatively little extra effect in the covalent case and dislocations are not so mobile. Thus in general, covalent materials are brittle, and the other materials are ductile and have a strength which is only a very tiny fraction of the theoretical value for a perfect crystal.

It was not until 20 years after their discovery that dislocations were directly seen, but with modern microscopy they are easily detected and their behaviour has been studied in great detail. Metals can be hardened and strengthened by precipitating out lots of little crystals containing impurities to trap the dislocations and prevent their movement within the pure host metal crystal.

Thin fibres can be manufactured that are essentially free of dislocations, and display almost all the possible strength of the solid crystal. Hence the use of carbon fibre composite materials for exceptional strength combined with exceptional lightness (and also exceptional cost!). Much of mechanical engineering depends upon the control of stress in materials. We do not have room in this book to pursue this important topic any further. Instead we now turn to a study of the thermal and electrical properties of solids, especially metals.

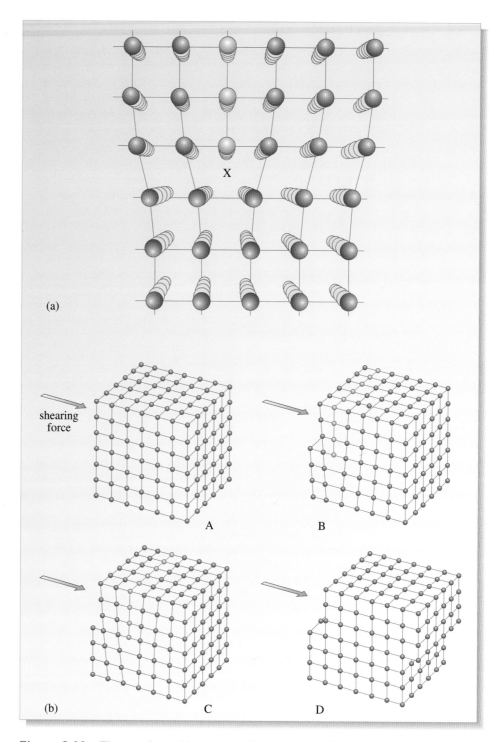

**Figure 2.10** The atomic positions near a dislocation. (a) The line of the dislocation runs perpendicular to the page at X. (b) Shear deformation of a small block of atoms, giving permanent plastic flow by the passage of a dislocation across the block. Starting with the perfect crystal A, the applied shearing force causes the dislocation seen in B to move through the crystal, as shown in C and D.

# 3 Free-electron models of conduction in solids

In Chapter 1 we introduced Drude's classical model of the free-electron gas in metals and showed how this was later replaced by Pauli's quantum-mechanical model. In this section we develop Drude's model a little further to see what it predicts for the thermal and electrical conductivities of metals. We then do the same for Pauli's quantum model. As you might expect, we find further failures of Drude's classical model and some more successes for Pauli's model. However, we shall find that Pauli's model too has its failures. These failures of Pauli's model set the stage for the band theory of solids in Section 4.

## 3.1 Survey of electrical and thermal conductivities

The relative abilities of different materials to conduct heat and electricity are indicated by the values of their *thermal conductivity* $\kappa$ and the *electrical conductivity* $\sigma$. The **electrical conductivity** is just the reciprocal of the resistivity $\rho$ of the material,

$$\sigma = \frac{1}{\rho}. \tag{2.2}$$

The *resistivity* of a material was introduced in *Static fields and potentials* (SFP). Resistivity has units $\Omega\,\text{m}$, and its value is numerically equal to the resistance between the plane faces of a cylindrical slab of the material of length 1 m and cross-sectional area $1\,\text{m}^2$ (Figure 2.11a). The units of the electrical conductivity $\sigma$ are seen from Equation 2.2 to be the reciprocal of the units of $\rho$, i.e. $\sigma$ is measured in $\Omega^{-1}\,\text{m}^{-1}$. The electrical conductivity $\sigma$ is numerically equal to the current that would flow across the above slab when a potential difference of 1 volt is maintained across its plane ends. This is illustrated in Figure 2.11b.

The electrical conductivity $\sigma$ (and the resistivity $\rho$) are characteristic of the material but depend on temperature. You'll find that it is sometimes convenient to work with the resistivity $\rho$, and sometimes with the electrical conductivity $\sigma$ — just remember that one is the reciprocal of the other.

Values of electrical conductivity $\sigma$ range from nearly $10^{8}\,\Omega^{-1}\,\text{m}^{-1}$ for the metals silver and copper, to very low values of about $10^{-20}\,\Omega^{-1}\,\text{m}^{-1}$ for insulators such as plastics. The ratio of maximum to minimum here is an astonishing 28 orders of magnitude, showing that solids differ far more widely in electrical conductivity than in any other property. The ratio of about a million between the mechanical strengths of steel and chewing gum is negligible by comparison. We can see in Figure 2.12 that the great majority of elements are metals with high electrical conductivities. Most chemical compounds are insulators.

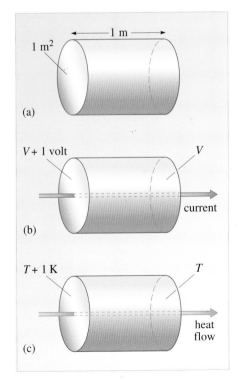

**Figure 2.11** (a) A cylindrical slab of material of cross-sectional area $1\,\text{m}^2$ and length 1 m. The electrical resistance $R$ from one plane face to the opposite plane face is numerically equal to the resistivity $\rho$ of the material from which it is made. (b) The current that flows through the slab when a potential difference of 1 volt is applied across its plane faces is numerically equal to the electrical conductivity $\sigma$ of the material. (c) The rate at which heat flows through the slab when there is a temperature difference of 1 K across its plane faces is numerically equal to the thermal conductivity $\kappa$ of the material. (The shape of the cross-section, taken to be circular for convenience, is arbitrary.)

71

**Figure 2.12** Electrical conductivities $\sigma$ at room temperature of solid elements with atomic numbers up to 54. The individual conductivities increase from the bottom of each column to the top. Thus silver (Ag) is the best elemental conductor at room temperature. Carbon appears both as diamond, an insulator, and as graphite which conducts almost as well as the metals. Insulators appear on the far left, and semiconductors in the middle of the conductivity range.

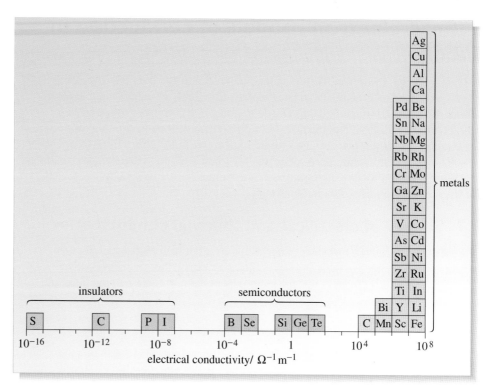

Although consistently high, the electrical conductivities for different metals at a given temperature vary by over an order of magnitude. This is shown for a wide range of metals in the second column of Table 2.3.

The ability of a solid to conduct heat is determined by its **thermal conductivity** $\kappa$ (units $W\,m^{-1}\,K^{-1}$). $\kappa$ is numerically equal to the rate at which heat would flow through a slab of the material of length 1 m and cross-sectional area 1 m$^2$, when the temperature difference across its ends is 1 K (Figure 2.11c). All materials give some modest heat conduction and so the range of thermal conductivities of solids is not as great as for electrical conductivities. However, if we restrict ourselves to metals we find that the $\kappa$ values vary from one metal to another in a similar way to the $\sigma$ values, a fact that we remarked upon in Chapter 1. This is illustrated for a selection of metals at room temperature in the fourth column of Table 2.3 where values of $\kappa/\sigma$ are seen to be approximately constant.

**Table 2.3**   The comparative conductivities of some metals at 293 K.

| Metal | $\sigma/10^7\,\Omega^{-1}\,m^{-1}$ | $\kappa/W\,m^{-1}\,K^{-1}$ | $(\kappa/\sigma)/10^{-6}\,W\,\Omega\,K^{-1}$ |
|---|---|---|---|
| silver | 6.21 | 429 | 6.9 |
| copper | 5.88 | 401 | 6.8 |
| aluminium | 3.65 | 237 | 6.5 |
| sodium | 2.11 | 141 | 6.9 |
| zinc | 1.69 | 116 | 6.9 |
| cadmium | 1.38 | 97 | 7.0 |
| iron | 1.02 | 80 | 7.8 |
| lead | 0.48 | 35 | 7.3 |

It is also found that whereas the thermal conductivities of metals are almost independent of temperature, electrical conductivities vary markedly with temperature. Figure 2.13 shows the temperature variation of the resistivity $\rho$ of copper. You can see that, above about 100 K the resistivity increases almost linearly with the absolute temperature.

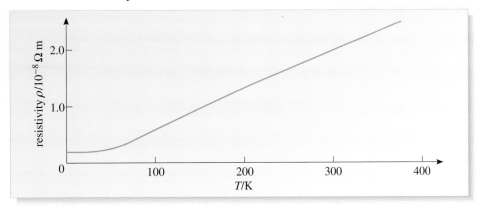

**Figure 2.13** The resistivity of copper as a function of absolute temperature.

Any theory of electrical conductivity of metals would need to explain this temperature variation as well as the actual values of conductivities such as those given in Table 2.3.

## 3.2 Free-electron models of electrical conduction

Here we develop Drude's classical free-electron model a little further, showing how it predicts Ohm's law for electrical conduction in a metal, and gives an expression for the resistance of a wire and hence the electrical conductivity $\sigma$ of the material. We shall then see how Pauli's quantum free-electron model improves on this but has troubles of its own.

Drude's model assumes that the valence electrons of the metal atoms are free to move throughout the volume of the metal but may collide with the fixed positive ions in the crystal lattice. You will recall from Chapter 1 that Drude treated these electrons as an ideal gas with a distribution of kinetic energies given by the Maxwell–Boltzmann distribution, with average thermal energy $\langle E \rangle = 3kT/2$. Because of the very small electron mass, the average speed of the electrons, $\langle v \rangle = (8kT/\pi m)^{1/2} \approx 10^5$ m s$^{-1}$ at room temperature, is two orders of magnitude larger than the average speed of gas molecules at the same temperature.

It is assumed for simplicity that when there are no external electric or magnetic fields applied to the metal, each electron moves along a straight path at constant speed between collisions with the fixed positive lattice ions. (We can neglect the effects of gravity.) You can picture the electrons rattling around at thermal speeds colliding with the fixed ions like balls on a pin-ball table, with each collision randomly changing the direction of motion. The resulting path of any electron over a long period of time, covering very many collisions, is therefore a random sequence of straight segments as shown in Figure 2.14a. It is important to note that during this random thermal motion, the average velocity component in any direction, $\langle v_x \rangle$ say, is zero. This is because electrons are equally likely to be moving in the positive or negative $x$-directions. For the same reason the average displacement of the electron along any direction during any long period of time is also zero.

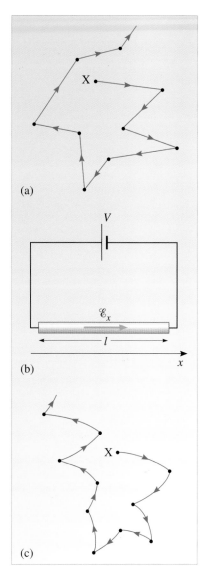

### The flow of electric current

How do the electron velocities change when an external electric field $\mathcal{E}_x$ is applied to the metal in the positive $x$-direction due to, say, a battery being connected across its ends, as in Figure 2.14b?

We can work this out using Newtonian mechanics since Drude's model is a classical one. The electric field exerts a force $F_x = -e\mathcal{E}_x$ on each electron. By Newton's second law, the electron will accelerate in the $x$-direction with acceleration

$$a_x = F_x/m_e = -e\mathcal{E}_x/m_e$$

where $m_e$ is the electron's mass.

How does this acceleration change the electron's velocity?

The electron's velocity components in the $y$- and $z$-directions will not change, but during any small interval $\Delta t$ between collisions the $x$-component of the velocity will change by an amount

$$\Delta v_x = a_x\Delta t = \left(\frac{-e\mathcal{E}_x}{m_e}\right)\Delta t.$$

Now the average distance that electrons travel between collisions is called the **mean free path** $\lambda$, and so the typical time between collisions is $\lambda/\langle v\rangle$, where $\langle v\rangle = (8kT/\pi m)^{1/2}$ is the average electron speed (see Chapter 1), and the typical velocity change between collisions is therefore

$$(\Delta v_x)_{\text{coll}} = \left(\frac{-e\mathcal{E}_x}{m_e}\right)\left(\frac{\lambda}{\langle v\rangle}\right).$$

You might think that the successive accelerations of the electrons between collisions would continually increase their speed and hence their kinetic energy. In fact, the *extra* kinetic energy given to the electrons by the electric field is lost at each collision and converted into thermal vibrational motion of the crystal lattice as a whole. Thus the lattice heats up and the electrons start each acceleration period afresh with their original thermal speed only. The net effect on the electron gas of the successive periods of acceleration between collisions is to give it an average 'drift' in the negative $x$-direction. The average drift velocity can be calculated using the methods of classical statistical mechanics. The result is in fact equal to the expression for the typical velocity change between collisions given above. Thus the drift is characterized by the **drift speed** $v_d$ given by

$$v_d = \left(\frac{e\mathcal{E}_x}{m_e}\right)\left(\frac{\lambda}{\langle v\rangle}\right). \tag{2.3}$$

This drift of the electron gas is superimposed on the rapid random thermal motion of the electron gas described previously, so that the electrons are found to move along slightly curved paths, as illustrated in Figure 2.14c.

The drift of the electron gas in the negative $x$-direction is the origin of the electric current in the positive $x$-direction produced by the battery, and the extra energy given up to the vibrational motion of the crystal lattice at each collision is the origin of the heating effect of the current.

### Drude's theory of electrical conductivity

We can now obtain an expression for the electrical conductivity $\sigma$ and compare it with measured values.

**Figure 2.14**  (a) In the absence of an applied electric field, the electrons in a metal move in straight-line paths at constant speed between collisions. These collisions occur at points indicated by dots. Starting at X, an electron bounces around randomly with no overall tendency to drift in any particular direction. (b) A metal wire of length $l$ connected to a battery of voltage $V$ provides an electric field $\mathcal{E}_x = V/l$ inside the metal. (c) Motion of electrons in a metal with the applied electric field $\mathcal{E}_x$. The curvature of the paths during the intervals of accelerated motion between collisions gives the electrons a tendency to drift to the left.

Consider a straight piece of wire of length $l$ and cross-sectional area $A$. See Figure 2.15. You have seen that when a battery is connected across a wire to produce an electric field $\mathcal{E}_x$ in the positive $x$-direction, the electrons will drift in the negative $x$-direction with drift speed $v_d$. This drift of electrons constitutes an electric current in the positive $x$-direction of magnitude

$$i = nev_dA \qquad (2.4)$$

where $n$ is the number density of valence electrons. This result was derived in Section 3.1 of *Static fields and potentials* and is briefly reviewed in the caption to Figure 2.15.

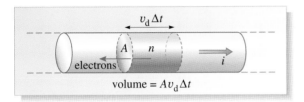

**Figure 2.15** The number of electrons that flow through a cross-section of wire in a time interval $\Delta t$ is equal to the electron density $n$ times the volume swept out, $n \times Av_d\Delta t$. Hence the magnitude of the total charge passing the cross-section in time $\Delta t$ is $|\Delta Q| = enAv_d\Delta t$, and the current flowing along the wire is $i = |\Delta Q|/\Delta t = neAv_d$.

Using Equation 2.3 for the drift speed, we obtain

$$i = \frac{ne^2\mathcal{E}_x\lambda A}{m_e\langle v\rangle}.$$

Now we make the connection with Ohm's law by putting $\mathcal{E}_x = V/l$, and so

$$i = \left(\frac{ne^2\lambda A}{m_e l\langle v\rangle}\right)V. \qquad (2.5)$$

If you compare this with Ohm's law, $i = V/R$, you can see that the resistance of the wire is

$$R = \left(\frac{m_e l\langle v\rangle}{ne^2\lambda A}\right) = \left(\frac{m_e\langle v\rangle}{ne^2\lambda}\right)\frac{l}{A}$$

where, for convenience, we have taken the factor $l/A$ outside the brackets. Now you know from *SFP* that the resistance $R$ of a wire is related to the resistivity $\rho$ of the material by

$$R = \rho\frac{l}{A}.$$

Hence, comparing the above two equations, we have the resistivity

$$\rho = \frac{m_e\langle v\rangle}{ne^2\lambda} \qquad (2.6)$$

and by using Equation 2.2, the electrical conductivity is

$$\sigma = \frac{1}{\rho} = \frac{ne^2\lambda}{m_e\langle v\rangle}. \qquad (2.7)$$

Thus we see that Drude's classical theory of the electron gas correctly predicts that the current in a metal obeys Ohm's law. It also gives an expression for the electrical resistivity $\rho$ and conductivity $\sigma$ that can be compared with experimental values.

## Comparison with experiment

The problem with using Equation 2.7 to predict a value for the conductivity, is knowing what value to substitute for $\lambda$, the mean free path. What we can do is to work backwards using the measured values of $\sigma$ to see what Equation 2.7 predicts for $\lambda$. Rearranging Equation 2.7 gives

$$\lambda = \frac{\sigma m_e \langle v \rangle}{ne^2}.$$

Using the values for copper on the right-hand side, put $\sigma = 5.88 \times 10^7\,\Omega^{-1}\,m^{-1}$ (see Table 2.3), $\langle v \rangle = (8kT/\pi m)^{1/2} = 1.08 \times 10^5\,m\,s^{-1}$ at room temperature, and $n = 8.45 \times 10^{28}\,m^{-3}$ (see Table 1.3). This gives

$$\lambda = 2.6 \times 10^{-9}\,m = 2.6\,nm.$$

This mean free path should seem suspiciously long to you. Copper has a close-packed structure of ions with a separation between ions of only about 0.3 nm. Our estimate of $\lambda$ is about ten times this length. It seems unlikely that an electron could travel through ten ions before suffering a collision.

There are more problems when we look at how the conductivities of metals depend on temperature and impurity content. You have seen from Figure 2.13 that the resistivity ($\rho = 1/\sigma$) of copper above about 100 K is found to increase *linearly* with the temperature $T$. If you look at Equation 2.6 you will see that the only factor in Drude's expression for $\rho$ that can depend on temperature is the average speed of electrons $\langle v \rangle = (8kT/\pi m)^{1/2}$, and this varies as $T^{1/2}$. Thus Drude's theory predicts wrongly that the resistivity $\rho$ should increase as the *square root* of the temperature $T$, a relatively slow variation.

These problems are even worse when the low-temperature resistivities of very pure forms of metals are considered. The resistivity of very pure copper, for example, decreases by a factor of 1000 or more when it is cooled to very low temperatures. These extremely low resistivities imply mean free paths greater than $10^{-6}\,m$, about 400 times greater than the spacing between ions.

We conclude that, although Drude's theory can describe the flow of electricity through metals in accordance with Ohm's law, it cannot explain the very long mean free paths of electrons or the measured temperature variation of electrical conductivities.

Furthermore, it is found that the very small low-temperature resistivities found for very pure metals are increased dramatically for samples containing quite small levels of impurities. Drude's model has no explanation at all for why small amounts of impurity should have such an effect.

In Chapter 1 you saw other failures of Drude's model. It was quite unable to explain why the electron gas does not make a significant contribution to the heat capacities of metals. There you saw how Drude's model was replaced by Pauli's quantum model of the electron gas. Pauli's model showed that the electron gas makes almost no contribution to the heat capacities of metals because only a tiny proportion of the electrons, those within a few $kT$ of the Fermi energy $E_F$, can move into higher unoccupied energy levels when heat is supplied. Can Pauli's theory correctly explain the electrical conductivities?

**Question 2.4**  Use Drude's theory, with parameters given above, to determine the drift speed of the electron gas when a current of 1 A flows through a copper wire of diameter 0.1 mm.  ■

## Pauli's theory of electrical conductivity

Recall that Pauli's quantum free-electron model (Chapter 1) takes full account of the exclusion principle by allowing only two electrons (in opposite spin states) to occupy any translational quantum state. This causes the valence electrons in a metal to pile up to very high translational energies. Pauli's distribution of free-electron energies is shown in Figure 2.16. The highest state that is occupied (at $T = 0$ K) is called the Fermi energy $E_F$, which is very much larger than $kT$. At normal temperatures only a tiny proportion of the electrons, those within a few $kT$ of the Fermi energy, are thermally excited into the empty higher states. The rest are trapped in the lower energy states. You saw in Chapter 1 how this explains the tiny contribution made by the free electrons to the heat capacities of metals. We now consider how Pauli's model accounts for the electrical conductivities of metals.

When an electric field in the positive $x$-direction, $\mathcal{E}_x$, is applied to a metal, the majority of the quantum states below the Fermi energy $E_F$ remain occupied and play no effective role in the conductivity. The net effect of the field is to change the distribution of occupied states near $E_F$. These states represent electrons with speeds approximately equal to the **Fermi speed** $v_F$, given by $m_e v_F^2 / 2 = E_F$. Electrons with velocities $v_x \approx -v_F$ are increased in number while those with velocities $v_x \approx v_F$ are decreased in number. This imbalance would increase in time while the electric field is applied were it not for the effects of electron scattering (to be discussed below) which tends to restore the original distribution. The net result is a constant drift of electrons in the negative $x$-direction creating a current in the same direction as the applied field, as in Drude's model.

When the details are worked out, the predicted expression for the electrical conductivity of a metal is

$$\sigma = \frac{ne^2 \lambda_F}{m_e v_F}. \tag{2.8}$$

This result looks similar to Drude's result, Equation 2.7, but there are important differences. The relevant speed is the Fermi speed $v_F$ rather than the average speed $\langle v \rangle$ of the Maxwell–Boltzmann distribution, and $\lambda_F$ is the mean free path of these electrons, i.e. the average distance they travel between scattering events. To fully understand the physical mechanisms behind Equation 2.8 and what makes it so different from Drude's result, Equation 2.7, we need to ask: which electrons can be scattered and what scatters them?

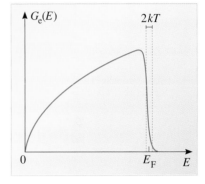

**Figure 2.16** Pauli's distribution $G_e(E)$ for an electron gas in a metal. Typically the Fermi energy $E_F$ is between 5 eV and 12 eV, and at room temperature $kT \approx 0.025$ eV. Only those electrons within a few $kT$ below $E_F$ are thermally excited into higher energy levels.

## Scattering of electron waves

In Pauli's model, the electrons are described by waves that represent the translational quantum states in the metal. Thus the classical notion of electrons undergoing collisions is replaced in the quantum-mechanical picture by electron waves being scattered. Now the Pauli exclusion principle allows electrons to be scattered only into *unoccupied* states, and the only unoccupied states are those near the Fermi energy and above. The energy change in a scattering event is very small, and so only those electrons close to the Fermi level can be scattered into unoccupied states. This is why the mean free path that you see in Equation 2.7 is $\lambda_F$, which characterizes the mean free path of electrons near the Fermi energy. But what scatters the waves?

An important property of waves is their ability to propagate through a perfect periodic structure without scattering. This is one reason why diamonds and other pure crystals are transparent. Light waves (in those cases) can propagate through the pure crystal without being scattered very much. The same is true of electron waves.

Scattering of electron waves can occur by reflections from planes of atoms when Bragg's law ($2d \sin \theta = n\lambda$) is satisfied, but this condition is so restrictive that only a tiny proportion of the electron waves can satisfy it. Pauli's model therefore predicts that the electron waves travel through a *perfect* lattice without any significant amount of scattering.

However, that idealized state of affairs ignores two things. Firstly, the crystal lattice is never perfect. You saw in Section 2.1 that there are crystal defects such as vacancies or impurities. These defects cause scattering of electron waves, and the extent of this kind of scattering is roughly independent of temperature and proportional to the defect density. For very impure alloys, the electrical conductivity is small and approximately constant as $T$ changes, due to this **defect scattering**.

The second thing that we have ignored is the thermal vibrations of the lattice ions. Thermal energy rattles the ions around near their defined sites in the lattice, like unruly children confined to their desks in a schoolroom, but never really sitting still in their places. This random thermal disorder also causes scattering of the electron waves, and the extent of this kind of scattering increases as the temperature rises and the thermal agitation of the lattice increases. This **thermal scattering** is also present in crystals with defects, but is often masked by defect scattering, unless the temperature is very high.

Pauli's model predicts that the resistivity of a metal increases linearly with temperature due to thermal scattering. A detailed derivation is beyond the scope of this course.

The case of copper shown in Figure 2.17a illustrates these effects. At temperatures below about 50 K, the thermal scattering is negligible and the resistivity is determined by defect scattering only. Because the defect density is independent of temperature, the resistivity is constant in this region. At temperatures above about 50 K the thermal agitation of the lattice becomes important and so the resistivity rises linearly with increasing temperature due to thermal scattering. Overall, the observed electrical conductivities of metals can thus be explained.

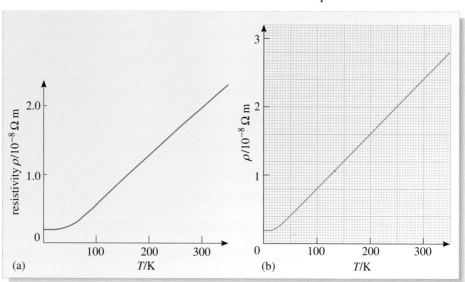

**Figure 2.17**   Resistivity as a function of temperature for a sample of (a) copper and (b) gold.

## 3.3  Free-electron models — successes and failures

You have seen how Pauli's quantum theory of the free-electron gas in metals can explain a variety of electrical properties. Like Drude's classical model, it successfully explains why Ohm's law ($V = iR$) is true for a metal and why a conductor heats up when a current flows. It gives a much better insight than the classical model into the origin of the electron scattering that is responsible for electrical resistance. According to the quantum model, the electrons are not scattered by a perfectly regular lattice of positive ions. The electron waves are scattered only by defects and random thermal disorder. The quantum model, with its vital ingredient of the Pauli exclusion principle, also makes clear the importance of electrons close to the Fermi energy. It is the scattering of these electrons by defects and thermal agitation that is responsible for the electrical resistance of a wire.

Pauli's free-electron model does, however, have its limitations and problems. For example, it cannot explain why some solids are insulators. There is nothing in the model that can explain why the four valence electrons in the carbon atom, for example, are not free to conduct electricity in diamond (an insulator) just as well as the valence electrons are in metals, or for that matter, why the valence electrons should be free in metals.

Another serious problem for any free-electron model is that some materials have anomalous Hall effects. The *Hall effect* was described in Chapter 4 of *SFP*, and is illustrated in Figure 2.18. A magnetic field $B$ is applied across a wafer of conducting material already carrying a current $i$ along its length. The magnetic force, $F_m = qv \times B$, pushes the charge carriers to one side thereby creating a voltage across the wafer at right-angles to the magnetic field. Figure 2.18 shows the magnetic force $F_m$ on electrons ($q = -e$) for the given directions of current flow and magnetic field. The electrons are pushed to the left, and so the left-hand side of the wafer becomes negatively charged and the right-hand side positively charged. Many metals, copper, silver and aluminium for example, show a Hall effect with this polarity. However, there are many other metals, zinc, cadmium and iron, for example, that show a Hall effect with the opposite polarity. To see the significance of this, suppose the current is carried by positively charged carriers ($q = +e$) moving in the positive $x$-direction instead of electrons moving in the negative $x$-direction. This changes the sign of $q$ but also reverses the velocity $v$, so the product $qv$ and hence the magnetic force $F_m$ are in the same direction as before. Thus the positive carriers are pushed to the left in the same direction as the electrons were, but the left-hand side of the wafer now acquires a *positive* charge and causes the Hall effect voltage to have the opposite polarity. The fact that many metals show this anomalous Hall effect suggests that the current in these metals is carried by positively charged carriers. There is no possibility of explaining this effect within a free-electron model.

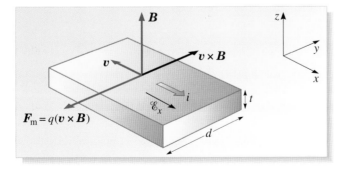

**Figure 2.18** The Hall effect.

In the next section we show how the limitations and problems associated with the free-electron model can be overcome by taking account of the ordered structure of the lattice ions in a crystal.

**Question 2.5**   Figure 2.17b shows the resistivity of a sample of gold as a function of temperature. Use the information in the figure to answer the following: (a) What mechanisms are responsible for the resistivity of the gold sample at (i) 10 K and (ii) 100 K? (b) What is the electrical conductivity of the sample at room temperature (300 K)?   ■

# 4  Band theory of solids

You have seen that Pauli's model assumes that the positive charge associated with the positively charged lattice ions is smoothed out across the whole volume of the metal. This means that the free electrons have a uniform potential energy (taken to be zero) throughout the metal. The electrons have a higher potential energy outside the metal, and so they are confined to a simple flat-bottomed potential energy well which has the size and shape of the metal specimen. This is the basis of the particle in a box model described in Chapter 1.

The quantum treatment of solids was deepened in the 1930s by considering more fully the interaction of the electrons with the lattice ions. Instead of assuming the positive charge of the lattice ions to be smoothly spread throughout the metal, account was taken of the locations of the ions in an ordered crystal structure. It was found that this ordered crystal structure created *gaps* in the distribution of energy levels, where no energy levels were to be found. This means that the energy distribution of electrons in a solid consists of *energy bands* containing the allowed states, separated by energy gaps. The energy band concept unlocked the whole riddle of insulator and semiconductor behaviour, with vast implications for technology, and was one of the major triumphs of twentieth-century physics.

## 4.1  Energy band formation in solids

There are two very different models for deriving energy bands in solids: the *nearly-free electron model* and the *tight-binding model*. The two methods ultimately give the same results and it is desirable that you gain some understanding of each, because they cast light on band structure concepts from different directions.

### Nearly-free electron model

In Pauli's model of the free electrons, the potential energy of an electron in the solid is assumed to be uniform. The **nearly-free electron model** corrects Pauli's model by adding a small spatially-periodic potential energy $E_{pot}$ to describe the effects of the ordered crystal structure of the positive ions. Thus $E_{pot}$ is lower than the average potential energy when the electron is near any positive ion and higher when it is between ions. This gives the bottom of the potential energy well a spatially periodic structure (Figure 2.19).

This additional potential energy $E_{pot}$ is then included in the potential energy term of Schrödinger's equation which is solved to obtain the allowed energies and wavefunctions.

The solutions show that the allowed energy levels are restricted to particular regions of the energy axis, separated by gaps where no energy levels are allowed. We can show these **energy bands** and **energy gaps** on a vertical energy axis $E$, as in Figure 2.20a. The electron dynamics that underlies these solutions of Schrödinger's equation is quite complex and it is difficult, within this model, to give a simple explanation of why energy gaps occur. It is useful to point out however that the gaps occur when the electron wavelength $\lambda$ satisfies the Bragg condition for normal

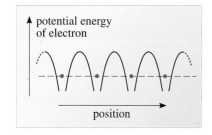

**Figure 2.19**   The bottom of the potential energy well for an electron inside a metal has a periodic structure with maxima of potential energy between the positive ions.

incidence (i.e. Equation 2.1 with $\theta = 90°$). Under these circumstances the electrons interact very strongly with the periodic lattice structure, and the Bragg reflections prevent electrons with energies within the energy gaps from travelling in the crystal. The banded energy level structure predicted by the nearly-free electron model is in marked contrast with the results of Pauli's model where the density of states function $D_e(E)$ is nonzero for all $E$. The absence of gaps in Pauli's model is illustrated in Figure 2.20b.

We shall make considerable use of energy band diagrams like those in Figure 2.20. Although the energy bands are shown as continuous blocks, you should keep in mind the fact that each band consists of a very large number of closely spaced energy levels.

## Tight-binding model

The alternative approach, which gives the same result but also considerable insight into why energy bands and gaps are formed, is the **tight-binding model**. This approach starts by considering the discrete energy levels of an isolated atom, as indicated schematically in Figure 2.21. This energy level diagram is the same for all isolated atoms of the same type. In the tight-binding model we consider what happens to these energy levels when a large number $N$ of atoms are brought together to form a crystalline solid.

A clue is to be found in Figure 2.1 where you saw how the total energy of two ground-state hydrogen atoms changes as the two atoms are brought together to form a hydrogen molecule. As the two hydrogen atoms come closer, their electron wavefunctions begin to overlap and the atoms begin to interact with one another via Coulomb forces between the charges. This interaction causes the 1s ground-state energy level of the isolated atoms to split into two separate energy levels. This splitting gave rise to the bonding and antibonding curves in Figure 2.1a. The splitting of the 1s level is shown Figure 2.22a, while Figure 2.22b indicates what would happen if a large number of hydrogen atoms were brought close together. You can see that the common 1s energy level splits into a band of closely separated levels, the number of individual levels being equal to the number of atoms.

Let's now return to a general unspecified atom whose energy levels are shown in Figure 2.21. This is more complicated than the hydrogen atom case because there is more than one electron per atom involved and some of the energy levels may be

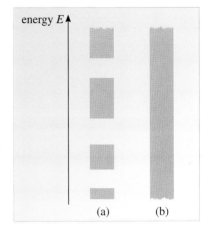

**Figure 2.20** The electron energy levels in a hypothetical solid are plotted on a vertical axis with energy increasing upwards. (a) In the nearly-free electron model the allowed energy levels are restricted to regions called energy bands. Between the energy bands there are energy gaps where no energy levels exist. Each energy band, shown as a grey region, consists in fact of a very large number of closely spaced energy levels. (b) In Pauli's model the distribution of energy levels is continuous with no gaps.

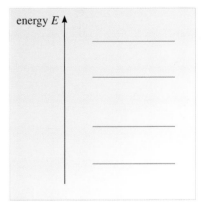

**Figure 2.21** Energy levels of an isolated atom.

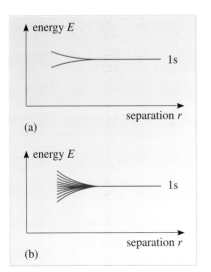

**Figure 2.22** (a) As two hydrogen atoms are brought close together, the common 1s energy level of each atom splits into two closely separated levels. (b) For $N$ hydrogen atoms, the common 1s ground-state level splits into a band of $N$ closely separated levels.

degenerate. Figure 2.23 shows what happens to the energy levels of the isolated atom when a large number of widely separated atoms are brought together to form a solid. On the right of the figure, where the separation $r$ between nearest neighbours is very large, you see the discrete energy levels common to the $N$ isolated atoms. As the atoms are brought closer, each discrete energy level splits into a band consisting of a large number of closely spaced levels. When $N$ is very large, the levels within a band are so crowded together that the band can be regarded as a continuum of levels. In the solid, where the separation of atoms has the equilibrium value, $r = r_0$, the wavefunctions of the high energy outer-shell electrons overlap considerably and each of these energy levels has become a wide band of levels.

The overlapping wavefunctions of the high energy levels extend across the whole solid. Electrons in these levels are no longer bound to particular atoms, but are free to wander from atom to atom, in some ways like the free electrons in Pauli's model. The lower energy levels also broaden out but by very much less. This is because the lower energy levels are those of the inner electrons, and these are more closely bound to the nuclei. These wavefunctions overlap very little and so these low energy electrons remain essentially bound to their particular atoms.

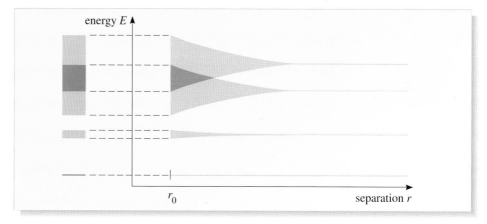

**Figure 2.23** Energy levels of a very large number $N$ of atoms of the same kind plotted against the separation $r$ between nearest neighbours. The solid is formed when $r = r_0$, the equilibrium separation. Each discrete level of the isolated atoms splits into a band of levels.

So what does the formation of energy bands tell us about the electrical properties of solids? You will see in the next section that the answer depends largely on whether the bands of energy levels are filled by the available electrons or only partially filled. This depends on the details of the particular type of atoms being considered, so let's first look at a particular example.

We'll consider sodium, which we know forms a solid metal at room temperature. The formation of sodium metal from $N$ widely dispersed sodium atoms is illustrated schematically in Figure 2.24a. The electronic structure of each isolated sodium atom is $1s^2 2s^2 2p^6 3s^1$. The inner subshells 1s, 2s and 2p of each atom are completely full with two, two and six electrons respectively. They remain largely bound to their respective nuclei and their energies are only slightly spread out into very narrow bands in the final crystalline solid. However, there is considerable overlap of the outer 3s wavefunctions. Consequently the 3s energy level evolves into the broad 3s band consisting of $N$ closely separated levels — this is similar to the splitting of the 1s level in hydrogen, illustrated in Figure 2.22b. Since the 3s subshell can accommodate two electrons per atom (with opposite spin states), the 3s band represents $2N$ quantum states. However, the 3s sublevel in each sodium atom contains only one electron, the single valence electron. Hence the 3s band in the solid contains only $N$ electrons and is therefore only half-full. The energies of the 3p and higher sublevels also develop into bands in the solid, but these bands are empty in sodium.

Figure 2.24b illustrates the conventional way of showing the energy band structure. The widths of the bands shown in Figure 2.24b correspond to the widths of the bands in Figure 2.24a at the equilibrium separation $r_0$. The fact that sodium metal has a band that is only half filled with electrons is extremely significant, as we now explain.

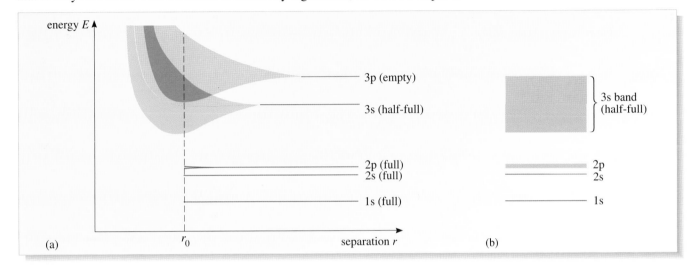

**Figure 2.24** (a) A schematic view of band formation in sodium metal. The stable crystal structure has minimum energy for the atomic spacing $r_0$. The formation of bands is seen, with the greatest width of spreading found for the higher bands, and starting, for them, at a greater separation. (b) The conventional way of showing the band structure of sodium. The blue colour shows bands or parts of bands that are occupied by electrons. At 0 K the $N$ 3s electrons occupy the $N/2$ lowest levels in the 3s band, and so the 3s band is half-full. The 3p band (not shown) is completely empty.

## 4.2 Conductors, insulators and semiconductors

The electrons in the broad 3s band of sodium are no longer localized on specific atoms but are nearly free and are reasonably well described by Pauli's free-electron model. This band is only half filled with electrons and so there are nearby empty states into which electrons can move if thermally or electrically excited, just as in Pauli's model. All the other alkali atoms (lithium Li, potassium K, rubidium Rb, caesium Cs) have a single valence electron in an s subshell and all the alkali solids have a similar half-filling of an s band, and so they are all metals.

Thus the band theory explains why sodium and the other alkali metals are conductors.

In general an electrical conductor is characterized by having a large number of electrons in a partly filled band (Figure 2.25a). But what happens if the highest occupied band is exactly full, as in Figure 2.25b? A material cannot conduct electricity by electrons in a totally *full* band, because there are no nearby empty states into which electrons can transfer in response to an electric field. Nor can we get conduction in a completely *empty* band, because there are no charge carriers at all to provide a current. So if the highest occupied band is exactly full, we have an insulator.

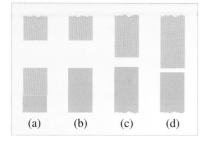

**Figure 2.25** A schematic diagram of energy bands and their occupation for (a) a conductor, (b) an insulator, (c) a semiconductor. Case (d) is also a conductor if the gap is small enough. (The blue colour shows bands or parts of bands that are occupied by electrons.)

However, the size of the gap between the full band and the higher empty one must be considered. If this is not too large compared with $kT$, some electrons will be thermally excited across the gap. These electrons then find themselves in a partially filled band and can give rise to a little conduction. This is the type of material we call a semiconductor (Figure 2.25c). If the gap is very small, like that shown in Figure 2.25d, we again have a conductor, since electrons are easily excited across a very small gap.

At 300 K, a gap of more than about 3 eV between the highest filled state and the first higher empty one will correspond to an insulator. A material with a gap of between about 0.1 eV and 3 eV is a semiconductor, since this gap can be crossed, though rarely. If the gap is less than about 0.1 eV, the gap is easily crossed at 300 K and we have a conductor. However, in general there are no sharp boundaries between conductors, semiconductors and insulators, and in most cases the degree of conduction depends strongly on the temperature. This is indicated in Figure 2.26. A gap between bands of 0.1 eV, for example, gives an insulator at 1 K, a semiconductor at 100 K and a relatively good conductor above 300 K.

**Figure 2.26** The conductivity of a material depends upon the size of the energy gap relative to $kT$. At 300 K a material is an insulator (white) for a gap over about 3 eV and a conductor (black) for a gap under about 0.1 eV. When the gap is between 0.1 eV and 3 eV the material is best described as a semiconductor (grey).

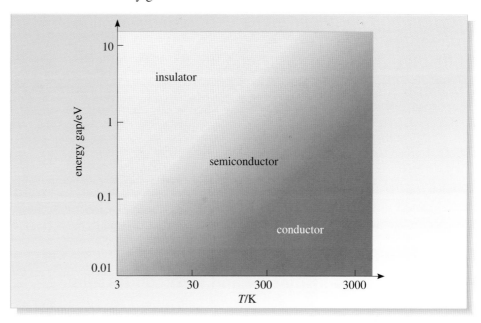

Thus the band theory of solids can explain the existence of electrical insulators and semiconductors, as well as metallic conductors such as sodium. Let's now see how these principles apply to some of the other solid elements in the Periodic Table. (You can find the Periodic Table in Figure 3.2 in the next chapter).

The next element after sodium (Na) in the Periodic Table is magnesium (Mg), with electronic structure $1s^2 2s^2 2p^6 3s^2$. Magnesium has a pair of 3s valence electrons exactly filling that subshell in the atom. Since the 3s band can hold two electrons per atom, this would seem to indicate that the 3s band in magnesium is full and that magnesium is an insulator, but we know it is a metal. What has gone wrong? We have forgotten the 3p band, ignored in our discussion of sodium as it is above the occupied energy range. In magnesium this 3p band actually overlaps the 3s band. The higher bands nearly always spread out so much in the solid that they share energy ranges and overlap with each other, hence the preponderance of conductors in the Periodic Table.

In Figure 2.27 we show the overlapping 3s and 3p bands for the first three elements (Na, Mg, Al) in the third row of the Periodic Table. Although the overlapping of the bands does not affect sodium, it plays a crucial role in magnesium, Figure 2.27b. Some of the 3s valence electrons, that would otherwise fill the 3s band, spill over into the overlapping 3p band, leaving both bands partially full. This makes magnesium a conductor. Similarly the other Group II elements (calcium Ca, strontium Sr, barium Ba) are all conductors because of band overlap.

Now consider aluminium, Figure 2.27c. The electronic structure of aluminium is $1s^2 2s^2 2p^6 3s^2 3p^1$. There are three valence electrons in each aluminium atom, two of them in 3s and one in 3p. In the solid, the 3s band is full with two electrons per atom, but the 3p band is only partially full since it can hold a total of six electrons per atom yet has only one. The partially filled 3p band allows the conduction of electricity and makes aluminium a metal.

**Question 2.6**   Figure 2.28 shows some energy bands and gaps for six materials A to F. The occupied states are shown in blue. Which materials are electrical conductors, which are semiconductors and which are insulators at (a) 300 K and (b) 10 K?   ■

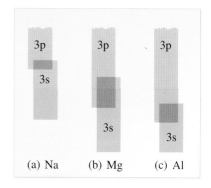

**Figure 2.27**   The overlapping 3s and 3p bands for the first three elements in the third row of the Periodic Table, (a) sodium, (b) magnesium, (c) aluminium, showing how the band occupation changes as the number of valence electrons per atom increases from one to three. The 3s band is shown displaced slightly to the right in order to show the overlap clearly. (The diagram is schematic and not to scale.)

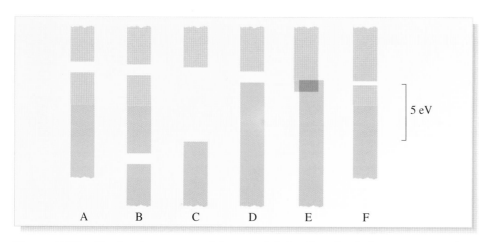

**Figure 2.28**   The band structures of six materials for Question 2.6.

## Valence and conduction bands in a semiconductor — the role of hybrid states

The next element in the Periodic Table after aluminium is silicon, with electronic structure $1s^2 2s^2 2p^6 3s^2 3p^2$. You might think that this element should be a metal similar to aluminium, but with two states per atom filled in the 3p band instead of one, leaving four of the six 3p states per atom unfilled. In fact, all the valence electrons are packed into the one band and fill it completely, while the next higher band is empty. An energy gap of 1.12 eV separates these bands, making silicon a semiconductor.

Why do the states in silicon sort themselves out so neatly into filled and unfilled bands? When discussing the bonding of methane in Section 2.1, the formation of hybrid bonds was mentioned. These are found for covalent bonding of all the Group IV elements (carbon C, silicon Si, germanium Ge and tin Sn in grey form), which have four valence electrons per atom. These hybrid bonds are found in all the important semiconductors used in electronic devices.

The band creation during the formation of a silicon crystal from $N$ isolated silicon atoms is illustrated in Figure 2.29. In each silicon atom there are 8 quantum states in

**Figure 2.29** The two highest energy bands of a silicon crystal are shown on the left. At 0 K the lower band (blue) is completely filled, the upper band (grey) is completely empty and the energy gap is 1.12 eV. Shown on the right is a schematic view of energy band formation as a crystal of silicon is formed from a large number $N$ of separated silicon atoms. As the atoms are brought together, the 3s and 3p wavefunctions mix to form hybrid wavefunctions. As this happens, the 3s and 3p energy levels form bands which merge and then reform as two hybrid bands, called the *valence band* and the *conduction band*, each of which represents $4N$ quantum states. The equilibrium separation of atoms in solid silicon is 0.235 nm. The diagram is not to scale vertically, particularly in relation to the depth of the full 1s, 2s and 2p states, which are near the indicated energy values.

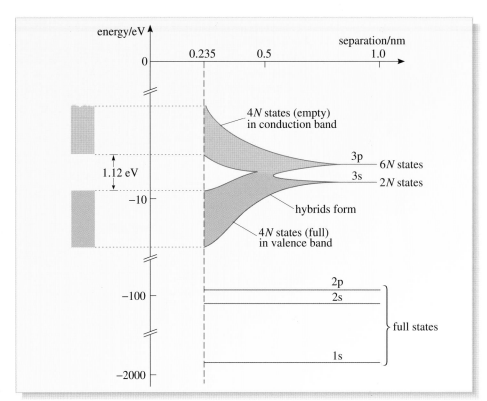

the $n = 3$ shell (two 3s states and six 3p states). In the solid these states mix to give two sets of hybrid wavefunctions and two wide energy bands, with $4N$ hybrid states for each band. The lower band is called the **valence band**, since in a pure and perfect crystal at low temperature it is completely filled with all $4N$ valence electrons. The upper band, called the **conduction band**, is totally empty in the same circumstances. At normal temperatures, some electrons are thermally excited from the top of the valence band, across the 1.12 eV gap to the bottom of the conduction band leaving empty states in the top of the valence band. This enables silicon to conduct electricity by *two mechanisms* when an electric field is applied.

- First, the thermally excited electrons in the bottom of the near-empty conduction band can move into accessible neighbouring states.
- Second, there are now empty states in the top of the nearly-full valence band into which other valence band electrons can move.

But there are some subtleties, as described below.

### Holes

When an energy band is half-full, as in sodium (Figure 2.24), the movement of electrons under the influence of an electric field is well described by Pauli's free-electron model. This is because the wavelengths of electrons near the surface of a half-filled band are very far from satisfying the Bragg condition. As a result, the electron waves, extended throughout the crystal, are affected by only the spatially-averaged effect of the positively charged ions, as assumed in Pauli's model. However, the situation is quite different for electrons in a band that is *nearly empty*, and for electrons near the top of a *nearly filled* band. These electrons have wavelengths $\lambda$ that nearly satisfy the Bragg condition for normal incidence and are therefore very strongly affected by the periodicity of the lattice ions. This has a profound effect on the complex dynamics of these electrons. We cannot discuss the details here, but we shall mention one important result.

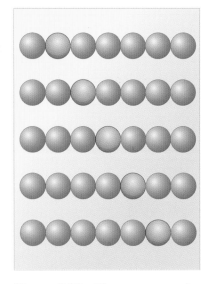

**Figure 2.30** The movement of a hole. The hole is coloured red to indicate the fact that it can be regarded as a positively charged particle.

Consider an energy band that is *nearly full* of electrons with a relatively small number of vacant states near the top. The overall movement of the electrons into the vacant neighbouring states in response to an electric field is the same as if particles with positive charge $+e$ were moving in the opposite direction to the electrons. These fictitious positively charged particles are called *positive holes*, or simply **holes**. A simple kinematic picture of a moving hole is illustrated in Figure 2.30. At the top of the figure is a line of spheres with one vacant space. Scanning down the diagram from one line of spheres to the next, you see the spheres shunting to the left into the vacant place. The overall effect is that the vacant space, or hole, moves to the right.

This situation occurs near the top of the valence band of a semiconductor such as silicon. At very low temperatures the valence band is completely full and the conduction band completely empty, as illustrated in Figure 2.29. At room temperature, a small number of holes exists near the top of the valence band because of the thermal excitation of a small number of electrons across the 1.12 eV gap into the conduction band. This is illustrated schematically in Figure 2.31a. A similar situation also occurs in some metals, notably boron, zinc and lead, because of the particular way in which the bands overlap in these metals.

When a battery is connected across the ends of a sample of silicon, an electric field $\mathscr{E}_x$ exists in the sample and a current flows. Figure 2.31b depicts the two contributions to this current in silicon: the flow of electrons that have been thermally excited into the conduction band and the flow of the resulting holes in the valence band. Note that while the electrons move in the direction opposite to the electric field direction, the flow of holes is in the same direction as the electric field just as you would expect from positively charged particles.

### The anomalous Hall effect explained

The fact that electric current is sometimes carried by positive holes explains the anomalous Hall effect (see Section 3.3) found in some materials. The Hall voltage appears to have the wrong polarity in metals such as boron and zinc and in some semiconductors because the current in these materials is largely carried by positive holes.

When the metal is melted into a liquid, it loses the regularity of atomic ordering in the crystalline lattice which is the basis for existence of holes and the band structure, and the Hall effect is normal. The explanation of the anomalous Hall effect was one of the major triumphs of the band theory of solids.

This is a good place to give a brief summary of the main results of the band theory of solids.

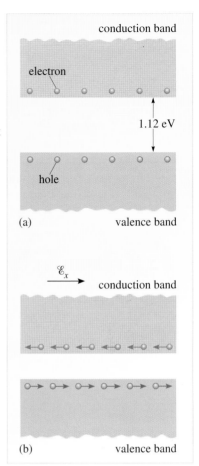

**Figure 2.31** (a) At room temperature a small number of electrons are thermally excited from the valence band into the conduction band, leaving an equal number of holes near the top of the valence band. (b) When a battery is connected across a sample of silicon, the flow of electrons in the negative $x$-direction and a flow of holes in the positive $x$-direction both contribute to an electric current flowing in the direction of the electric field $\mathscr{E}_x$.

### Results from the band theory of solids

1   Metals have good electrical and thermal conductivity due to the presence of an energy band which is only partially full. This happens in sodium because each constituent atom has a single 3s valence electron, and in magnesium because there is a large band overlap.

2   Insulators have a completely full valence band holding all the valence electrons below a completely empty conduction band separated by a very large energy gap relative to $kT$.

3   Semiconductors are qualitatively similar to insulators, but the size of the energy gap is not so great and a few electrons are excited into the conduction band from the valence band. This gives some electrical conductivity due to the movement of electrons in the conduction band and holes in the valence band.

4   The polarity of the Hall effect voltage is determined by whether the current is mostly carried by electrons or by holes.

## 4.3 Types of semiconductors

You have seen that in a semiconductor such as silicon, the current is carried by electrons in the conduction band and holes in the valence band. We'll now look at some of the details of how electrons and holes are thermally generated, and how the electron and hole concentrations can be dramatically changed by 'doping' the semiconductor crystal with tiny amounts of suitable impurity atoms. Doping is of huge importance in modern technology, and since nearly all integrated circuits are made from silicon, we shall choose that as our prime example.

### Intrinsic semiconductors

In silicon, the valence band is essentially full, holding 4 valence electrons per atom. Using Equation 1.26 we find that this corresponds to an electron density in the valence band of approximately $2 \times 10^{29}\,\text{m}^{-3}$. By thermal agitation, all of these are attempting to leap up to the conduction band, but the 1.12 eV barrier is almost impossible to cross at room temperature. The success rate is very small, but not zero, and occasionally an electron near the top of the valence band does succeed in reaching the bottom of the conduction band. When it does, a pair of potential current carriers is produced, namely an electron in the conduction band and a hole in the valence band.

Each electron in the conduction band has an average lifetime of a few seconds before it meets a hole and undergoes **recombination**, the inverse process to excitation, returning the electron to the valence band and destroying a hole there. Electron–hole pairs are being created by thermal excitation all the time, and in thermal equilibrium the rate of the two processes, excitation and recombination, must balance. This dynamic equilibrium results in an average electron–hole density of about $10^{16}\,\text{m}^{-3}$ at room temperature. This density is extremely small compared with the density of valence band electrons given above ($2 \times 10^{29}\,\text{m}^{-3}$). It is estimated that a typical electron is excited into the conduction band once every million years!

A pure crystal of silicon, containing a relatively small density of electron–hole pairs produced by the process of thermal excitation described above, is an example of an **intrinsic semiconductor**. The presence of electron–hole pairs in an intrinsic semiconductor is shown schematically in diagrams like Figure 2.32a.

An important characteristic of an intrinsic semiconductor is that the electrical conductivity increases rapidly with temperature as the rate of thermal excitation of electron–hole pairs increases (Figure 2.33a).

### Extrinsic semiconductors

You have seen that in an intrinsic semiconductor the number of free electrons and the number of free holes must be the same. This is because all the electrons and holes are created in pairs by thermal excitation of electrons from near the top of the valence band. However, that equality can be dramatically upset by a process known as **doping**, which is the adding of a tiny amount of suitable impurity atoms to take the place of silicon atoms in the crystal structure. This creates what is called an **extrinsic semiconductor**, which is the type of semiconductor normally used in practice.

Solid silicon has the diamond structure with four covalent bonds per atom. **Donors** are impurity atoms like arsenic or phosphorus with five valence electrons — too many electrons for bonding to neighbouring silicon atoms. When some arsenic atoms are put into the crystal to replace some of the silicon atoms, only four electrons are used to bond each arsenic atom into the lattice. The extra electron is so loosely bound to the arsenic atom that at room temperature it is thermally excited into the conduction band, leaving a positive arsenic ion in the crystal lattice. We illustrate this in Figure 2.32b by showing the extra so-called *donor electrons* in the conduction band and drawing empty *donor energy levels* just below the edge of the conduction band inside the energy gap of the semiconductor. The doping density is sufficiently large that the number of donor electrons in the conduction band far exceeds the number of electrons thermally excited into the conduction band from the valence band. These donor electrons are available for carrying a current, and so at room temperature the doped material has a much higher electrical conductivity than the pure silicon. A semiconductor material doped with donor atoms, such as arsenic, is referred to as an **n-type semiconductor**.

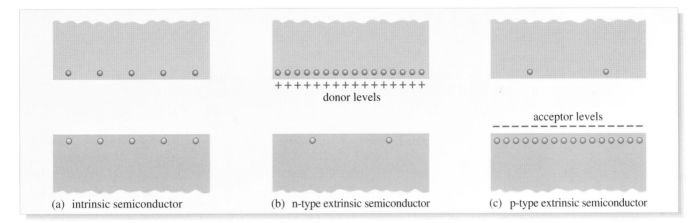

(a) intrinsic semiconductor      (b) n-type extrinsic semiconductor      (c) p-type extrinsic semiconductor

There is also a tiny density of holes in the valence band due to thermal excitation of electrons from there into the conduction band. However, there is now a huge number of donor electrons in the conduction band, about $10^{22}\,\mathrm{m}^{-3}$ at typical doping concentrations, compared to only about $10^{16}\,\mathrm{m}^{-3}$ in pure silicon at room temperature. As a result, the recombination rate is greatly increased — a hole cannot exist for very long in such an environment with so many conduction-band electrons ready to fall into it. This greatly reduces the number of holes from about $10^{16}\,\mathrm{m}^{-3}$ in pure intrinsic silicon at room temperature to about $10^{10}\,\mathrm{m}^{-3}$, as illustrated schematically in Figure 2.32b.

A **p-type semiconductor** is created by adding **acceptors**, which are impurity atoms with too few electrons for bonding. Examples of acceptor atoms are boron and aluminium, which have three valence electrons, one less than silicon. When a boron atom replaces a silicon atom in the lattice, it forms bonds with four silicon atoms in the usual way. This requires an extra electron taken from the valence band. This leaves a positive hole in the valence band free to move and carry a current, and a negative boron ion fixed in the lattice. The energy required to do this is extremely small, and so the acceptor energy levels, all filled with bonding electrons, are shown in Figure 2.32c just above the edge of the valence band inside the energy gap of the semiconductor.

When the density of acceptor atoms is sufficiently large, there is a large number of positive holes in the valence band available for carrying a current, and the p-type material, like the n-type material, has a much larger electrical conductivity than pure silicon where only a relatively small number of thermally excited electron–hole pairs is available.

There are of course some electrons that have been thermally excited into the conduction band from the valence band, but their number is very small because the relatively large number of holes in the valence band makes the recombination rate large.

**Figure 2.32** (a) Thermally excited electron–hole pairs in an intrinsic semiconductor. (b) An n-type extrinsic material. There are extra electrons in the conduction band. The empty donor energy levels close to the conduction band are indicated by the line of plus signs which is also indicative of the positive impurity ions left in the lattice due to thermal excitation of the donor electrons. There are very few holes in the valence band because the recombination rate is high. (c) A p-type extrinsic material, with acceptor levels close to the valence band and negative impurity ions in the lattice. There is a high density of holes in the valence band but only a few electrons in the conduction band due to the high recombination rate.

Although doping has a dramatic effect on the electrical conductivities of semiconductors, doping concentrations in both n-type and p-type materials are very dilute, typically about 1 impurity atom per 5 million silicon atoms. This indicates the high level of purity required for the silicon crystals used in the manufacture of extrinsic semiconducors.

### Visualizing the numbers involved

It is very hard to grasp the size of the numbers quoted above. Let us try a series of demonstrations to illustrate them. (These are intended as imaginary 'thought' experiments, but feel free to try them at home if you have lots of spare time and money!)

Clear all the furniture from a medium sized room with a floor area of about $12\,m^2$. Buy some apples of a similar size, say about 6 cm diameter, and cover the whole floor with these in a single hexagonal close-packed layer, like Figure 2.5a. You will need about 4000 apples. Replace one apple by a lemon. That single lemon on the floor of the room represents the level of accidental contamination by undesirable impurities, which was considered negligible in chemical supplies in 1960 before the semiconductor processing industry got started.

Now let us model the deliberate doping of n-type silicon with donors. The number density of silicon atoms in a crystal is about $5 \times 10^{28}\,m^{-3}$, and typically we put in a donor density of about $10^{22}\,m^{-3}$ for n-type material. So for each donor there are about 5 million silicon atoms. Borrow a large sports stadium like Wembley, and spread out a close-packed layer of 5 million apples, which will approximately cover the whole area. Replace one apple in that vast array by an orange, and you will see what one in 5 million looks like! Of course the *accidental* impurity contamination must be much less than this, in most cases well below 1% of that of deliberate doping. The equivalent is no more than one lemon on about 1000 sports fields, which represents the improved purity required today for electronics-grade materials.

Finally, let's try to picture the tiny hole density in the valence band for an n-type sample in a similar way. This, remember, is much lower than in pure silicon because of the much higher recombination rate. When doped as above, the silicon will have an electron density in the conduction band of $10^{22}\,m^{-3}$ and a hole density in the valence band of about $10^{10}\,m^{-3}$. This represents one hole for about $5 \times 10^{18}$ silicon atoms. A cube of silicon with an edge of length 1 mm has a volume of $10^{-9}\,m^3$ and contains $5 \times 10^{19}$ atoms. So lay out $5 \times 10^{19}$ apples to represent the silicon atoms in a 1 mm cube, select $10^{13}$ of these apples at random and replace them by oranges to represent the electrons in the conduction band. Finally take ten apples at random and substitute lemons to represent the ten holes in that 1 mm cube of volume $10^{-9}m^3$. If you were to try this in practice, however, you would have great difficulty in finding so many apples, and the area needed to lay them out in is about $1.6 \times 10^{17}\,m^2$, which is over 300 times the total surface area of the Earth (land and sea)! Now you may have an idea of how rare the holes are in n-type material — each like one lemon lying somewhere on 30 planets the size of the Earth!

As you might expect, the electrical conductivity of extrinsic semiconductors is dominated by the impurity density, which determines the density of electrons in the conduction band of n-type materials, and the density of holes in the valence band of p-type materials. This is illustrated in Figure 2.33b for n-type silicon doped with varying amounts of arsenic. At room temperature, indicated by the line AB, the conductivity $\sigma$ increases steadily with increasing arsenic number density. There are other features of interest in this graph. The sharp rise in conductivity for temperatures greater than about 400 K (to the left of C) is due to the intrinsic conductivity taking over as more and more electrons are excited from the valence band into the conduction band. The fall-off at temperatures below about 80 K (to the right of D) is due to a drop in the number of electrons excited into the valence band from the donor levels.

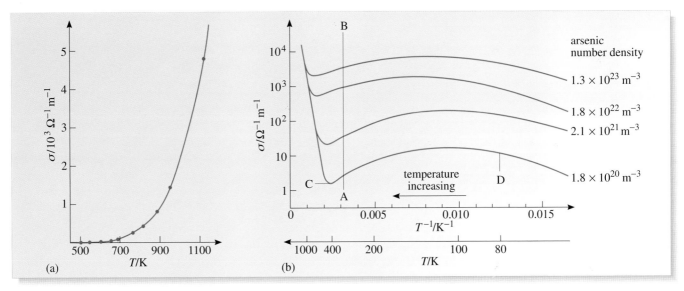

**Figure 2.33** (a) The electrical conductivity of pure silicon increases rapidly with temperature. (b) Electrical conductivities of four silicon specimens doped with different number densities of arsenic atoms. Note that the $\sigma$-scale is logarithmic and $\sigma$ is plotted against the reciprocal of the temperature. This choice of scale allows the huge range of conductivities to be displayed efficiently.

Figure 2.33b reveals the wide range of conductivities that can be obtained from doping. By selecting semiconductors with suitable energy gaps and doping with suitable amounts of impurities, semiconductor engineers can produce materials for a huge variety of devices. Generally the conductivities of semiconductors are very much less than for metals (compare the values in Figure 2.33 with those in Table 2.3). This means that much higher electric fields can be applied in semiconductors, since the much lower conductivity restricts the current to values giving negligible power dissipation and heating until huge fields are applied. This implies that the current carriers in semiconductors can have very high drift speeds, comparable even with the thermal speed in extreme cases. This enables semiconductor circuits to be made that can operate at very high speed and very high frequency, which is extremely important in computers.

**Question 2.7** Fill in the blanks in Table 2.4, selecting the number densities from: $5 \times 10^{28}\,\text{m}^{-3}$, $10^{22}\,\text{m}^{-3}$, $10^{16}\,\text{m}^{-3}$, $10^{10}\,\text{m}^{-3}$. ■

**Table 2.4** Typical number densities in pure silicon at room temperature and in silicon doped with arsenic or boron atoms at doping densities of $10^{22}\,\text{m}^{-3}$.

| Material | Electron number density in conduction band/m$^{-3}$ | Hole number density in valence band/m$^{-3}$ | Number density of silicon atoms/m$^{-3}$ |
|---|---|---|---|
| pure silicon | | | |
| n-type silicon | | | |
| p-type silicon | | | |

**Question 2.8** (a) Explain why the electrical conductivity of a metal such as sodium decreases with temperature, while that of a pure semiconductor such as silicon increases with temperature. (b) Explain why the electrical conductivity of pure silicon increases rapidly with temperature while that of n-type and p-type silicon depends only weakly on temperature over a wide temperature range. ■

# 5 The p–n junction and semiconductor devices

The previous section introduced you to the band theory of solids and some of the basic physics of semiconductors. We begin this section with a brief overview of the materials used in semiconductor devices. We then discuss the p–n junction diode. This is the principal component in almost all semiconductor devices and so we look at the principles of its operation in some detail. We can then go on to see the p–n junction in action in a variety of devices including the transistor, laser diodes and solar cells.

## 5.1  Semiconducting materials

For most of history, the science of a topic has arrived a very long time after its technology. In the past the effect of science (if any) has been to explain and sometimes to improve the empirical results already achieved by the engineers, and to suggest new areas for invention. For example, electricity was used to light theatres and send signals across the Atlantic Ocean long before the electron was discovered.

However, in the past 50 years that primacy of technology has been lost. Industry now is based on semiconductor electronics where science comes first in all cases. Only in the 1950s was there the simultaneous maturation of solid-state physics, optics, chemical processing, materials technology, electron microscopy, circuit design, etc. which made this industry possible. In fact, each stimulated the others into a rapid advance which is still continuing.

Silicon is an excellent compromise material for the construction of solid-state circuits, mainly because it is an element, making processing much easier, and because of the very high quality of silicon oxide, used in various ways within circuits. Development of crystal growth methods has resulted in 80 kg single crystals of silicon up to 3 m in diameter with remarkable quality and purity. These are sliced into thin wafers, and then used to manufacture integrated circuits.

Various semiconducting compounds, such as gallium arsenide GaAs, gallium phosphide GaP, and many others (such as InSb, CdS, CdTe, ZnSe) are also used for specialized applications. A problem with these is the partial ionicity of the bonding, which reduces the electrical conductivity and has many other deleterious effects. Crystals of the compounds are available at up to about half the size of the silicon ones, and are very much more expensive. The processing into devices is more difficult and often the resulting products are less stable and reliable.

Because the semiconductor compound materials are expensive and hard to use, silicon is preferred for most applications. A recent development is the investigation of new conducting and semiconducting polymers. None are yet suitable for use, but many show very interesting properties.

There is increasing use of very thin layers of material, grown as single crystals on the surface of the main semiconductor wafer. The sophisticated use of such *epilayers* has made possible many new applications. These often use *mixed crystals*, e.g. a controlled intermingling of aluminium and gallium in a single crystal. By varying the relative proportions of aluminium to gallium, intermediate semiconducting properties can be designed into the material. These layers can be made extremely thin, down to only a few atoms thick. Electrons can be trapped within these ultra-thin layers, known as *quantum wells*, with special effects on their energies. It is now possible to perform engineering at the atomic scale. Such *nanotechnology* is one of the most exciting growth areas of semiconductor technology.

## 5.2 The p–n junction in equilibrium

The p–n junction is the basic component of almost all semiconductor devices. If you understand the physics of the p–n junction you are half-way towards understanding the operation of transistors, light-emitting-diodes, laser diodes, solar cells and many other achievements of solid-state technology, so you should study this section carefully.

A **p–n junction** is a region in a semiconductor where the doping changes from p-type to n-type. Normally the transition takes place over a very thin region typically about 0.5 μm wide, and can be accomplished by converting the surface layer of a p-type slice of semiconductor into n-type through the introduction of a greater number of donors than existing acceptors in that region. To understand the electrical effects at the junction, however, it is helpful to consider the results of an imaginary process of joining a piece of n-type semiconductor to a piece of p-type, even though this is impractical.

### The p–n junction in equilibrium

Figure 2.34a shows the starting materials, n-type silicon and p-type silicon, and their valence and conduction bands. Now let's imagine joining them together and seeing how an equilibrium situation develops.

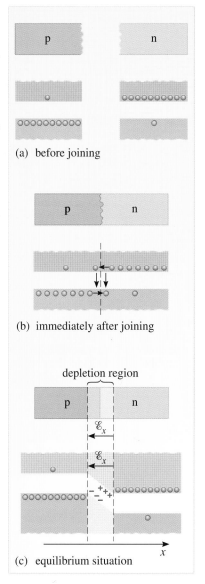

(a) before joining

(b) immediately after joining

(c) equilibrium situation

As soon as the materials come into contact (Figure 2.34b), electrons in the conduction band begin to diffuse from the n-region to the p-region in an attempt to equalize the electron pressure throughout the material, like a gas expanding to fill an empty space. Similarly, holes in the valence band diffuse from the p-region into the n-region.

However, most of the electrons and holes don't get very far. They meet in the junction region and recombine — electrons fall down into holes. This gives rise to a very narrow layer of semiconductor at the junction which is depleted of electrons and holes. This thin layer at the junction is called the **depletion region**, and it's normally no more than 1 μm ($10^{-6}$ m) thick.

You might think that this two-way diffusion process would continue until all the holes and electrons have recombined or have spread uniformly throughout the material. This is not so. The absence of electrons and holes in the depletion region exposes fixed positively charged donor ions on the n-side of the depletion region and fixed negatively charged acceptor ions on the p-side. These fixed charges set up an electric field $\mathcal{E}_x$ in the depletion region, directed from the positively charged n-region to the negatively charged p-region. This is illustrated in Figure 2.34c which shows the equilibrium situation obtained very shortly after the junction is made.

The electric field $\mathcal{E}_x$ in the depletion region opposes further diffusion of electrons and holes. This is because it exerts a force on the negatively charged electrons pushing them back into the n-region, and a force on the positively charged holes pushing them back into the p-region.

**Figure 2.34** An imaginary way to make a p–n junction. (a) Two isolated pieces of semiconductor, and their valence bands and conduction bands. (b) The formation of a p–n junction. The valence and conduction bands are shown *immediately* after the p- and n-materials come into contact. Electrons begin to diffuse from the n-region into the p-region and holes diffuse from the p-region to the n-region. This is indicated by the horizontal arrows. The downward arrows indicate electron–hole recombinations near the junction region. (c) The final equilibrium condition. The donor ions (+) and acceptor ions (−) in the depletion region set up an electric field $\mathcal{E}_x$ directed from the n-region to the p-region that opposes further diffusion. (Note that if the positive *x*-direction is chosen to be from left to right, as indicated, then $\mathcal{E}_x$ is negative.)

We can see this another way. Figure 2.34c also shows the effect of the electric field $\mathcal{E}_x$ on the energy levels of the electrons in the p- and n-regions. You know that an electric field is always directed from a region of high electrostatic potential towards a region of low electrostatic potential. Thus the n-side of the junction is at a higher electrostatic potential than the p-side. This means that the electrons have a *lower* potential energy on the n-side than on the p-side, because electrons are negatively charged. This lowering of electron energies in the n-region is shown in Figure 2.34c by a lowering of the energy bands. Thus electrons trying to diffuse across the junction from the n- to the p-region are inhibited from doing so by a potential energy gradient. Similarly, holes trying to diffuse the other way, from p to n, are also faced with a potential energy gradient; this is because holes are positively charged and have a higher potential energy in the n-region than in the p-region.

Thus we conclude that in equilibrium, the electric field in the depletion region presents an electrostatic potential energy gradient opposing further diffusion of both electrons and holes across the junction.

### The equilibrium diffusion current

The equilibrium however is a dynamic one. The electrons and holes have a range of energies, and there are always a few electrons and a few holes with enough energy to climb the hill before eventual recombination on the other side of the junction. This continual passage of a small number of energetic electrons from n-type material to p-type and a small number of energetic holes from p-type to n-type, represents a small electric current called the **diffusion current**.

What is the direction of the diffusion current?

A flow of electrons moving from the n-type material to the p-type represents an electric current in the opposite direction — from p-type to n-type. A flow of holes moving from the p-type material to the n-type also represents an electric current in that direction. Thus both electrons and holes contribute to the diffusion current which flows from the p-type to the n-type material.

If the semiconductor material is not connected to an external voltage source, there can be no net current flow through the junction in equilibrium. It follows that, in equilibrium, the small diffusion current must be cancelled by a current of the same small magnitude flowing in the opposite direction.

### The pair current

This opposing current comes from thermally excited electron–hole pairs discussed in connection with intrinsic semiconductors in Section 4.3. In equilibrium, electrons and holes are thermally excited throughout the material, but we shall follow the fate of electrons in the p-region, and holes in the n-region.

Consider first the thermally generated electrons in the conduction band of the p-region. Some of these will survive recombination in this hole-rich environment and wander into the depletion region. Here they come under the influence of the electrostatic field $\mathcal{E}_x$ directed from the n-region to the p-region. This field sweeps them across the junction into the n-region, resulting in a small electric current in the opposite direction — from the n-region to the p-region.

Now consider holes in the valence band of the n-region. Some of these will survive recombination in this electron-rich environment and wander into the depletion region. Consider the following questions.

● How are these holes affected by the electrostatic field in the depletion layer? Does this process give rise to an electric current flow through the junction? If so, in what direction is this current?

○ The electric field in the depletion layer is directed from the n-type material to the p-type material. The holes behave like positively charged particles and so the electric field sweeps them in the direction of the field, giving rise to an electric current flowing across the junction from n to p. ■

We conclude that some of the thermally excited electrons and holes are swept across the junction by the electrostatic field producing a small electric current across the junction from the n-type material to the p-type. This current, called the **pair current**, is in the opposite direction to the diffusion current, and in equilibrium these currents balance exactly. The magnitude of these opposing equilibrium currents is extremely small, typically less than $1\,\mu A$ ($10^{-6}\,A$).

## 5.3 The p–n junction diode

A p–n junction can act as a *rectifying diode*, a two-terminal device that allows a large current to flow in one direction only. To understand this behaviour we consider how an external voltage applied across the junction disturbs the balance between diffusion current $i_d$ and pair current $i_p$.

### The p–n junction with reverse bias

Figure 2.35a depicts the equilibrium state of the p–n junction, as described above, before any external voltage is applied. The diffusion current and the pair current are in balance so no net current flows.

Suppose we now connect the positive terminal of a battery to a contact on the n-material and the negative terminal to a contact on the p-material. This is shown in Figure 2.35b. The junction is now said to have a **reverse bias**.

**Figure 2.35** A p–n junction and energy band diagrams. (a) No bias. The diffusion current and pair current cancel. (b) Reverse bias stops the flow of the diffusion current $i_d$ without changing the pair current $i_p$. (c) Forward bias greatly increases the diffusion current flow without changing the pair current. The downward arrows indicate recombinations of electrons and holes in the depletion region.

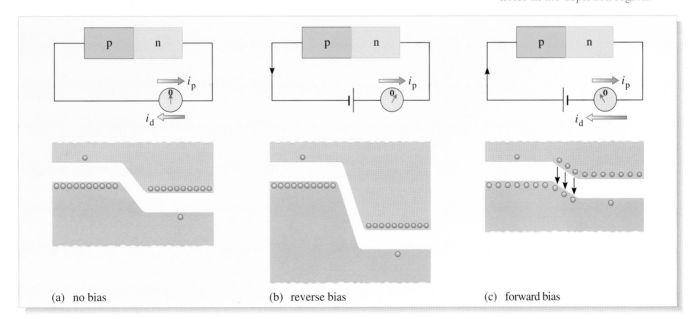

(a) no bias (b) reverse bias (c) forward bias

●   What effect do you think the reverse bias has on the electric field in the depletion layer and electron energy levels in the n-material relative to those in the p-material?

○   Connecting the positive terminal of the battery to the n-material and the negative terminal to the p-material raises the electrostatic potential in the n-material relative to the p-material, thereby increasing the electric field in the depletion layer. This causes a further *lowering* of the electron energy levels in the n-region relative to the p-region, as illustrated in Figure 2.35b. Thus the effect of the battery is to *increase* the height of the potential energy hill for the equilibrium diffusion current, completely stopping it. ■

What effect does the reverse bias have on the pair current in the opposite direction? None! As before, some of the thermally excited electrons in the p-region will wander into the junction region where they are all swept across the junction by the electrostatic field. The magnitude of the field has no effect on the magnitude of the electron current, and similarly for holes in the n-region. (This is analogous to the fact that you will end up on the ground whether you jump out of a third-floor window or a tenth-floor window.)

Thus we conclude that in reverse bias, the junction passes a very small pair current (of order 1 μA, as before) directed across the junction from the n-type to the p-type, i.e. an anticlockwise current in the circuit shown in Figure 2.35b.

### The p–n junction with forward bias

Now suppose the two battery connections are interchanged, as in Figure 2.35c. The negative terminal of the battery is connected to the n-material and the positive terminal to the p-material. The junction now is said to have a **forward bias**.

The electrostatic potential in the n-region is now lowered relative to the p-side, reducing the magnitude of the electrostatic field in the depletion region and raising the electron energy levels in the n-material relative to the p-material, as illustrated in Figure 2.35c. This allows more electrons to climb the potential energy hill from n to p and more holes to climb from p to n, thereby increasing the diffusion current. The pair current, on the other hand, remains unchanged; those thermally generated electrons in the p-material and holes in the n-material that reach the junction region are still swept across. (You will still hit the ground even if you only jump out of a first-floor window).

Thus in a forward-biased p–n junction the diffusion current exceeds the pair current and there is a net current flow across the junction from the p-side to the n-side i.e. a clockwise current in the circuit shown in Figure 2.35c. This net current can be quite large since the electron density in the n-material and hole density in the p-material can be made large by the doping.

### Characteristics of the junction diode

Figure 2.36 shows a plot of net current through the junction against battery voltage applied across it. You can see that when the applied voltage is negative (reverse bias) only a very small current flows — the pair current; but when the applied voltage is positive (forward bias) the current is dominated by a large diffusion current. This plot summarizes the above discussion of the p–n junction with reverse and forward bias. Clearly the p–n junction does not obey Ohm's law.

The characteristics shown in Figure 2.36 are in fact those of a **diode rectifier**, a two-terminal device that can pass a high current in one direction only. This is useful in itself. The first solid-state devices were semiconductor diodes, just a rectifying junction with a wire attached to each side. The very earliest of these were the crystal

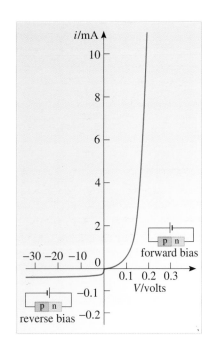

**Figure 2.36** Current against voltage for a p–n junction. Note the change in scale on both axes at the origin. This is why the graph appears to have a discontinuity in slope at the origin.

detectors used in the first radio sets, but those were old technology, based on trial-and-error development. Reliable devices were produced only after the concepts described here had been developed and used in manufactured crystals of high quality and uniformity. Nowadays diode rectifiers are commonly used in circuits for converting AC (alternating current) power supplies to DC (direct current) supplies.

**Question 2.9**   The depletion region in a certain p–n junction is 1.0 μm wide. The electrostatic field in the region is directed in the negative $x$-direction (see Figure 2.34c) directly across the junction from the n-material towards the p-material, and has a uniform magnitude $|\mathscr{E}_x| = 1.5 \times 10^6 \text{ V m}^{-1}$.

(a) Determine the potential energy changes that occur when: an electron moves (i) from the n-region to the p-region, and (ii) from the p-region to the n-region.

(b) Repeat part (a) for a hole instead of an electron.

(c) State the direction of (i) the force acting on an electron in the electrostatic field of the depletion layer, and (ii) the force acting on a hole in the same layer.  ■

## 5.4  Semiconductor devices

You have seen one application of a p–n junction, the diode rectifier. There are many other common applications of p–n junctions, such as solar cells, light-emitting-diodes, laser diodes and transistors, which we describe below.

### Light-detectors and solar cells

Consider a p–n junction connected to an external circuit containing only a resistor. You have seen that when the junction is in equilibrium with no bias the thermally generated pair current is very small and is balanced by the diffusion current, so no net current flows. This situation is shown in Figure 2.37a.

Now suppose some process creates extra electron–hole pairs in the region of the junction. The strong electric field in the depletion region would sweep the electrons into the n-region and the holes into the p-region, thereby increasing the pair current. The small diffusion current would remain unchanged since no bias has been applied. Thus a net current would flow through the circuit with electrons leaving the n-region, flowing through the external resistor and then into the p-region where they combine with holes moving the other way.

An example of such a pair-creation process is the absorption of photons of sufficient energy to excite electrons from the top of the valence band to the bottom of the conduction band. This is illustrated in Figure 2.37b. Some of these electrons are excited in the depletion region where the strong electric field sweeps them across the junction. The net electric current that flows in the circuit when light falls on the p–n junction (Figure 2.37c) is called a *photocurrent*. Thus the p–n junction can be used as a light-detector.

**Figure 2.37**   (a) A p–n junction with no bias. (The electric field in the junction region is not shown.) No net current flows: the small diffusion current $i_d$ and pair current $i_p$ balance. (b) Photons of sufficient energy can excite electrons from the valence band to the conduction band leaving behind an equal number of holes. (c) When light falls on the junction region a photocurrent flows through the resistor $R$.

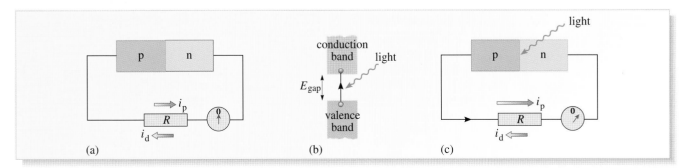

Electron–hole creation by light is similar to the *photoelectric effect* (see *QPI* for details) where photons eject electrons from the surface of a metal. In the p–n junction, the photon energy is used to excite an electron from the valence band to the conduction band, leaving behind a hole. To be effective, the photon energy $hf$ must be equal to or greater than the energy gap $E_{gap}$. That is,

$$hf \geq E_{gap}.$$

Also, the absorption of light should take place near the junction so that the electrons and holes are not lost by recombination before the electric field can sweep them across the junction.

The same principles can also be used to detect not only light but beams of subatomic particles, which can generate electron–hole pairs by losing kinetic energy in the junction region. This is the basis of many modern particle detectors used to detect the nuclear radiation and high energy particles that are created in collisions in particle accelerators (see Chapter 4).

A **solar cell** is a device that makes use of the power dissipated by the photocurrent in a resistance in the external circuit. The basic construction of the p–n junction in a solar cell is illustrated in Figure 2.38. Light falls on the n-type top layer. The surface of this layer has an anti-reflection coating so as to absorb as much light as possible, and the layer is very thin (about $0.5\,\mu\text{m}$) so that the light can penetrate into the vicinity of the junction where it is absorbed to create electrons and holes and hence a pair current that flows in the external circuit.

One major problem with the solar cell is that solar radiation reaching the ground covers a broad range of wavelengths, stretching from 2000 nm in the infrared to 350 nm in the ultraviolet. To see why this is a problem, try Question 2.10.

**Question 2.10** Silicon, which has an energy gap $E_{gap} = 1.12\,\text{eV}$, is commonly used for solar cells. A typical spectrum for solar radiation reaching the Earth's surface is shown in Figure 2.39. Which region of the spectrum is ineffective in generating electrical power from a silicon solar cell? ■

For an ideal silicon solar cell about 21% of the solar power falling on the device can be converted into electrical power; the rest goes mainly into lattice vibration which uselessly heats the cell itself. This maximum efficiency assumes that all of the incident light gets into the crystal and all the electrical power is dissipated in the external circuit and not inside the cell. Neither of these assumptions is valid in practice. For a start, a certain proportion of the light is reflected from the surface of the silicon. These losses can be only partially reduced by using anti-reflection coatings. Light is also lost by absorption in the metal electrodes on the top surface. These electrodes are needed to collect the electrons and take them to the external circuit. Typically about 10% of the surface has to be covered by metallic electrodes. Of course, in many applications the surface of the cells can become further obscured by the droppings from inconsiderate birds!

In small-scale terrestrial applications, single silicon crystals give an overall efficiency of about 15%. This provides a good compromise between efficiency and expense for powering garden lights, pond fountains, car park ticket dispensers, pocket calculators, etc. (Often this silicon is from wafers intended for integrated circuit manufacture but rejected for being of too low a quality.) In any large-scale terrestrial application, such as the use of solar cell arrays on rooftops for domestic power supply, polycrystalline silicon or even amorphous silicon is used, and efficiencies are down to about 5%.

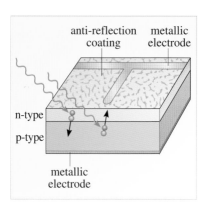

**Figure 2.38** Photons create electron–hole pairs near the p–n junction of a solar cell. The metallic electrodes are connected to an external circuit (not shown) where heat is generated in a resistor.

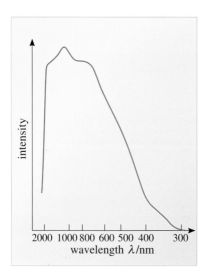

**Figure 2.39** The solar spectrum (solar power as a function of wavelength) at sea-level when the Sun is overhead.

For applications like power for space satellites, where the cost of the solar cell is less relevant, the p–n junctions are made using semiconducting compounds such as gallium arsenide (GaAs) which have a higher ideal efficiency than silicon.

## Light-emitting-diodes and laser diodes

You have seen that light falling on a p–n junction can produce an electric current. Conversely, an electric current passing through a p–n junction can sometimes produce light through the mechanism of electron–hole recombination. A forward-biased p–n junction passes a current by the diffusion of electrons from the n-region to the p-region and by the diffusion of holes from the p-region to the n-region. Electrons and holes meet near the junction region and recombine with the release of energy. Each recombination event releases an amount of energy approximately equal to the energy gap $E_{gap}$. Some of this liberated energy is dissipated uselessly as increased lattice vibrations, but some recombinations are accompanied by the emission of photons (Figure 2.40). The proportion of the recombination energy converted into light depends strongly on the precise details of the energy band structure of the semiconductor and other factors such as temperature and the impurities and defects that are present.

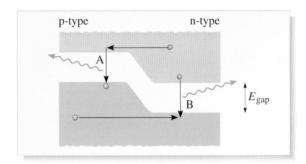

**Figure 2.40** In a forward-biased p–n junction, electrons in the conduction band diffuse from n to p and recombine with holes, emitting photons (process A). Holes diffuse from p to n, recombine with electrons and, again, photons are emitted (process B).

A forward-biased p–n junction used to generate light is called a **light-emitting-diode**, or LED. One of the most efficient materials used for LEDs is the semiconductor compound gallium arsenide (GaAs). In a forward-biased p–n junction made from GaAs, practically 100% of the recombinations are accompanied by photon emission. The energy gap in GaAs is $E_{gap}$ = 1.43 eV. Knowing the relationship between photon energy and frequency, $E = hf$, you can show (by the method you used in Question 2.10) that this corresponds to a photon frequency of $f = E_{gap}/h$, and a wavelength of $\lambda = c/f = ch/E_{gap} = 867$ nm, which lies in the near infrared. Light-emitting-diodes made from GaAs have been widely used in optical-fibre communication systems.

**Question 2.11** LEDs are often used for visual displays and warning devices. Would GaAs be a suitable material for this purpose? The semiconducting material gallium phosphide (GaP) has an energy gap of 2.27 eV. Would this be a suitable material for making an LED for visual display?  ■

Particular semiconductor compounds have fixed energy gaps, giving light of fixed wavelengths. This inflexibility can be overcome by the use of *mixed crystals*. For example, a mixture of GaAs, which has an energy gap too small to produce visible light, and GaP which has an energy gap corresponding to green light, can be used to create mixed crystals that can be described by the chemical formula $GaAs_xP_{1-x}$. By altering the relative proportions (the fraction $x$ in the formula) the colour can be selected for particular applications. For example, when LEDs are used in gas detectors, a mixed crystal can be selected where the LED wavelength is equal to one that is strongly absorbed by a specified gas. The amount of light lost within the gas on its way to a suitable light-detector indicates the concentration of the gas.

Light from an LED originates from the *spontaneous* recombination of electrons and holes in a p–n junction. Each photon is emitted in a random direction and so the light from an LED is given out in all directions. Furthermore, the emitted light does not have an exactly defined wavelength, since the electrons in the conduction band and the holes in the valence band occupy a small but finite range of energy levels.

The nature of the light emitted by a forward-biased junction is dramatically altered in the **laser diode**. Laser light is obtained by *stimulated emission*. In a laser diode, the stimulated emission is obtained by electron–hole recombinations that are *stimulated* by the presence of *existing photons*. As Figure 2.41 illustrates, a photon in the junction region can stimulate an electron to recombine with a hole, emitting a second photon in the process. This second photon has the same wavelength and travels in the same direction as the stimulating photon.

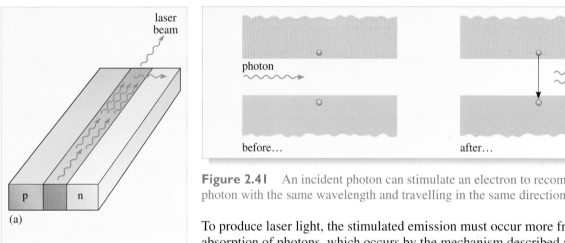

**Figure 2.41**  An incident photon can stimulate an electron to recombine with a hole. A photon with the same wavelength and travelling in the same direction is emitted.

To produce laser light, the stimulated emission must occur more frequently than the absorption of photons, which occurs by the mechanism described above in connection with light-detectors and solar cells. To achieve this, there must be a *population inversion*, i.e. the number of electron–hole pairs ready to be stimulated to recombine and emit photons, must be greater than the number of electrons near the top of the valence band ready to absorb photons. This *population inversion* can be achieved by passing a *very large current* through a forward-biased p–n junction, a much greater current than is normally used in an LED. This large current produces a huge density of electrons and holes in the junction region where stimulated recombinations occur.

In the laser diode the p–n junction region is long and thin, with flat polished end-faces (Figure 2.42a). Photons travelling at an angle to the junction have little chance of stimulating other photons. Photons travelling along the axis of the junction, however, are reflected back and forth by the polished end-faces and so stimulate an intense beam in this direction, a small fraction of which is transmitted by one of the polished end-faces to give the laser beam.

The stimulated photons that make up the laser beam have the same wavelength as the photons that stimulate their emission. This concentrates the emission into a much narrower range of wavelengths than in an LED, (Figure 2.42b).

A simple p–n junction is not very suitable for a laser because the large currents required to produce the population inversion burn out the device. In practice, devices with three layers of semiconducting materials (called *double heterostructures*) are used to achieve the required population inversion with smaller currents.

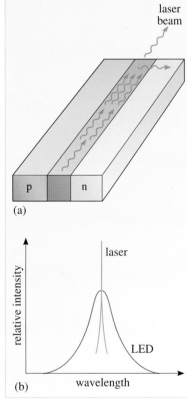

**Figure 2.42**  (a) In a laser diode the junction region is long and thin. A high density of stimulated photons builds up along the axis of the junction and is reflected back and forth by the polished ends. A proportion of the photons is transmitted by one end-face and forms the laser beam. (b) The light from a laser diode covers a much narrower range of wavelengths than the light from an LED.

Laser diodes provide very small, reliable, robust and relatively cheap lasers for the infrared and red regions of the spectrum. They are used in the technology of optical communications as well as in CD-players and supermarket checkouts. There is much current research into developing suitable materials with wider energy gaps so that laser diodes operating in the green, blue and ultraviolet regions may become available.

### The transistor

The most spectacular of the many achievements of semiconductor physics is certainly the invention of the **transistor**, for the development of which Bardeen, Brattain and Shockley were awarded the Nobel Prize for Physics in 1956. The transistor today is used to perform a variety of essential operations in electronic equipment, especially the amplification of weak signals.

Transistors are semiconducting devices, usually with three electrodes, in which a current flowing between one pair of electrodes can be controlled by varying the current or voltage applied to another pair of electrodes. We shall describe one of the common types of transistor, the *junction transistor*, but you should bear in mind that this is one of a family of different types of transistor and we do not have room in this book to discuss others.

A junction transistor has three semiconductor regions and it can be thought of as made up of two p–n junctions back to back. The transistor shown schematically in Figure 2.43 has a p-type region, known as the *base*, sandwiched between two n-type regions called the *emitter* and *collector*.

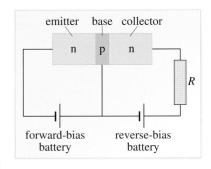

**Figure 2.43** An n–p–n bipolar junction transistor showing the operating voltage bias conditions. The emitter–base junction is forward-biased and the base–collector junction is reverse-biased.

There are two important design features:

- the p-type base region is very thin, of order 1 μm ($10^{-6}$ m);
- the n-type doping of the emitter region is much greater than the p-type doping of the thin base.

In operation the emitter–base junction is forward-biased and the base–collector junction is reverse-biased, as shown in Figure 2.43. The effect of these biasing voltages on the energy bands in the three regions of the transistor is shown in Figure 2.44.

With the emitter–base junction forward-biased, there is a large diffusion current flowing from the p-type base into the n-type emitter, as described in Section 5.2. However, because of the stronger n-type doping of the emitter region, this diffusion current is mainly carried by electrons diffusing from the n-type emitter to the p-type base, with hardly any holes diffusing the other way. Now because the p-type base region is lightly doped and very thin, most of these electrons pass straight through it, with minimal loss due to recombination with holes, and reach the base–collector junction. This junction is reverse-biased, and so the electrons are swept across into the collector region by the large electrostatic field. These electrons reach the collector, leave the collector terminal, pass clockwise through the circuit in Figure 2.43 and re-enter the transistor via the emitter terminal. Thus we have a steady current flowing through the transistor. Now let's see how the transistor is used as an amplifier.

● Suppose the forward bias of the emitter–base junction were to be increased slightly. What effect would this have on the current through the circuit?

○ Look at the graph of the current in a p–n junction versus applied voltage in Figure 2.36. In the forward-bias region, the graph is very steep. This means that a small increase in forward-bias voltage can result in a large increase in the current flowing across the p–n junction. Now the emitter–base junction in the transistor is a forward-biased p–n junction. It follows that a small increase in the forward bias of the emitter–base junction results in a large increase in the current flowing across the emitter–base junction, into the collector and around the circuit. ■

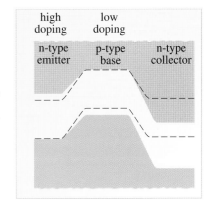

**Figure 2.44** The energy bands in an n–p–n junction transistor with bias voltages applied. The broken lines show the edges of the bands with no bias voltages applied.

Now suppose we have a very weak time-varying signal voltage that we wish to amplify. We can add this time-varying signal voltage to the steady forward bias of the emitter–base junction by connecting it in series with the forward-bias battery, as illustrated in Figure 2.45. The current through the transistor will then increase and decrease as the signal voltage increases and decreases. Thus a voltage will appear across the resistor $R$ that varies in the same way as the signal voltage and is an amplified version of it.

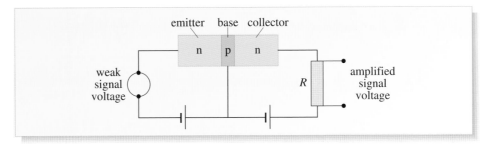

**Figure 2.45**   The source of a weak time-varying signal voltage $V(t)$ is put in series with the forward-bias battery. The voltage across the resistor $R$ is an amplified version of $V(t)$.

In 1959 a total of 125 million transistors were produced in the USA. Nowadays that number of transistors can be found in a single integrated circuit on a silicon chip the size of your thumbnail. This has been the fastest and greatest technological change in history. There is no scope to discuss the transistor and its applications any further in this course, even though arguably it is the most significant invention in the past 2000 years!

> Open University students should leave the text at this point and use the multimedia package *Electrons in Solids*. This activity will occupy about 1 hour.
>
> Open University students should then view Video 8 (*The Incredible Shrinking Chip*). You should return to the text when you have finished viewing.

# 6   Superconductivity

This section introduces the phenomenon of superconductivity, gives a brief outline of its quantum-mechanical origins and discusses some of its applications.

## 6.1   Introducing superconductivity

In 1908, Kamerlingh Onnes (1853–1926) succeeded in liquefying helium. At atmospheric pressure, helium liquefies at 4.2 K and will cool to below this temperature by evaporation. Using liquid helium as a coolant, it is possible to study the properties of materials over a range of temperatures down to about 1 K. Many interesting results were obtained, but none more startling than the observation by Kamerlingh Onnes himself in 1911 that the electrical resistance of solid mercury suddenly vanishes as it is cooled below 4.2 K. His original results are shown in Figure 2.46a. We now know that this **superconductivity** is a property of about 30 elements, thousands of metallic alloys, and many compounds. It is believed that, at a temperature below the **superconducting transition temperature** $T_C$, the electrical resistance is exactly zero for direct current flow in a superconductor (Figure 2.46b). Just as a wheel on frictionless bearings would spin for ever without slowing down, the absence of resistance allows the current in a superconducting loop to circulate eternally without diminishing, and this current needs no voltage source to sustain it. Somehow the electrons inside a solid are able to flow in perpetual motion with no scattering at all due to defects or to thermal agitation.

Of course it is impossible to confirm experimentally that the current is eternal or that the resistance is precisely zero. What has been shown is that no current change can be detected after a year of careful observation with the most sensitive equipment, proving that the current will last for at least 100 000 years. Although the conductivity may not be infinite, it is certainly greater than $10^{25}\,\Omega^{-1}\,\text{m}^{-1}$, which is very large compared with typical values of about $10^{8}\,\Omega^{-1}\,\text{m}^{-1}$ for normal metals.

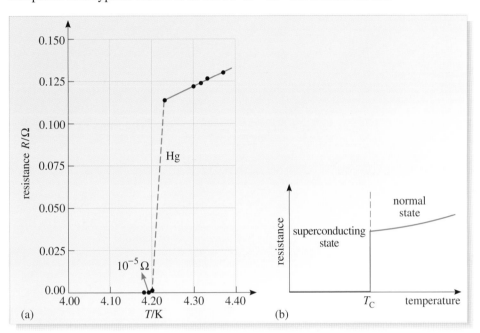

**Figure 2.46** (a) The resistance $R$ in ohms of a mercury specimen versus temperature $T$, as recorded by Kamerlingh Onnes, showing the first observation of superconductivity. The black dots are the experimental points. (b) The superconducting transition temperature $T_C$ marks the boundary between the superconducting state and the normal state of the material.

The perfect conduction is extraordinary enough, but equally unusual effects also occur in the magnetic properties of a superconductor. You know from *Dynamic fields and waves* that transient currents are induced in a circuit while the magnetic flux through the circuit is changing, as described by *Faraday's law of induction*. The induced current flowing in normal conductors dissipates its energy in the circuit resistance and disappears rapidly once the magnetic flux stops changing, but in a superconductor the induced current will persist indefinitely. It flows in the direction given by *Lenz's law*, i.e., in the direction that produces a magnetic field that opposes any change in the applied magnetic field. As there is no resistance, the supercurrent has a high enough value to *fully* cancel out any magnetic field that we may try to apply to the sample. Thus magnetic flux cannot enter a superconductor, as the induction of a large loss-free current, flowing in the surface of the superconductor, prevents this.

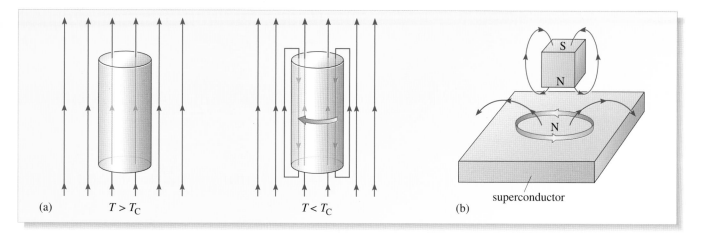

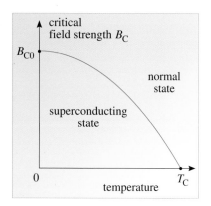

**Figure 2.47** The Meissner effect, in which magnetic flux is excluded from a superconductor as it is cooled below the transition temperature $T_C$. (a) Above $T_C$, the flux enters the material. Below $T_C$, a surface current (red arrow) is induced circulating around the sample (towards the left at the front) creating an internal field that fully cancels the original field inside and reinforces it outside, in effect moving the magnetic flux to outside the superconductor. (b) A schematic view of the induced surface current loop that neutralizes the flux inside the superconducting slab and levitates the magnet above the surface.

**Figure 2.48** The temperature-dependence of the critical magnetic field strength $B_C$. The higher the magnetic field at the surface of the superconductor the lower the temperature must be to maintain the superconducting state.

More surprising still is the fact that any magnetic flux *already* present in a normal material is expelled when it goes through the superconducting transition on cooling. This is the **Meissner effect** illustrated in Figure 2.47a. The Meissner effect similarly involves the induction of a loss-free current in the surface of the superconductor to create the internal field that cancels the applied field.

The effects of magnetic flux exclusion can be large and dramatic. A popular demonstration of these effects involves a magnet initially resting on the surface of a material that is above its superconducting transition temperature $T_C$. Under these conditions the material is an ordinary conductor and the magnetic field of the magnet pervades the space around it including the body of the material. The material is then cooled to below its transition temperature $T_C$. As the material becomes superconducting, an electric current builds up rapidly on the surface of the superconductor, producing a magnetic field that neutralizes the flux inside the material and levitates the magnet above the surface, as illustrated in Figure 2.47b.

In fact, the external magnetic field does enter the superconductor surface to some small extent within a thin layer in which the induced supercurrent is flowing. The thickness of this current layer is called the **penetration depth**; it is only about 10 nm thick near 0 K, but rises to about 1 µm at $T_C$. There is also a limit to the flux exclusion property. No magnetic field enters into the superconductor up to a **critical magnetic field strength** $B_C$ at the surface, but if that limit is exceeded the material reverts to the normal condition. The value of $B_C$ depends upon temperature, being zero at $T_C$ and rising steadily to a maximum of $B_{C0}$ at zero temperature. A good description is given by the equation

$$\frac{B_C}{B_{C0}} = 1 - \left(\frac{T}{T_C}\right)^2 .$$

It is clear from this equation and from its illustration in Figure 2.48 that superconductors should be cooled well below the transition temperature $T_C$ to maximize the magnetic field strength at which they will operate.

## 6.2 The origin of superconductivity

The discovery of the Meissner effect in 1933 stimulated even more interest in superconductivity, and a great many complex observations were accumulated. Lots of new superconductors were found and studied, but in spite of a great deal of research the *reasons* for the existence of superconductivity remained a mystery. It was clear that quantum-mechanical effects must be involved, as the behaviour of the conduction electrons is greatly altered from the usual state discussed earlier in this chapter and the superconducting state is found only at very low temperatures where quantum physics becomes dominant.

Finally in 1957 the answer came in what is called the **BCS theory** from the surnames of its authors, John Bardeen (one of the inventors of the transistor), Leon Cooper and Robert Schrieffer. The BCS model is beyond the level of this course, but we can give you a highly simplified outline of the main ideas. The essence of the BCS model is an interaction between each electron and another electron in a related quantum state, which causes them to become bound in a **Cooper pair**. It is the circulation of the negatively charged electrons in Cooper pairs that gives rise to the supercurrents. The difficulty of breaking up the electron pair is the reason for the absence of normal resistance via scattering.

The binding of the Cooper pair arises from the way an electron very slightly alters the positive ion lattice around it. This small lattice distortion results in an attractive interaction between the electron and a matching electron in a quantum state with a similar wavefunction, but propagating in the opposite direction and having the opposite spin state. These bound Cooper pairs are not localized at one place in space, but are represented by wavefunctions that spread over a considerable range within the metal, as much as 1 μm in a superconductor. This is more than 1000 times greater than the average distance between individual electrons in the superconductor. Because each pair is made up of two fermions it may be thought of as a boson, and so the Cooper pairs can undergo a kind of Bose–Einstein condensation in which they all have the same energy and momentum.

You have seen that resistance in a normal conductor is caused by the scattering of electron waves by defects and thermal agitation. When two electrons are bound in a Cooper pair, scattering of either electron would change its direction of propagation and destroy the Cooper pairing. Thus a scattering process requires the supply of at least the binding energy of a Cooper pair. As a result of the Bose–Einstein condensation, the effective binding energy per Cooper pair is substantial, about $3.5kT_C$ at 0 K, where $T_C$ is the superconducting transition temperature.

This effective binding energy is called the **superconducting energy gap**, expressing the fact that a minimum amount of energy is required to destroy the Cooper pairs and stop the superconductivity. The superconducting energy gap is a function of temperature (Figure 2.49). As the temperature rises towards the superconducting transition temperature $T_C$, the Cooper pairs gradually break up, the superconducting energy gap falls and becomes zero at $T_C$ where the material reverts to the normal (non-superconducting) state.

The Cooper pairing is strongest when the interactions between individual electrons and the lattice are largest, a condition which in the normal state gives strong

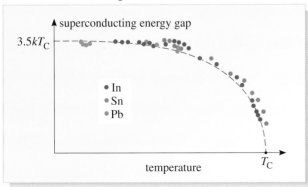

**Figure 2.49** The temperature-dependence of the superconducting energy gap for three materials: indium (In), tin (Sn) and lead (Pb).

scattering of electron waves and a low conductivity. Thus the best superconductivity arises in materials normally of little use as conductors, like lead. The best normal conductors, such as copper, are the worst superconductors. In fact some, including copper, never become superconducting no matter how far they are cooled, at least in the range of temperatures, down to a few μK, that we can reach in practice.

## 6.3  Superconductors and their applications

One of the major uses for superconductors is in the creation of large magnetic fields. In principle an electromagnet is easy to make, since you can get any magnetic field you want by passing a large enough current through sufficient turns of copper wire. However, it is not at all that simple in practice! Lots of turns of wire means lots of resistance, even for good quality copper. Although magnetic fields of 35 T have been achieved, conventional electromagnets reach a practical limit at a field of about 10 T, perhaps needing a current of 12 000 A to produce that field and 250 V to cause the current to flow. The power used is the product of those two quantities, $3.0 \times 10^6$ W. Not only must the supply of power at such inconvenient levels as 3 MW be specially negotiated with the electricity supplier, and paid for of course, but all that expensive energy is converted to heat in the copper wire and must be disposed of in a safe and environmentally friendly way. Typically water is used for cooling, in view of its huge heat capacity. However, around a thousand gallons a minute of purified cooling water is needed to absorb 3 MW, normally recirculated until near to boiling, when work finishes for the day!

By contrast, a superconducting solenoid cooled by liquid helium may have many more turns of wire (as the length of the conductor may be any value at all without increasing the resistance) and so may need a current of only about 60 A to produce a 10 T field. Once it is started, the current flows without energy loss and with no need for a power supply, removing all the problems listed above. In a small well-designed system, the liquid helium consumption by boil-off is so low that topping up is needed only a few times a year. However, the sophisticated superconducting wire is very expensive, and the capital cost of the magnet could be £300 000 (this is in year 2000 prices). Even the liquid helium usage costs can be significant in large systems, perhaps £20 000 per year for the huge magnets needed in hospital scanners for MRI, (Nuclear) Magnetic Resonance Imaging. (The word 'nuclear' is usually avoided because it is open to misinterpretation.)

However, superconductors do not allow the creation of magnetic fields more intense than about 10 T, and even that requires very special superconductor materials. The maximum magnetic field in a superconducting electromagnet is limited by the need to keep the field value at the superconductor surface below the critical magnetic field strength $B_C$ (Figure 2.48). The magnetic field strength at a radial distance $r$ from a wire carrying current $i$ is $B(r) = \dfrac{\mu_0 i}{2\pi r}$ where $\mu_0$ is the permeability of free space.

Putting this field equal to $B_C$ and setting $r$ equal to the radius $a$ of the wire, we see that the current in the wire should not exceed the value of $i_C = \dfrac{2\pi a B_C}{\mu_0}$. This maximum current is usually expressed in terms of a *critical current density*. The current density $J$ is simply current per unit cross-section of the wire, i.e. $J = i/(\pi a^2)$.

Hence the **critical current density** is

$$J_C = \frac{i_C}{\pi a^2} = \frac{2 B_C}{\mu_0 a}$$

and should be as large as possible for most applications.

**Question 2.12** At its operating temperature of 4.2 K, a superconductor has a critical magnetic field strength of 0.20 T. What is the maximum current density and maximum current if it is used to make: (a) a wire 0.10 mm in diameter; (b) a filament 1.0 μm in diameter? ■

The superconducting materials we have discussed so far are called **type I superconductors**. There is a class of materials, mainly very impure metals, alloys and compounds known as **type II superconductors** which can tolerate a certain amount of internal magnetic flux and higher magnetic field strengths at their surface, greatly increasing their usefulness for solenoids and allowing 10 T fields to be generated from them.

Amongst possible future application of superconductors is in the distribution of electrical energy to industry and households. In existing power distribution systems, which are not superconducting, some of the transmitted energy is converted into heat within the conductors that are used to connect the load to the power station, such as the overhead wires between the pylons. Some 5% to 10% of total electrical power produced is wasted in this way, costing nearly £2 billion each year in the UK alone. It should be noted that superconducting cables would not save all of that loss, because exactly zero resistance applies only when a constant current flows (DC) and there are some energy losses for the alternating current (AC) used in power distribution. Nevertheless, at least half the present transmission losses could be prevented. Many feasibility studies have been performed on superconducting power cables, but it was always found that the refrigeration costs of cooling with liquid helium, and other difficulties, are too great for superconducting power distribution to be viable.

There are many highly specialized applications of superconductors, but we have space to outline only one of these. If two superconductors are separated by an oxide layer only 1 nm or so thick, the Cooper pairs in one material can get through the insulating barrier into the other by the process of *quantum-mechanical tunnelling* (see *Quantum physics: an introduction*). This was first suggested in 1962 by Brian Josephson (then a 22-year-old research student) and the structure is called a *Josephson junction*. A magnetic field can be applied to switch off the supercurrent extremely quickly, making the device a very small fast switch that could, in principle, be used in logic circuits in computers.

The speed and size advantage of Josephson junctions prompted work on superconducting microelectronic circuits for computers, in particular by IBM from 1968 until 1983. However, the advances in the competing silicon technology kept coming, and the IBM effort was abandoned after huge research costs. The technology developed is still of interest in some laboratories and may be revived for the powerful computers expected in 20 years time.

Various applications of superconducting devices came out of this and other related work, including a Josephson junction device called a SQUID (*Superconducting Quantum Interference Device*) for the measurement of tiny changing magnetic fields. A SQUID can detect the magnetic field caused by a current flow of about ten electrons per minute and is a thousand times more sensitive than any other magnetic field sensor. This has made possible whole new areas of research such as the detailed study of magnetic fields produced by the tiny electric currents associated with thought processes in the brain.

## 6.4 High-temperature superconductivity

A good superconductor needs a high transition temperature $T_C$ and a high critical magnetic field strength $B_C$. There have been steady improvements in both of these.

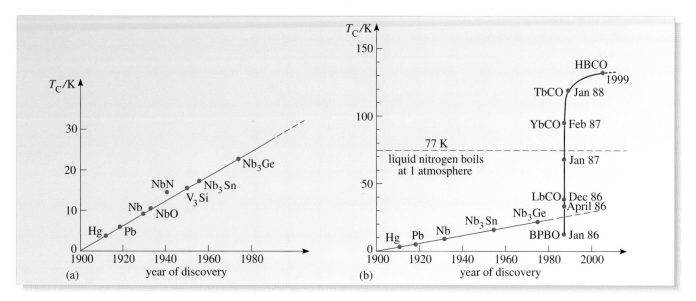

**Figure 2.50** (a) The slow rise in critical temperature $T_C$ from 1910 to 1980. (b) The discovery of high-temperature superconductors. (You do not need to remember the chemical formulas or shorthand names of these materials.)

The best superconductor discovered by 1985 was a compound of niobium Nb and germanium Ge ($Nb_3Ge$) with a $T_C$ of 23.2 K (Figure 2.50a). A further large increase in $T_C$ was thought unlikely.

The situation was transformed when J. Georg Bednorz and K. Alex Müller began work on an obscure oxide with strange metallic-like electrical properties. In January 1986 they found that it was a new kind of superconductor with a $T_C$ of 12 K. At once the hunt was on for other oxides with even better properties. By April 1986 their best $T_C$ had passed the previous record and by December 1986 the maximum $T_C$ had reached a record 35 K. These and later dramatic developments are illustrated by Figure 2.50b. For their work on superconductivity in oxides Bednorz and Müller won the 1987 Nobel Prize in Physics. This was the first time the physics prize was awarded for the discovery of a material.

Others pushed the superconductor boundaries onwards, reaching a $T_C$ of 70 K in January 1987, 90 K in February 1987 and 120 K in January 1988. The crucial step was getting a $T_C$ above the boiling point of liquid nitrogen, 77 K. In commercial terms that opened up whole new areas for the possible application of superconductors without the use of expensive liquid helium, which costs about a hundred times more than liquid nitrogen which is cheaper than beer. In 1999 a superconductor was found with a $T_C$ of 134 K, namely $HgBa_2Ca_2Cu_2O_8$ (referred to as HBCO in Figure 2.50b).

These materials are all type II superconductors in extreme form with very high critical magnetic field strength $B_C$ as well as a high critical temperature $T_C$. They are described as **high-temperature superconductors**, even though to most people 134 K is still rather cold! Sometimes they are called 'warm superconductors'.

These new high-temperature superconducting materials are brittle oxides with very difficult chemistry, and challenging mechanical properties. In the publicity and excitement surrounding the discoveries, the difficulties of transforming such difficult materials into useful forms of hardware (like flexible wires) were greatly underestimated. In spite of the problems, a long period of development since 1987 has gradually transformed dreams into realities. Wires capable of carrying 100 A at 77 K have been made in lengths approaching 1 km. It is possible that we shall soon witness the introduction of high-temperature superconducting cables, cooled by liquid nitrogen, for large-scale power distribution. This would not only save on energy losses in transmission but greatly reduce the size and mass of the cables, allowing replacement within existing ducts at higher capacity and saving on the usage of increasingly scarce copper. Power transformers and electric motors using high-temperature superconducting windings have been successfully operated. Applications are sure to become widespread as quality rises and costs fall.

At the start of the new millennium, room temperature superconductivity still remains a dream. But in 1985 an increase by a factor of over 12 was required to raise $T_C$ to 300 K from 24 K, and now we need a factor of only just over 2 on the 134 K already achieved. Though such a further rise is perhaps unlikely, it is by no means impossible. Even if liquid nitrogen cooling is always required, it may be that high-temperature superconducting connectors and devices will become standard within domestic equipment. One day the supermarkets may stock liquid nitrogen, or the milkman might deliver it each day in a vacuum flask for us to top up our computers.

# 7 Closing items

## 7.1 Chapter summary

1   Individual atoms combine to form molecules by ionic and covalent bonding involving the transfer or sharing of valence electrons in the outer subshells of the atoms.

2   Solids form by condensation of atoms or molecules. They are held together by ionic bonding, covalent bonding, metallic bonding, or by the much weaker van der Waals force and hydrogen bonding in the case of molecular crystals. Often bonding is of mixed type and not fully predictable by present day theory.

3   Solids can condense into single crystals where the atoms, molecules or ions are arranged in a regular lattice structure. Most solids are polycrystals (i.e. lots of small crystals joined at the crystal boundaries) and some are amorphous (i.e. there is little or no ordered structure).

4   Most crystals have various kinds of defects, such as vacancies or impurities or other imperfections such as dislocations.

5   Covalent solids tend to be brittle and to break under load due to stress concentration at holes and cracks. Pure metals are normally ductile due to the ease of movement of dislocations in the crystal structure.

6   Solids exhibit a vast range of electrical conductivities. The conductivities of good conductors differ from those of insulators by a factor of up to $10^{28}$. Until the band theory of solids was developed using quantum mechanics, there was no real explanation for these differences.

7   Drude's free-electron model was developed early in the twentieth century. It is based on classical mechanics and is similar to the kinetic theory for gases, with the electrons colliding with the lattice ions. Due to their small mass, the electrons have a high thermal speed of order $10^5 \, \text{m s}^{-1}$ at room temperature.

8   When an electric field exists inside a metal due to a voltage being applied across its ends, a small drift of the free electrons in a direction opposite the electric field direction is superimposed on the thermal motion. The electric current represented by this drift is related to the applied voltage by Ohm's law. In Drude's theory the resistivity of a metal is due to collisions of the free electrons with the lattice ions. The theory gives a wrong prediction for the temperature variation of resistivity.

9    Pauli's quantum free-electron model treats the free electrons as a gas of fermions. Because of the Pauli exclusion principle, the electrons have energies and speeds very much greater than in Drude's classical model. Only those electrons with energies close to the Fermi energy can move into nearby empty levels in response to an electric field.

10   The electron waves in Pauli's model are not scattered in a perfect crystal, but are scattered by crystal defects and by the thermal agitation of the crystal lattice. Thus the observed mean free path values and the temperature variation of resistivity can be explained.

11   The free-electron models of Drude and Pauli are unable to explain why the valence electrons are free or why some solids are insulators; nor can they explain the anomalous Hall effect which suggests that in some materials electric current is conveyed by positively charged carriers.

12   In the band theory of solids there are energy gaps between bands of allowed electron energy levels. Metals are materials where the highest occupied band is only partly filled, or where there are bands overlapping so that both are only partly filled.

13   In insulators the lowest occupied band, the valence band, is full and contains all the valence electrons, and the next higher band, the conduction band, contains no electrons. In silicon, the valence band and conduction band form hybrid states produced by mixing the 3s and 3p wavefunctions. When the energy gap $E_{\mathrm{gap}}$ between the valence and conduction bands is relatively small, the material is a semiconductor, and electrons can be thermally excited from the valence band into the conduction band, leaving behind an empty state in the valence band.

14   An empty state in the valence band constitutes a positive hole, and can give rise to conduction, as if by a positively charged particle. The existence of positive holes accounts for the anomalous Hall effect.

15   Pure semiconductor materials, that conduct electricity by the movement of the thermally generated electrons in the conduction band and holes in the valence band, are called intrinsic semiconductors.

16   The electrical properties of semiconductors can be strongly modified by doping with donor or acceptor impurities to introduce additional electrons into the conduction band or additional holes into the valence band, making respectively, n-type or p-type extrinsic material. This can greatly increase the conductivity of the material. A relatively small number of thermally generated electrons and holes are also always present.

17   In a p–n junction the doping changes from p-type to n-type in a thin layer called the depletion region where electrons and holes have recombined. The depletion region contains fixed acceptor and donor ions that produce an electric field and hence a potential energy hill which inhibits the diffusion of electrons from the n-to the p-material and holes from p to n. In equilibrium there is only a very small diffusion current due to very high energy electrons and holes that can climb the potential energy hill. This small diffusion current is exactly cancelled by the pair current which arises from thermally generated electrons and holes being swept across the junction by the electric field.

18   In reverse bias, the magnitude of the electric field in the depletion region is increased, totally stopping the diffusion current but allowing the small pair current to flow unchanged. In forward bias, the magnitude of the electric field is reduced, allowing a large diffusion current to flow, which is much greater than the small pair current. Thus the p–n junction acts as a diode rectifier, allowing a large current to flow in one direction, from p to n, only.

19   In some semiconductor materials additional electron–hole pairs can be produced by exposure to light or to beams of subatomic particles. If these electrons and holes are produced near the depletion layer, the electric field there can sweep them across the junction thus producing an electric current, called a photocurrent in the case of light. To be effective, the photon energy needs to be equal to or greater than the energy gap between the valence band and the conduction band. Applications include light-detectors and solar cells for electricity generation.

20   When a current flows in a forward-biased p–n junction, the recombination of electron–hole pairs in the junction region can, in some materials, produce photons of energy approximately equal to the energy gap. Light emission from a forward-biased junction is used in light-emitting-diodes (LEDs).

21   If the current through the p–n junction is increased, stimulated electron–hole recombinations can be increased to such a level that laser light is produced. In a laser diode the p–n junction region is long and thin with polished end-faces. This allows an intense build-up of stimulated photons by multiple reflections, with an output laser beam emitted from one end-face.

22   A bipolar junction transistor is a three-terminal device consisting of two p–n junctions back to back, one junction being forward-biased and the other reverse-biased. A signal in the forward-biased emitter–base circuit is then amplified in the reverse-biased base–collector circuit. Millions of transistors and other circuit components can be incorporated into integrated circuits on a single silicon chip.

23   Below the superconducting transition temperature $T_C$, many materials become superconductors with infinite conductivity for direct current flow. As superconducting material is cooled to below its transition temperature, any magnetic flux initially present in its interior is expelled. If the magnetic field at the surface of the superconductor exceeds a critical magnetic field strength $B_C$, the superconducting state is destroyed and the material reverts to an ordinary conductor. $B_C$ is temperature dependent; it is zero at $T_C$ and becomes large upon cooling well below $T_C$.

24   Superconductivity is a low-temperature quantum-mechanical effect. According to the BCS theory it involves the formation of bosonic Cooper pairs of electrons. The binding energy of the Cooper pairs constitutes an energy gap between the superconducting and normal states of the material. The energy gap gets smaller as the temperature rises and is zero at $T_C$.

25   Superconductors can be used to make electromagnets that produce large magnetic fields without power loss. They are also used in other applications, including Josephson junction devices like the SQUID which can measure extremely small magnetic fields.

26   Recently metal oxide materials have been discovered with transition temperatures $T_C$ up to 134 K. These are known as high-temperature superconductors and applications of these materials usually involve only liquid nitrogen cooling. Difficulties in preparing these materials for practical use in large-scale situations are severe, but are now being overcome.

## 7.2 Achievements

After you have completed your work on this chapter, you should be able to:

A1   Explain the meaning of all the newly defined (emboldened) terms introduced in this chapter.

A2    Discuss the various bonding mechanisms and crystal defects in solids and the role these play in determining the structure and properties of solids.

A3    Discuss Drude's classical free-electron model in metals, and use it to obtain an expression for the electrical conductivity and resistivity of metals.

A4    Describe the failures of Drude's free-electron model in relation to mean free paths of electrons, the temperature-dependence of resistivity, and Hall effect results.

A5    Discuss Pauli's quantum free-electron model in metals and indicate how it describes electrical conductivity. Explain how it accounts for the long mean free paths of electrons and the temperature-dependence of resistivity.

A6    Discuss the failures of Pauli's quantum free-electron model, and describe how the assumptions in Pauli's model are modified in the band theory of solids.

A7    Outline the origin of energy bands and energy gaps in crystalline solids.

A8    Describe how the band theory of solids explains the existence of metals, insulators and semiconductors.

A9    Discuss the concentrations of electrons and holes in intrinsic semiconductors and in n-type and p-type extrinsic semiconductors.

A10   Describe the motion of electrons and holes in a p–n junction in equilibrium, and how this leads to a rectifying effect when forward- and reverse-bias voltages are applied.

A11   Describe the operation of devices based on the p–n junction, such as light-detectors, solar cells, light-emitting-diodes, laser diodes and the bipolar junction transistor.

A12   Describe the phenomenon of superconductivity, its causes and applications.

## 7.3  End-of-chapter questions

**Question 2.13**   Explain, in terms of bonding properties, the following: (a) Diamond is very much harder than grey tin, even though diamond and grey tin both have the same diamond-like crystal structure. (b) Graphite is not as hard as diamond, even though both are pure carbon. (c) Most metals can be hammered into new shapes but glass cannot.

**Question 2.14**   Use Pauli's equation (Equation 2.8) for the electrical conductivity of a metal, $\sigma = ne^2\lambda_F/m_e v_F$, to confirm that the units of $\sigma$ are $\Omega^{-1}\,m^{-1}$.

**Question 2.15**   Suppose the electric field is held constant within a metal wire. During the following (imagined) changes, does the electric current rise or fall or stay the same? Give reasons for your answers, based on Pauli's quantum free-electron model where appropriate.

(a) The length of the wire is increased, (b) the conductivity of the material is reduced; (c) the mean free path of electrons is increased, (d) the temperature is increased substantially, (e) the free-electron density is increased.

**Question 2.16**   Very briefly state how the tight-binding model of solids introduces energy bands and gaps.

**Question 2.17**   Explain how the band theory of solids accounts for the existence of insulators, semiconductors and metals.

**Question 2.18**   Briefly describe the movement of electrons and holes in the equilibrium state of a p–n junction, and how the junction can act as a rectifier.  ■

# Chapter 3  Nuclear physics

## 1 Energy puzzles

In 1898 Marie Curie (Figure 3.1) discovered the existence of a mysterious new element which she called *radium*. Present in minute amounts in ores of the radioactive metal uranium, radium is nevertheless responsible for a large part of the ore's radioactivity. In a remarkable endeavour, Madame Curie set about the task of chemically separating the radium from the uranium ore. After many years of work with many tons of the uranium ore *pitchblende*, she eventually isolated a few milligrams of pure radium.

Radium is radioactive to an extraordinary degree, millions of times more so than uranium. As a result of its radioactivity, radium generates heat. A sample of radium is warm to the touch and glows in the dark and, amazingly, maintains its heat output for months or even years without any apparent change of composition. In fact a sample of radium can heat its own weight of water from zero degrees to boiling in about 46 minutes, and this power output falls by only 1% in every 23 years! This means that a sample of radium can generate nearly a million times more heat than an equal mass of burning coal. Where can this energy come from? Energy conservation is one of the fundamental pillars of physics so you can imagine how puzzled the physics community was at the time of Marie Curie's discovery.

**Figure 3.1**  Madame Curie (1867–1934).

In fact, the radium problem was just a variation of an energy puzzle known from the middle of the nineteenth century. Where does the Sun get *its* energy from? Chemical energy can be discounted: the burning of a lump of coal of the same mass as the Sun would produce enough energy to keep the Sun going for a mere 6000 years or so. A much more promising contender is gravitational contraction. A slow contraction of the Sun under the pull of its own gravitational forces, by say 10% in radius, would result in the conversion of gravitational potential energy into heat sufficient to sustain the Sun's present output for about $3 \times 10^8$ years. For a time it was believed that this was the answer, but in the early part of the twentieth century it became clear from geological and astronomical evidence that the Earth was at least $4 \times 10^9$ years old and the Sun has maintained its present power output for at least this long.

The radium puzzle and the Sun puzzle have essentially the same solution: the energy comes from reactions occurring inside the nuclei of atoms. About 99.9% of the mass of everything that you can see lies in the nuclei of atoms, yet apart from our dependence on the Sun's heat, the energy locked up in atomic nuclei does not directly affect our lives. This is because most of our activities, such as the metabolism of our bodies and the burning of fossil fuels, etc., are driven by chemical energy which has its origin in changes in the arrangements of the outermost electrons in atoms and molecules. The atomic nuclei remain uninfluenced by all but the most violent environments, such as those found in stars.

So why is nuclear physics important? Firstly, most of the known universe consists of violent stellar environments (if you discount the vastness of empty space) in which nuclear reactions not only provide the energy source but have also led to the synthesis of all the naturally-occurring elements, starting from a primeval soup of protons, neutrons and electrons. The mercury in your tooth fillings, for example, was created by nuclear reactions in a supernova explosion sometime before the Solar System was formed. In fact nuclear physics provides the answers to some of the most basic questions concerning the world about us. Why, for example, is there more carbon than gold? Why are there just 90 naturally occurring elements? And at the practical level: how can the energy of the atomic nucleus be safely extracted and harnessed?

This chapter begins with a description of some of the basic properties of atomic nuclei including the phenomenon of radioactivity. The emphasis is on an overview of all known nuclei, seeing how their properties and behaviour fall into a pattern. Sections 3 and 4 attempt to explain this pattern using models of the nucleus. In fact, the structure of the nucleus is still not fully understood but you will see that there are a number of different models which describe different aspects of the nucleus. Of course, quantum mechanics lies at the heart of any description of nuclear phenomena, and nuclei provide an arena where quantum mechanics performs some of its most dazzling acts. Section 5 is devoted to the decisive role played by quantum-mechanical tunnelling in a wide range of nuclear reactions, including radioactivity, power generation in the Sun, and nuclear fusion and fission.

# 2 Characteristics of atomic nuclei

We begin by describing the *composition* and *size* of atomic nuclei and introducing the terminology and symbols used to specify individual nuclei. Some nuclei are stable, meaning that they last forever, while others are *unstable* meaning that they are subject to *radioactive decay*. Most of the known nuclei are, in fact, unstable. Section 2.3 describes the different modes of radioactive decay of unstable nuclei, while Section 2.4 considers the rate at which radioactive decay occurs and introduces the *exponential decay law* and *half-life*. A survey of all known nuclei, stable and unstable, is given in Section 2.5. Much of the rest of this chapter is concerned with understanding why nuclei have the properties and behaviour revealed by this survey.

## 2.1 Nuclear composition

Nuclei are composed of *protons* and *neutron*s held together by the so-called *strong nuclear force*, a short-range force operating only on the nuclear scale. The single proton is, in fact, the lightest nucleus, that of ordinary hydrogen. The proton has an electric charge equal in magnitude and opposite in sign to that of the electron, but is about 1850 times heavier than the electron. The neutron is electrically neutral and a little heavier than a proton.

The number of protons in a nucleus is called the **atomic number** and is denoted by the symbol $Z$. The number of neutrons, denoted by $N$, is called the **neutron number**. The total number of particles in the nucleus is called the **mass number** $A$. Thus

$$A = Z + N. \tag{3.1}$$

We use the term **nucleon** to mean a particle, proton or neutron, in an atomic nucleus. Thus $A$ is equal to the total number of nucleons in the nucleus, and is sometimes called the *nucleon number*. Since the neutron and proton have similar masses, the mass number $A$ is *approximately* equal to the total mass of the nucleus relative to the mass of a single proton or neutron.

The number of atomic electrons in the electron shells surrounding the nucleus is also equal to the atomic number $Z$, since the atom as a whole is electrically neutral. Thus the atomic number determines what chemical element we are dealing with. For example, $Z = 1$ specifies hydrogen, $Z = 2$ is helium and $Z = 26$ is iron. The Periodic Table of the elements (see Figure 3.2) shows the atomic numbers and chemical symbols of the elements.

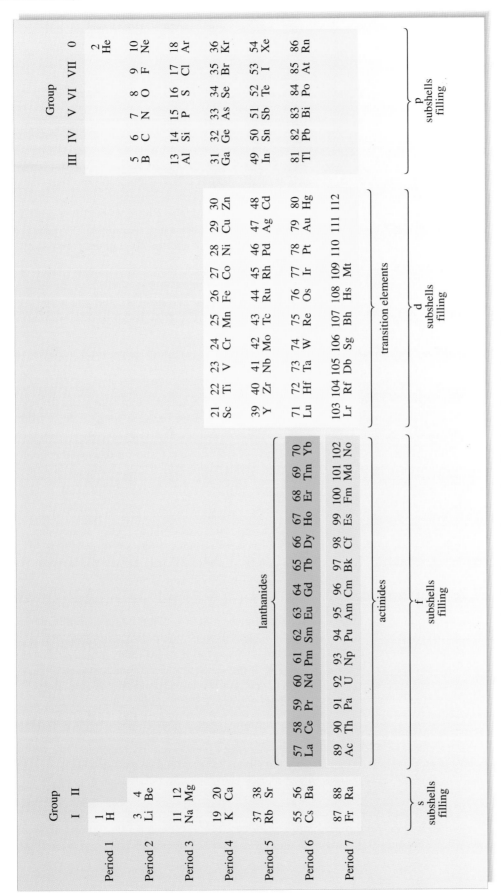

**Figure 3.2** The Periodic Table of the elements showing the atomic numbers Z and the chemical symbols.

It turns out that for any given atomic number $Z$, nuclei exist which contain various different numbers of neutrons, $N$, and thus have correspondingly different mass numbers $A$. Nuclei with the same $Z$ but different $N$ are called **isotopes**. Since $Z$ determines the chemical element, we speak of the isotopes of such and such an element. The chemical properties of different isotopes of the same element are almost identical.

The full identification of the nucleus of an element with chemical symbol Q is given symbolically as $^A_Z Q_N$. For example, a nucleus of the most abundant isotope of iron (Fe) has $A = 56$, $Z = 26$, $N = 30$ and is specified symbolically as $^{56}_{26}\text{Fe}_{30}$. This clearly contains redundant information. First of all, since $A = N + Z$, one can omit the subscript specifying $N = 30$. Moreover, since the chemical symbol Fe contains the same information as $Z = 26$ (if you can remember it!) the $Z$ number is often omitted too. For example, we can specify the iron nucleus above more simply as $^{56}\text{Fe}$ and we refer to it as 'iron 56'. Whether or not we omit the $Z$ and the $N$ numbers depends on the context.

You are not expected to remember $Z$ values for elements but you will find it useful to know, at least, that $Z = 1$ for hydrogen (H), $Z = 2$ for helium (He), and you will probably pick up a few more like $Z = 6$ for carbon (C), $Z = 8$ for oxygen (O) and $Z = 92$ for uranium (U).

Hydrogen nuclei with $A = 2$ are called *deuterons* and those with $A = 3$ are called *tritons*.

**Question 3.1**   (a) Hydrogen, $Z = 1$, has three known isotopes, with $A = 1$, 2 and 3. Write down the full (redundant) symbols $^A_Z H_N$ for these three isotopes.

(b) Repeat this for the two isotopes of oxygen, $Z = 8$, that have the most and the least numbers of neutrons, given that the mass numbers $A$ for oxygen isotopes range from 12 to 26.

(c) For hydrogen, the stable isotopes have $N = 0$ and $N = 1$ and for oxygen the stable isotopes have $N = 8$, 9 and 10. List these stable isotopes in the non-redundant form $^A\text{H}$ and $^A\text{O}$. (By *stable* we mean non-radioactive. We discuss radioactivity in Section 2.3.)   ■

Several of the unstable oxygen isotopes were discovered in the 1990s, and by the time this book is printed the number of known isotopes for oxygen may have been extended beyond $A = 26$ with the discovery of some new very unstable isotopes.

## 2.2  Nuclear sizes

$\alpha$-*particles* are nuclei of the common helium isotope $^4_2\text{He}_2$.

What is meant by the size of an atomic nucleus and how can it be measured? Like the electrons in atoms, the nucleons in a nucleus must be described by quantum mechanics. Thus we must think of the nucleus in terms of nucleon wavefunctions and nucleon energy levels in analogy with the description of atomic electrons in *Quantum physics: an introduction*. The nucleon wavefunctions fall off smoothly with distance from the centre of the nucleus, hence the boundary of the nucleus is not well defined and a nucleus does not have an unambiguous size. However, it is possible to measure the distribution of nuclear charge and mass across the nuclear volume. One way of doing this is by *scattering experiments*. By about 1913 Rutherford (Figure 3.3) had already put rough limits on the size of the atomic nucleus from the way $\alpha$-*particles* scattered from nuclei in the experiments of Geiger and Marsden. The scattering of $\alpha$-particles of sufficiently low energy is governed by the Coulomb repulsion between the $\alpha$-particle and the target nucleus, as described in *Quantum physics: an introduction*. For higher energies and for low-$Z$ nuclei, the two particles can come closer together and interact by the strong nuclear force as well as by the Coulomb repulsion. Thus deviations from pure Coulomb scattering give information about the nuclear size. Roughly speaking, such experiments implied that nuclear radii were no more than about $10^{-14}$ m, that's about $10^4$ smaller than the radius of the atom.

Although much information about nuclear charge sizes has been obtained from modern forms of these experiments, by far the most detailed information has been obtained by *high-energy electron scattering* experiments. You know from *Quantum physics: an introduction* that electrons have a wave nature, as demonstrated by the Davisson–Germer experiments in which electron diffraction patterns were obtained by scattering electrons from a crystal target. This famous experiment established the wave nature of electrons as described by the relationship $\lambda_{dB} = h/p$ where $\lambda_{dB}$ is the de Broglie wavelength, $p$ the magnitude of the electron's momentum and $h$ is Planck's constant. The electron energies used in the Davisson–Germer experiment were typically about 50 eV which gives de Broglie wavelengths of the same order as atomic dimensions.

**Figure 3.3** Lord Rutherford (1871–1937).

What we are discussing now is rather similar except that the electrons are scattered from nuclei rather than from atoms. Because nuclei are much smaller than atoms, the de Broglie wavelengths need to be correspondingly shorter to resolve details of nuclear sizes, and so the electrons must be accelerated to much greater energies, typically about $500 \times 10^6$ eV or 500 MeV. The MeV is a very convenient unit of energy in nuclear physics and we shall use it a lot.

Nuclear energies are measured in MeV: 1 MeV = $10^6$ eV.

An electron scattering experiment is depicted in Figure 3.4. Electron detectors arranged around the target count the numbers of electrons that have been scattered though different angles $\theta$ by the nuclei in the target.

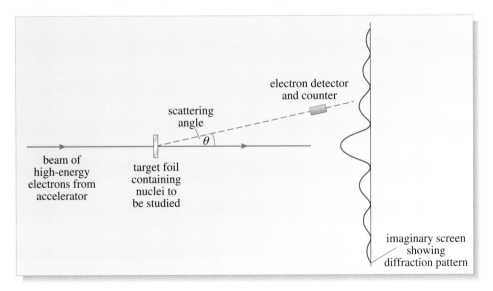

**Figure 3.4** Essential elements of an electron scattering experiment. The beam of high-energy electrons from the accelerator enters from the left and strikes a target foil containing the nuclei to be studied. Many electrons go straight through, but some will be scattered by the nuclei in the target. The diffraction pattern, indicated on an imaginary screen, is actually measured by electron detectors and counters. The observed diffraction pattern is made up of contributions from the large number of nuclei where the beam passes through the target. This region is kept as small as possible.

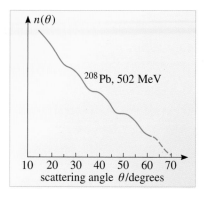

**Figure 3.5**  Diffraction of 502 MeV electrons from $^{208}$Pb nuclei. $n(\theta)$ is proportional to the number of electrons diffracted through an angle $\theta$.

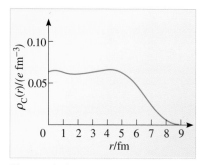

**Figure 3.6**  Nuclear charge densities $\rho_c(r)$, in $e\,\text{fm}^{-3}$ (i.e. elementary charges per fm$^3$) for lead $^{208}$Pb.

Muons will be discussed in Chapter 4 of this book.

These numbers depend on the angle $\theta$ in an oscillatory way characteristic of a diffraction pattern. Figure 3.5 shows the diffraction patterns for 502 MeV electrons scattering from lead nuclei, $^{208}$Pb.

The diffraction minima in Figure 3.5 are not very deep, partly because nuclei have ill-defined edges, unlike a sharp edged hole in a screen which would create deep minima in an optical diffraction pattern (see *DFW*). Figure 3.6 shows the nuclear charge density as a function of radial distance $r$ from the centre obtained from diffraction data of this kind using detailed theoretical analysis of electron scattering. The $r$-axis is marked out in femtometres (fm), where 1 fm = $10^{-15}$ m, a convenient unit of distance on the nuclear scale.

Nuclear dimensions are measured in femtometres (fm): 1 fm = $10^{-15}$ m.

Note that the nuclear charge density $\rho_c(r)$ is fairly uniform inside the lead nucleus, and falls off gradually over the outer 3 fm or so. We can define the **radius of a nucleus** to be the value of $r$ at which $\rho_c(r)$ falls to one-half the value it has at the nuclear centre ($r = 0$). You can see that with this definition the radius of the $^{208}$Pb nucleus is about 6.5 fm.

Nuclear sizes can also be deduced from high-resolution measurements of optical and X-ray spectral lines emitted by atoms. You may find this surprising since atomic spectral lines are produced, not by changes in the nucleus, but by the atomic electrons jumping from one electronic energy level to another. However, you will recall (*QPI*) that some electron states, especially $s$-states (those with quantum number $l = 0$) have large probability densities inside the nuclear volume. This nuclear penetration influences the energies of these electron states and the frequencies of the emitted spectral lines. Modern laser techniques make this a very sensitive method for measuring the sizes of unstable nuclei that have lifetimes that are too brief for other methods to work.

Much larger penetration effects can be obtained using *muonic atoms*. The heavier cousin of the electron, known as a *muon*, can be captured into orbits about nuclei in place of electrons. Because muons are about 200 times heavier than electrons and carry the same electric charge, their probability distributions are about 200 times smaller than those of electrons. Hence they can probe the nucleus much more deeply than atomic electrons. The spectra of muonic atoms has been an important source of information on nuclear sizes.

In the electron scattering and spectroscopic techniques described above, the electrons (or muons) interact with the protons in the nucleus through the Coulomb force, and so they give information only about the nuclear *charge* distribution $\rho_c(r)$. To obtain information about the *matter* distribution, i.e. the distribution of nucleons, irrespective of whether they are neutrons or protons, we must use techniques that involve the *strong nuclear force*, the short-range force which holds nucleons together. Electrons and muons are *unaffected* by the strong nuclear force, and so the experiment must use some other particle that *is* affected by the strong nuclear force. One technique involves a modern form of the Geiger and Marsden experiments mentioned above, in which high-energy $\alpha$-particles get close enough to the target nuclei to be scattered by the strong nuclear force. Such experiments show that the nuclear matter distribution is generally very similar to the charge distribution and hence that neutrons are distributed in much the same way as protons. Figure 3.7 shows the nuclear matter densities $\rho_m(r)$ in six stable nuclei.

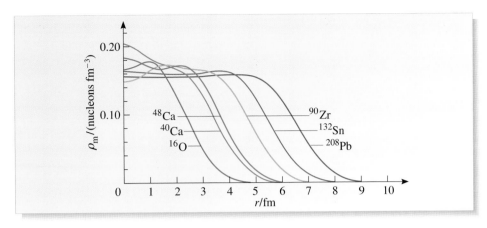

**Figure 3.7**  Nuclear matter densities $\rho_m(r)$ for six stable nuclei.

The vertical axis gives the number of nucleons per $fm^3$. You can see that the matter density in the central region, i.e. everywhere except the edge where $\rho_m(r)$ falls with $r$, is much the same for all but the lightest nuclei; those with $A$ less than about 50 show the most variation. A characteristic value for nuclear matter density is about 0.17 nucleons $fm^{-3}$.

**Question 3.2**    (a) Calculate the characteristic nuclear matter density in $kg\,m^{-3}$. What mass would a 5 ml spoonful of nuclear material have? (1 ml = 1 millilitre = $10^{-6}\,m^{-3}$.) (b) Evaluate the radius of a spherical nucleus of mass number $A$ and uniform density 0.17 nucleons $fm^{-3}$.  ■

As a consequence of the constant nuclear density, the volume occupied by a nucleus is proportional to its mass number $A$, and so the nuclear matter radius $r$, as defined above, is proportional to $A^{1/3}$. The usual value taken for the nuclear radius is

$$r \approx (1.2\ fm)A^{1/3}$$

which takes into account the diffuse surface, and agrees fairly well with your answer to Question 3.2(b).

Here we meet one of the dramatic differences between atoms and nuclei: nuclei have more or less the *same density* whereas atoms have more or less the *same size*. As the nuclear charge $Z$ increases, more and more outer electron shells are filled, but the atoms do not get larger because the electrons are more strongly attracted towards the nucleus. On the other hand, nuclear matter cannot be compressed, and even the centres of neutron stars (pulsars) have a density comparable to that at the centre of a lead nucleus.

Atomic sizes in fact vary by a factor of two or so in a periodic fashion with increasing $Z$, from the inert gases (smallest) to the alkali metals (largest), but there is no significant progressive increase in atomic size with $Z$.

## 2.3  Radioactive decay

In 1896, Henri Becquerel (Figure 3.8) discovered a wholly unexpected new phenomenon-radioactivity: the emission of ionizing 'radiations' by certain substances. Unlike the then recently-discovered X-rays, this emission occurred with no apparent external source of energy. Marie and Pierre Curie and others soon showed that radioactivity is a property of certain elements, such as uranium, and not related to chemical composition, i.e. whether the radioactive substance is in the form of the pure element, or a compound, such as an oxide. When the Curies discovered previously unknown elements, especially radium and polonium which are radioactive to an extraordinary degree and release considerable amounts of heat, radioactivity seized the public imagination.

**Figure 3.8**  Henri Becquerel (1852–1908).

Within five or six years Rutherford had shown that there were two distinct kinds of radiation emitted by radioactive substances which he called *α-radiation* and *β-radiation* (a third kind, *γ-radiation*, was discovered later). He also discovered that radioactivity involved the transmutation of one element into another. This transmutation of elements demolished the cherished chemical dogma that elements cannot be converted from one to another. It was for this discovery that Rutherford won the Nobel Prize for chemistry in 1908. Rutherford also discovered that radioactivity exhibited statistical behaviour, later recognized by Einstein as an indication that quantum physics was at work.

We now know that **radioactivity** is caused by the spontaneous disintegration of unstable nuclei. There are three main modes of radioactive decay: α-decay, β-decay and γ-decay, and these are discussed below. Other modes of decay can also occur, such as the spontaneous fission of heavy elements, but these are less common.

### α-decay

In α-**decay**, a nucleus spontaneously emits an *α-particle*, i.e. a helium nucleus, $^4_2\text{He}_2$. The α-particles are emitted with very high energy, usually in the MeV range. These high-energy α-particles are what Rutherford called α-rays. Since the helium nucleus carries two protons, the atomic number $Z$ of the emitting nucleus decreases by two.

**Question 3.3**   What are the changes in $A$ and $N$ when a nucleus undergoes α-decay?  ■

An example is the α-decay of the commonest isotope of uranium which is thereby transmuted into an isotope of thorium. We write this decay symbolically as

$$^{238}_{92}\text{U} \rightarrow \ ^{234}_{90}\text{Th} + \alpha$$

or, since the α-particle is a helium nucleus,

$$^{238}_{92}\text{U} \rightarrow \ ^{234}_{90}\text{Th} + \ ^4_2\text{He}.$$

In the latter form you can easily check that the total number of nucleons hasn't changed (left superscripts: 238 = 234 + 4) and that charge is conserved (left subscripts: 92 = 90 + 2). We have not troubled to show the (redundant) information that the total number of neutrons also remains the same.

α-decay occurs mainly in heavy nuclei with $Z$ greater than 80 or so; we shall explain why this is so in Section 4.

**Question 3.4**   Write out the α-decay process for Marie Curie's famous radium nucleus $^{226}_{88}\text{Ra}$, leading to an isotope of radon, Rn.  ■

Emitted α-particles have discrete energies. For example, $^{226}_{88}\text{Ra}$ emits α-particles of energies 4.78 MeV and 4.60 MeV. A decay is usually accompanied by the emission of γ-rays, which also have discrete energies. As you might guess, the discreteness of the energies of emitted particles reflects the fact that nuclei, like atoms, exist in discrete energy levels, a topic we shall discuss in some detail in Section 4.

In any decay process it is common to call the initial nucleus the **parent nucleus** and the final nucleus the **daughter nucleus**. Thus in the decay process $^{238}_{92}\text{U} \rightarrow \ ^{234}_{90}\text{Th} + \alpha$, the nucleus $^{238}_{92}\text{U}$ is the parent nucleus and $^{234}_{90}\text{Th}$ is the daughter nucleus.

## β-decay

There are three distinct kinds of decay processes that are classed as β-**decay**. They are negative β-decay, positive β-decay and electron capture:

### Negative β-decay (β⁻-decay)

β⁻-**decay** is effectively the transformation of a neutron into a proton, an electron and a wonderfully elusive, almost massless, electrically neutral particle called a *neutrino*. Thus we write

$$n \rightarrow p + e^- + \overline{\nu}_e$$

where the symbol n denotes a neutron, p is a proton (or hydrogen nucleus $^1_1H$), $e^-$ is an electron and the symbol $\overline{\nu}_e$ denotes a neutrino, in fact an *antineutrino of electron type* to give it its full title. A typical β⁻-decay is the one used in the process of *carbon dating*: the decay of $^{14}_6C$ into an isotope of nitrogen,

$$^{14}_6C \rightarrow {}^{14}_7N + e^- + \overline{\nu}_e.$$

The electron and antineutrino are ejected from the nucleus, sharing the discrete amount of energy released in the process, in this case 156 keV.

In Chapter 4 we shall distinguish the neutrino $\nu_e$ of β⁺-decay from the antineutrino of β⁻-decay.

The electrons ejected in β⁻-decay are the *β-rays* discovered by Rutherford in 1898. The neutrinos interact so weakly with matter that they are almost undetectable. In fact neutrinos were detected for the first time in 1953, some 55 years after Rutherford's discovery of β-rays. However, their presence in β⁻-decay was inferred long before that by the fact that the β-rays were ejected with a spread of energies rather than discrete energies, and so the existence of an accompanying particle was postulated in order to maintain conservation of energy, momentum and angular momentum.

An important point to notice is that β⁻-decay always *increases* the proton number $Z$ of the nucleus by one unit but does not change the mass number $A$; of course it also reduces the neutron number by one unit since $A = Z + N$.

The β⁻-decay process occurs in nuclei that are unstable because they contain too many neutrons for stability, as we shall explain in Section 4. For carbon, $Z = 6$, only the isotopes with $N = 6$ and $N = 7$ are stable, whereas all carbon nuclei with $N \geq 8$ undergo β⁻-decay.

But what about nuclei that have too *few* neutrons for stability, such as carbon isotopes with $N < 6$? This question leads us to β⁺-decay.

### Positive β-decay (β⁺-decay)

β⁺-**decay** occurs in nuclei that have too many protons for stability; it is effectively the transformation of a proton into a neutron, a positron and a neutrino. A **positron**, denoted by the symbol $e^+$, is the *antiparticle* of an electron; it has the same mass as the electron but carries a unit of positive electric charge. Thus we can express β⁺-decay by writing $p \rightarrow n + e^+ + \nu_e$, but this process can occur only inside a nucleus. The neutron remains inside the nucleus while the positron and neutrino are ejected, sharing the released energy.

Free protons are stable. Energy conservation does not allow a free proton to decay to a neutron since the neutron has a slightly greater mass.

Note that in β⁺-decay the atomic number $Z$ of the nucleus *decreases* by one unit while the mass number $A$ remains the same. Here is an example of β⁺-decay that has application in a medical imaging technique called *positron emission tomography* (to be discussed in Section 2.6)

$$^{13}_7N \rightarrow {}^{13}_6C + e^+ + \nu_e.$$

**Question 3.5**   Two other examples of β$^+$-decay (also used in positron emission tomography) are the decays of $^{15}$O and $^{11}$C. Write out the transformations following the form given above. Specify the parent and daughter nuclei in each case. (You may need to consult Figure 3.2 to determine the nuclei produced.)   ■

*Electron capture*

**Electron capture** is made possible by the fact that some atomic electrons (particularly those with quantum number $l = 0$) penetrate into the nucleus to some extent, allowing a proton to absorb one of them. Thus we write $p + e^- \rightarrow n + \nu_e$. No particles are emitted by the nucleus, apart from the almost undetectable neutrino which carries off all the released energy. Electron capture can be detected by observing the *characteristic X-rays* that are emitted when atomic electrons in outer shells jump into the vacancies created by the capture. For heavy nuclei, electron capture is by far the dominant decay process for nuclei with too many protons for stability. Like β$^+$-decay, electron capture *decreases* the atomic number $Z$ of the nucleus. An example is electron capture in beryllium

$$^7_4\text{Be} + e^- \rightarrow {}^7_3\text{Li} + \nu_e.$$

Electron capture in $^7_4$Be produces a significant part of the neutrino flux from the Sun.

*γ-ray emission*

α-decay and β-decay are usually accompanied by the emission of γ-rays. These consist of very short-wavelength electromagnetic radiation similar to X-rays, but whereas X-rays are commonly produced by atomic electrons falling into a vacancy in an inner shell, γ-rays have their origin in energy changes inside the nucleus itself. The emission of γ-rays from radioactive nuclei is often referred to as γ-**decay**.

We have mentioned that nuclei are described by quantum mechanics and exist in discrete energy levels. When a parent nucleus undergoes α-decay, or any form of β-decay, the daughter nucleus may be left in an excited nuclear energy level, from which it can make a transition into a lower nuclear energy level. This transition is accompanied by the emission of a photon of energy equal to the difference in the energies of the levels involved. Thus the emission of γ-rays from an excited nucleus is the nuclear equivalent of the emission of light by excited atoms described in *Quantum physics: an introduction*. You saw there that the frequency $f$ of the emitted electromagnetic radiation is related to the energy change by *Planck's law*, $hf = E_2 - E_1$. The separation of nuclear energy levels, $E_2 - E_1$, is typically in the MeV range rather than just a few eV in the case of atomic levels, so the emitted photons have energies typically some million times greater than the energies of the photons of light emitted from excited atoms. These high-energy photons emitted when excited nuclei decay to lower nuclear energy levels are called γ-**rays**.

**Question 3.6**   Electron capture in a certain isotope of barium is described by $^{133}_{56}\text{Ba} + e^- \rightarrow {}^{133}_{55}\text{Cs} + \nu_e$ . The daughter $^{133}$Cs nucleus is left in an excited energy level 0.437 MeV above its nuclear ground state. See Figure 3.9. Use Planck's law to determine the wavelength of the γ-ray emitted when the $^{133}$Cs nucleus makes a transition to the lower energy level 0.081 MeV above the ground state.   ■

As shown in the answer to the above question, γ-ray wavelengths can be similar to those of X-rays, although they can also be even shorter, making γ-radiation extremely penetrating, much more so than α- and β-rays.

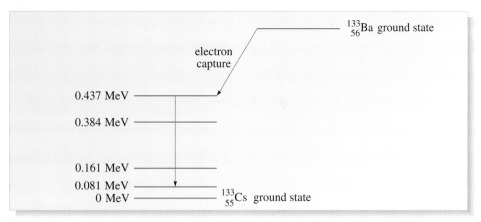

**Figure 3.9**  Some energy levels of the caesium nucleus $^{133}_{55}\text{Cs}$. When the $^{133}_{56}\text{Ba}$ nucleus undergoes electron capture the daughter Cs nucleus can be left in the fourth excited state, as indicated by the sloping arrow. The Cs energy levels are indicated by horizontal lines labelled with the energies above the Cs ground state in MeV. The transition referred to in Question 3.6 is indicated by the vertical arrow.

## 2.4  The exponential decay law

Radioactivity, like everything else at the atomic and nuclear level, is governed by the laws of quantum mechanics. Consequently only statistical predictions about the decay process are possible. Suppose, for example, you have a sample containing barium nuclei $^{133}_{56}\text{Ba}$. The barium nuclei are indistinguishable particles (they are fermions in fact) and are therefore identical in all respects. Yet they will not all undergo radioactive decay at the same time; in fact it is not possible to predict exactly when any particular nucleus will decay. All we can say is that if you have $N$ unstable (parent) nuclei, then the *average number* that will decay during a short interval of time, $\Delta t$, is proportional to both $N$ and $\Delta t$. That is,

$$\Delta N \propto -N\,\Delta t$$

where the minus sign comes from the fact that $\Delta N$, the average change in the number of parent nuclei, is a negative number, i.e. $|\Delta N|$ is the average number that decay in the interval $\Delta t$.

We can write the proportionality above as $\Delta N/\Delta t \propto -N$. If we let $\Delta t$ become indefinitely small and then take the limit $\Delta t \to 0$, $\Delta N/\Delta t$ becomes the derivative $\mathrm{d}N/\mathrm{d}t$, the *instantaneous rate of change* of parent nuclei; it is of course negative since $N$ is decreasing with time. Introducing a constant of proportionality $\lambda$, we have

By average number in this context we mean that if you had a very large number of samples each containing exactly $N$ parent nuclei, then $|\Delta N|$ is the number that decay in the interval $\Delta t$, averaged over all samples.

$$\frac{\mathrm{d}N}{\mathrm{d}t} = -\lambda N. \qquad (3.2)$$

The constant $\lambda$ is called the **decay constant**. Its value characterizes the particular nuclear species and decay process. For example, the decay constant for $\alpha$-decay of the most common uranium isotope $^{238}\text{U}$ is $\lambda = 1.53 \times 10^{-10}$ year$^{-1}$.

Equation 3.2 is a differential equation and has a solution

$$N(t) = N_0 \exp(-\lambda t) \qquad (3.3)$$

where $N(t)$ is the average number of parent nuclei at time $t$, and $N_0$ is the number of unstable nuclei at $t = 0$. This equation describes the **exponential decay law** that governs radioactivity.

The reciprocal of $\lambda$ is $\tau = 1/\lambda$, and is known as the *time constant* of the decay. The time constant is the time taken for the average number of parent nuclei to fall to $1/e$ of the initial number.

In the context of radioactive decay it is conventional to refer to the **half-life**, represented by the symbol $T_{1/2}$, rather than the time constant $\tau$. The half-life is the time taken for the number of parent nuclei to fall to one half of the initial number.

We can see how the half-life is related to the decay constant $\lambda$ and to $\tau$ by making use of the fact that when $t = T_{1/2}$, $N(t) = N_0/2$. Substituting these values of $N(t)$ and $t$ into Equation 3.3, gives

$$N_0/2 = N_0 \exp(-\lambda T_{1/2}).$$

The $N_0$ cancels and we are left with $1/2 = \exp(-\lambda T_{1/2})$. Taking natural logarithms of both sides of this equation, and remembering that $\log_e(\exp(x)) = x$, we obtain the result $\log_e(1/2) = -\lambda T_{1/2}$. Remembering also that $\log_e(1/2) = -\log_e(2)$ we obtain the result

$$T_{1/2} = \frac{\log_e(2)}{\lambda} = \frac{0.6931}{\lambda} = 0.6931\tau \quad \text{(to 4 significant figures).} \quad (3.4)$$

We can now express the exponential decay law as

$$N(t) = N_0 \exp\left(-\frac{0.6931}{T_{1/2}} t\right). \quad (3.5)$$

Since the parameters $\lambda$, $\tau$ and $T_{1/2}$ are simply related to one another by Equation 3.4, any one of them is sufficient to characterize a decay process. We shall refer mainly to the half-life.

### Example 3.1

Suppose you are given $N_0$ radioactive nuclei characterized by a half-life $T_{1/2} = 1$ year. Use the exponential decay law, given in Equation 3.5, to determine: (a) How many are left after 2 years? (b) How many are left after 3 years?

### Solution

(a) Using Equation 3.5, the average number after 2 years is found by putting $t = 2$ years and $T_{1/2} = 1$ year. Thus

$$N(2 \text{ years}) = N_0 \exp\left(-\frac{0.6931 \times 2 \text{ years}}{1 \text{ year}}\right)$$

$$= N_0 \exp(-1.3862)$$

$$= 0.2500 N_0 = N_0/4.$$

(b) After 3 years we have

$$N(3 \text{ years}) = N_0 \exp\left(-\frac{0.6931 \times 3 \text{ years}}{1 \text{ year}}\right)$$

$$= N_0 \exp(-2.079)$$

$$= 0.1250 N_0 = N_0/8.$$

You can see from the above example that,

> During any interval equal to the half-life, $T_{1/2}$, the average number of undecayed nuclei halves (see Figure 3.10).

We have stated that the radioactive decay is a statistical process and the exponential decay law gives only the *average* number of undecayed nuclei. However, you can regard it as exact when $N$ is very large. Obviously it cannot be exact when $N$ is very small; you only have to think of the extreme case of $N_0 = 1$. Then, the number can never halve since you cannot have half a nucleus!

Nuclear half-lives cover a remarkably large range. For the common isotope of uranium, $^{238}$U, the half-life for $\alpha$-decay is $4.51 \times 10^9$ years, whereas for the polonium isotope $^{212}$Po the half-life for $\alpha$-decay is $3 \times 10^{-7}$ seconds, a factor of about $10^{16}$ smaller. The explanation of this huge range relies on quantum-mechanical tunnelling, as we shall show in Section 5.

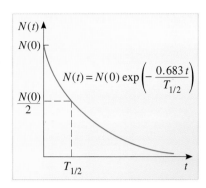

**Figure 3.10**  The exponential decay law. For any interval of time equal to $T_{1/2}$, the population halves.

**Question 3.7**  The half-lives of the two naturally-occurring uranium isotopes $^{238}$U and $^{235}$U are respectively $4.51 \times 10^9$ years and $7.04 \times 10^8$ years. The age of the Earth is approximately $4.5 \times 10^9$ years. (a) Determine what fractions of the original amounts of each isotope are still here today. (b) Natural uranium consists currently of 99.28% $^{238}$U and 0.72% $^{235}$U. (There is also 0.0057% $^{234}$U which we are ignoring in this question.) Determine the original proportions of these two isotopes.  ■

As Question 3.7 suggests, the knowledge of half-lives of radioactive nuclei is an important source of information concerning the ages of rocks, as recognised by Rutherford as early as 1905.

## 2.5  A survey of known nuclei

There are 90 naturally-occurring elements representing all integer values of atomic number $Z$, from $Z = 1$ (hydrogen) to $Z = 92$ (uranium), with the exception of $Z = 43$ (technetium) and $Z = 61$ (promethium) which do not exist naturally but have been made artificially in laboratories. In addition, many other nuclei, with $Z > 92$, have been artificially produced.

Each element exists in several or many isotopic forms, typically between 10 and 20 known isotopes per element, most of which are unstable, i.e. radioactive. All this is dramatically illustrated by Figure 3.11 from which it can be seen that many new isotopes were discovered between 1995 and the date of publication. No doubt by the time you read this many more will have been discovered.

In this figure the nuclei that are naturally abundant on Earth are indicated with a black and white box. Generally speaking these are stable, but a few, like $^{238}$U, are radioactive yet have sufficiently long half-lives to have survived since they were produced. Figure 3.11 contains many noteworthy features and patterns:

- The stable and very long-lived nuclei, i.e. those naturally abundant on Earth, do not cover the diagram uniformly but are confined to a narrow curving path called the **path of stability** (see inset to Figure 3.11). Some signposts along the path are $Z = N = 8$ ($^{16}$O); $Z = 26$, $N = 30$ ($^{56}$Fe); $Z = 82$, $N = 126$ ($^{208}$Pb). The path of stability follows a more-or-less straight line, $N = Z$, until about $Z = 20$. For larger $Z$, the path curves downwards as there is a rapidly increasing tendency for $N$ to exceed $Z$. The most abundant isotope of tin, for example, is $^{120}_{50}\text{Sn}_{70}$.

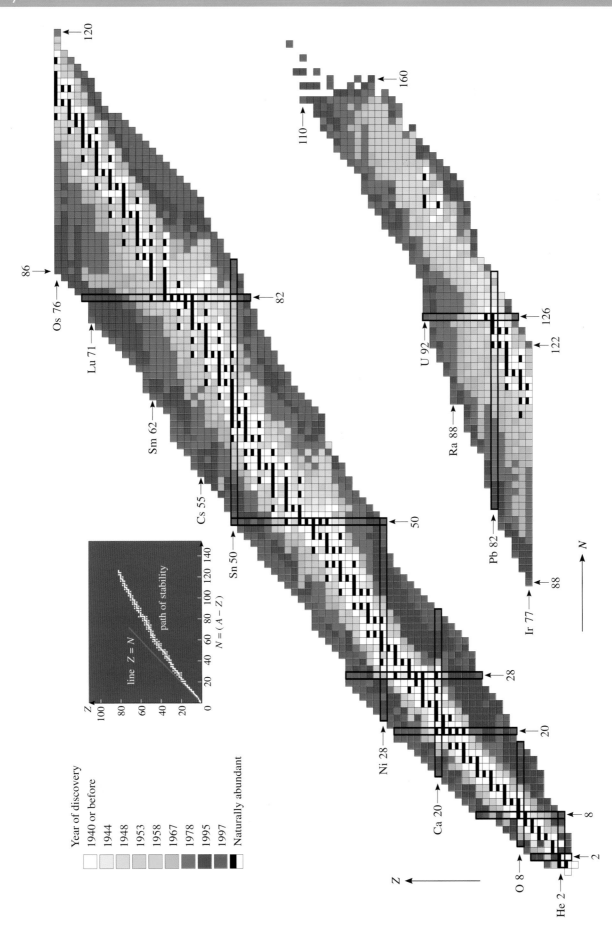

Year of discovery

1940 or before
1944
1948
1953
1958
1967
1978
1995
1997

Naturally abundant

**Figure 3.11**  Table of isotopes. $Z$ is plotted vertically and $N$ horizontally. (In some other versions of this table, $N$ is plotted vertically and $Z$ horizontally.) Each square corresponds to a specific combination of $Z$ and $N$. Those marked in black and white are stable or at least sufficiently long-lived to be naturally abundant. The figure shows how the number of known isotopes has been extended outward as new methods for exploring unstable nuclei have been developed.

- The further a nucleus is from the path of stability, the more recently it is likely to have been discovered.

- There is a marked odd-even effect. There are far more stable nuclei for even $Z$ than for odd $Z$. For example, for $Z = 50$ (tin) there are 10 stable isotopes, but for $Z = 51$ (antimony) there are just two. Tin has the largest number of stable isotopes. Furthermore, the numbers, 2, 8, 20, 28, 50, 82 and 126 are favoured, there being more stable nuclei with either $Z$ or $N$ equal to one of these numbers than might be expected from looking at nearby values. These numbers are called **magic numbers**. The corresponding rows and columns have thick black borders on Figure 3.11.

- There are no stable nuclei for $Z > 83$ (bismuth), although thorium ($Z = 90$) and uranium ($Z = 92$) are shown as stable because their very long half-lives enable them to be naturally present on the Earth. All nuclei with $A > 209$ are radioactive.

- Apart from the proton ($^1$H) there is only one stable nucleus with $Z > N$, namely $^3$He.

You may be wondering, in view of the fact that new nuclei seem to be discovered each year, whether there are any limits to possible combinations of $Z$ and $N$. There certainly are. What is not evident from Figure 3.11 is that the further out from the path of stability one goes, the shorter the half-lives become, i.e. the more unstable the nuclei are. Eventually the half-lives become so short that nuclei cannot be said to exist at all. We shall return to Figure 3.11 many times in the remainder of this chapter.

**Question 3.8**   Refer to Figure 3.11. How many naturally-abundant isotopes are there of lead ($Z = 82$)? Identify them. Is $^{205}_{82}$Pb one of them?  ■

Sequences, or chains, of radioactive decays starting from the long-lived isotopes of the heavy elements uranium ($Z = 92$), thorium ($Z = 90$) and actinium ($Z = 89$) are responsible for the natural occurrence of relatively short-lived elements with $Z$ between 83 and 90 (including radium, $Z = 88$ and polonium, $Z = 84$, discovered by Marie Curie). These decay chains end with the production of stable isotopes of lead ($Z = 82$). The first six and the last two decays of the uranium decay chain, starting from $^{238}_{92}$U, are shown in Figure 3.12.

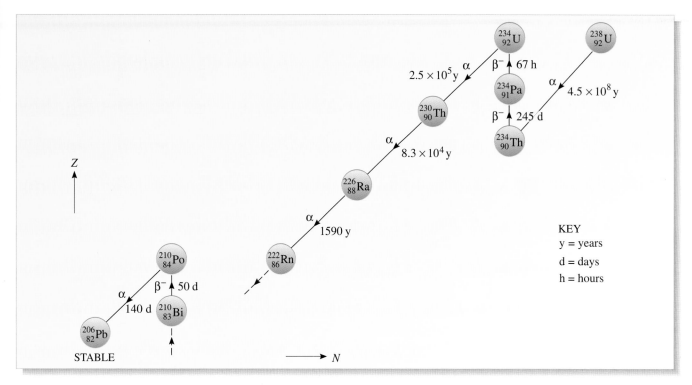

**Figure 3.12**   The uranium radioactive decay chain. The half-life of each parent and the decay mode are shown. For example, $^{226}_{88}$Ra has a half-life of 1590 years for α-decay to $^{222}_{86}$Rn (that is $^{226}_{88}$Ra $\rightarrow$ $^{222}_{86}$Rn + α).

## 2.6  Two applications of radioactivity

### Positron emission tomography (PET)

You will know that radioactive isotopes, or *radioisotopes*, are used in various ways in medicine. Tomography is a technique for imaging a chosen two-dimensional section, or slice, of an internal organ or tumour etc., using penetrating radiations, traditionally beams of X-rays. Here we discuss **positron emission tomography**, or PET, where the source of penetrating radiation is a β⁺-emitting isotope introduced into the region to be imaged. Obviously the radioisotope should have a short half-life, no longer than the time taken to do the scan, so as to minimise the harmful effects. Commonly-used isotopes in PET are $^{15}$O ($T_{1/2} \approx 2$ minutes), $^{13}$N ($T_{1/2} \approx 10$ minutes) and $^{11}$C ($T_{1/2} \approx 20$ minutes).

The positron is the antiparticle of the electron. Particles and their anti-particles tend to annihilate one another, with the conversion of their combined mass $m$ into an equivalent amount of energy $E$, calculated using $E = mc^2$. When a positron is emitted in β⁺-decay it rapidly slows down by ionising the material through which it passes, and finally encounters an electron and annihilates with it. The combined mass of the two particles, $m = 2m_e$, is thus converted into an amount of energy $2m_e c^2$ in the form of two 511 keV γ-rays. Momentum conservation requires that the two γ-ray photons go off in opposite directions from the point of annihilation. PET is a medical imaging technique that exploits the fact that the detection of the two oppositely-directed γ-rays defines a line on which the point of annihilation occurs, as illustrated in Figure 3.13a. A positron emitter, such as radioactive oxygen $^{15}$O, is introduced into the region (an organ or tumour, etc.) to be imaged, and that region is positioned at the centre of a circular array of photon detectors as in Figure 3.13b. Each coincident detection of two γ-rays defines a line, and after a very large number of events have been recorded, the density of the β-emitter can be mapped. Figure 3.13c shows a PET image of the brain of a patient who was given a small amount of oxygen $^{15}$O to inhale. The regions of low density indicate low oxygen levels and therefore poor blood flow.

### Carbon dating

Another well-known application of radioactivity is the *dating* of various geological and archaeological samples. If you measure the amount of a particular radioactive isotope in a sample, and you know how much of that isotope was present in the sample when it was formed, then you can use the half-life of the decay process to determine the age of the sample.

Atmospheric carbon, in the form of carbon dioxide, consists of stable carbon isotopes (mainly $^{12}$C) with tiny amounts of the radioactive isotope $^{14}$C which has a half-life of 5730 years. The $^{14}$C is produced by cosmic ray neutrons bombarding atmospheric nitrogen ($^{14}$N + neutron $\rightarrow$ $^{14}$C + proton). Since it is known that the cosmic ray intensity has remained constant over at least the last 50 000 years or so, we can be sure that the amount of $^{14}$C in the atmosphere, about one part in $10^{12}$, has remained steady over this period. (Cosmic rays are discussed in Chapter 4.)

All living organisms continuously take in carbon from the atmosphere, directly during photosynthesis in the case of plants, or indirectly through the food chain in the case of animals. The tissues of living organisms therefore contain radioactive $^{14}$C in the same proportion as in the atmosphere from whence it came. After death however, no new carbon is taken in and the existing $^{14}$C in the tissues decays with the half-life of 5730 years. Given a sample of dead bone or plant to be dated, the first step is to chemically extract and weigh the total carbon content. The amount of $^{14}$C present is then measured using very sensitive radiation counters, and so the ratio of $^{14}$C to stable carbon in the sample is found. Assuming that the ratio of $^{14}$C to stable carbon at the time of death was the same as that in the atmosphere today, we can then calculate the age of the sample using the exponential decay law.

Because the $^{14}$C content is so small, giving rise to perhaps only a few tens of decays per minute, there are many precautions necessary for reliable results. To begin with, it is essential to know that the sample has not been contaminated by fresh carbon since the organism died. Then it is necessary to carefully shield the radiation counters from cosmic rays and other sources of radioactivity that may be more intense than the sample itself, such as the natural radioactivity in the surroundings, including yourself if you are doing the experiment.

## 3 Nuclear stability and binding energies

You have seen that very few of the mathematically-possible combinations of $Z$ and $N$ correspond to stable nuclei. A more substantial number correspond to radioactive (unstable) nuclei but most do not correspond to even fleetingly-existent nuclei. In this section we shall explore, from an energy point of view, the stability of the naturally-occurring nuclei lying along the path of stability of Figure 3.11. This will require an introduction to the *binding energy* of a nucleus. We shall then begin to understand the pattern of stable and unstable nuclei and the principles behind processes of immense social importance: the release of nuclear energy by the *fission* of heavy nuclei and the *fusion* of light ones.

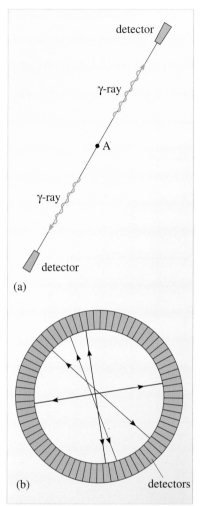

(a)

(b)  detectors

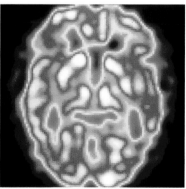

(c)

**Figure 3.13**  (a) A positron and electron annihilate one another at point A. Detection of the two oppositely-directed γ-rays defines a line. (b) Coincident events recorded in detectors arranged in a circle define lines from which the density of the β-emitter can be mapped. (c) PET image of the brain of a patient who has been given radioactive oxygen.

## 3.1 Nuclear binding energies

To introduce the concept of binding energy we first consider atomic binding energies. You know from your study of atomic structure in *QPI* that the *total energy E* of the hydrogen atom in its ground state (i.e. the sum of the electrostatic potential energy and electron's kinetic energy) is $E = -13.6$ eV. This means that a minimum energy of 13.6 eV is needed to pull the electron out of a ground-state hydrogen atom and allow it to escape to infinity. Conversely, if a ground state hydrogen atom were to be formed from a proton and an electron, both initially at rest and remote from one another, then an energy of 13.6 eV would be released, probably in the form of optical photons.

One way of expressing this is to say that the hydrogen atom in its ground state has a *binding energy B* = 13.6 eV, but remember, the *total energy E* is *minus* the binding energy: $E = -B = -13.6$ eV. The same ideas apply to nuclei, except that due to the nature of the strong nuclear force, the binding energies of nuclei are very much larger than atomic binding energies, typically tens or hundreds of MeV. For example, the binding energy of the deuteron (i.e. the nucleus of the hydrogen isotope $^2_1$H) is $B = 2.22$ MeV. This is the minimum energy needed to pull the proton and neutron apart. In practice the disintegration of the deuteron could be brought about by a sufficiently energetic γ-ray photon. The photon would be destroyed in the process and any energy in excess of 2.22 MeV would appear as kinetic energy of the proton and neutron. Conversely, under certain circumstances a neutron can be captured by a proton (the neutron might tell the story the other way round!) with release of the 2.22 MeV binding energy as a γ-ray photon.

The idea of binding energy applies to any nucleus, and we have the following definition:

> The **binding energy** *B* of a nucleus is the least energy needed to completely disassemble the nucleus into its constituent protons and neutrons.
>
> The binding energy is also equal to the energy that would be released if the nucleus were to be created from its disassembled constituents.

You will not be required to calculate binding energies from the tabulated atomic masses. We shall always give you the binding energies when you need them.

It follows from Einstein's mass–energy relation, $E = mc^2$, that the mass of a nucleus is less than the sum of the masses of its constituent nucleons by $B/c^2$. For example, the mass of the deuteron is less than the sum of the masses of the proton and the neutron by 2.22 MeV/$c^2$. Binding energies of nuclei can be calculated from the measured atomic masses given in standard tables.

Binding energies are very useful for discussing the stability of nuclei and determining whether particular processes are energetically possible. For example, the binding energy of the triton, the nucleus of the hydrogen isotope $^3$H, is 8.48 MeV. A 12 MeV photon, say, would have enough energy to break up a triton into a proton and two neutrons (i.e. $^3_1$H + γ → p + 2n) or into a deuteron and a neutron ($^3_1$H + γ → $^2_1$H + n). According to quantum mechanics each of these energetically-allowable processes would have some probability of occurring. The photon would be completely destroyed in the process, so that in the former case there would be 12 MeV − 8.48 MeV = 3.52 MeV left over as kinetic energy of the three constituents, while in the latter case there would be 12 MeV − 8.48 MeV + 2.22 MeV = 5.74 MeV left over as kinetic energy of the deuteron and the neutron.

**Question 3.9** (a) A γ-ray photon of energy 8.00 MeV 'photodisintegrates' a deuteron, i.e. splits it into its component proton and neutron. What is the total kinetic energy of the proton and neutron? (b) Could such a photon photodisintegrate the triton into a proton and two neutrons? Could it photodisintegrate the triton into a deuteron and a neutron? ■

## 3.2 Binding energy per nucleon

In order to compare binding energies of different nuclei and to obtain insight into nuclear stability and the process of radioactive decay, we now introduce the **binding energy per nucleon**. If the binding energy of a nucleus is denoted by $B$, then the binding energy per nucleon is simply $B/A$ where $A$ is the mass number i.e. the number of nucleons in the nucleus. $B/A$ provides us with a good measure of the relative stability of different nuclei: a large binding energy per nucleon indicates a stable nucleus.

**Question 3.10** The binding energy of the triton, $^3$H, is 8.48 MeV. What is the binding energy per nucleon for the triton? ■

Figure 3.14 shows a plot of $B/A$ values for some of the stable and long-lived nuclei near the path of stability, as a function of $A$.

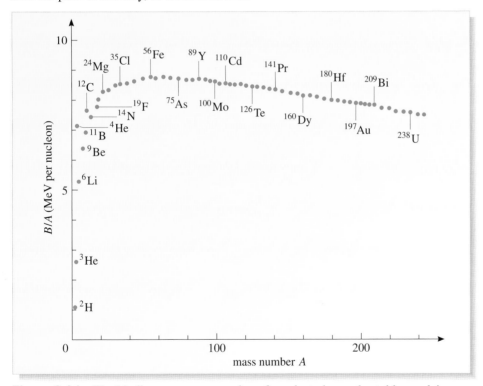

**Figure 3.14** The binding energy per nucleon for selected, mostly stable, nuclei.

The general pattern is that except for the very lightest nuclei, $B/A$ is in the range 7.5 MeV per nucleon to 8.7 MeV per nucleon for all nuclei. It has a maximum value of 8.795 MeV per nucleon for $^{62}$Ni, but $^{56}$Fe is close behind at 8.790 MeV per nucleon. For heavy nuclei, near uranium, $B/A$ falls to about 7.5 MeV per nucleon. This might not seem a large drop but it represents a large amount of energy when you remember that the total binding energy of a nucleus is $A \times (B/A)$, a fact with momentous consequences, as you will see later.

At the small-$A$ end of the figure, you can see a rapid fall-off. You have seen that the deuteron, $^2_1\text{H}$, has binding energy $B = 2.22$ MeV, so $B/A$ for the deuteron is only about 1.11 MeV per nucleon. What is remarkable is that for the $\alpha$-particle, $^4_2\text{He}$, the binding energy is 28.30 MeV, so that $B/A$ is more than 7 MeV per nucleon, which is exceptionally large compared to that of its nearest neighbours.

> The $\alpha$-particle $^4_2\text{He}$ is an exceptionally stable nucleus with a binding energy of 28.30 MeV.

Indeed, light nuclei that are made up of three, four, five or six $\alpha$-particles have exceptionally large binding energies. For example, $^{12}_6\text{C}$ and $^{24}_{12}\text{Mg}$ have the same nucleons as three $\alpha$-particles and six $\alpha$-particles respectively, and have larger $B/A$ values than their nearest neighbours.

We can now begin to understand why many heavy nuclei undergo $\alpha$-decay. Since $\alpha$-particles are exceptionally stable combinations of nucleons, we can think of them as existing, in some sense, inside heavy nuclei. If a heavy nucleus can expel an $\alpha$-particle its $A$ value will fall by four units. This represents a move on the binding energy per nucleon curve (Figure 3.14) towards higher values of $B/A$. As a result the nucleus becomes more stable and energy is released in the form of kinetic energy of the ejected $\alpha$-particle and possibly also the emission of $\gamma$-radiation.

**Question 3.11**  Consider the $\alpha$-decay of the common isotope of uranium,

$$^{238}_{92}\text{U} \rightarrow {}^{234}_{90}\text{Th} + {}^4_2\text{He}.$$

Given that the binding energy per nucleon of $^{238}_{92}\text{U}$ is about 7.6 MeV per nucleon, that of $^{234}_{90}\text{Th}$ is 0.027 MeV per nucleon greater, and the total binding energy of the $\alpha$-particle is 28.3 MeV, determine the kinetic energy of the ejected $\alpha$-particle. (Assume that all the available released energy is in the form of kinetic energy of the $\alpha$-particle. There are in fact no $\gamma$-rays emitted in this decay and the heavy thorium nucleus takes a negligible proportion of the kinetic energy.) ∎

There are other departures from smooth behaviour in Figure 3.14 as well as the exceptional binding energy of the '$\alpha$-particle' nuclei and we shall return to discuss these in Section 4.

## 3.3  Fission and fusion

The release of nuclear energy by the fission of uranium nuclei and the fusion of hydrogen nuclei is of immense social importance. The binding energy per nucleon curve, Figure 3.14, reveals why these processes release energy. Let's do some simple energy accountancy. Suppose a $^{235}_{92}\text{U}$ nucleus were to divide into a krypton nucleus and a barium nucleus:

$$^{235}_{92}\text{U} \rightarrow {}^{94}_{36}\text{Kr} + {}^{141}_{56}\text{Ba}.$$

Is this process *endothermic* or *exothermic*, i.e., does it require a source of energy in order to take place, or is energy released? You can do a rough estimate based on reading values of $B/A$ from Figure 3.14. The three nuclei have $B/A$ values of: 7.4 MeV per nucleon (U), 8.5 MeV per nucleon (Kr) and 8.2 MeV per nucleon (Ba), so the actual binding energies are approximately 1700 MeV, 800 MeV and 1200 MeV, respectively. The total binding energy of the Kr and Ba nuclei is thus about $(800 + 1200)$ MeV $= 2000$ MeV, which is 300 MeV greater than the binding energy of the $^{235}\text{U}$ nucleus. This implies that approximately 300 MeV is released for every $^{235}$U nucleus that divides in this way. The released energy appears mainly in the form of kinetic energy of the two fragment nuclei, with smaller amounts in the form of γ-ray photons and radioactive decay of the two fragments. Roughly speaking, this release of energy comes about because the process takes us from a region of low $B/A$ to a region of higher $B/A$ in Figure 3.14. The process we are discussing is, of course, **nuclear fission**. Fission is the source of the energy released in atomic bombs and nuclear power stations. In fact fission reactions in power stations and bombs are induced by the presence of neutrons. For example, the fission reaction

$$^{235}_{92}U + n \rightarrow {}^{93}_{37}Rb + {}^{141}_{55}Cs + 2n$$

releases nearly 200 MeV as well as an additional neutron. In a large mass of $^{235}_{92}U$, the neutrons go on to induce more fissions in other $^{235}_{92}U$ nuclei thereby setting off a *chain reaction.*

We shall discuss fission more fully in Section 5, but first we do some more energy accountancy, this time at the other end of the curve in Figure 3.14.

You have seen that the binding energy per nucleon for the deuteron is $B/A = 1.11$ MeV per nucleon. For the helium isotope $^3$He, $B/A = 2.57$ MeV per nucleon (see Figure 3.14). On the basis of this, would you expect that the absorption of a proton by a deuteron to produce $^3$He would release energy? Note that it involves moving on the curve of Figure 3.14 in the direction of the maximum $B/A$ region. In fact, the following so-called *proton capture* reaction does release energy, mostly in the form of a γ-ray photon:

$$p + {}^2H \rightarrow {}^3He + \gamma.$$

The γ-ray produced in this reaction has an energy of about 5.5 MeV.

**Question 3.12**  Given that the binding energy per nucleon of the helium isotope $^3$He is 2.57 MeV per nucleon and that of the deuteron $^2$H is 1.11 MeV per nucleon, confirm that the energy of the γ-ray in the *proton capture* reaction above is 5.5 MeV. (You may neglect the kinetic energies of the nuclei.) ■

All processes in which two nuclei 'fuse' together to form a heavier nucleus are called **nuclear fusion** processes, and if the nucleus produced has $A \leq 56$ (i.e. iron or anything to the left of it on Figure 3.14) or thereabouts, the process is likely to be exothermic, i.e. it will release energy. Fusion is crucial in stars for both energy production, and for the production of the lighter elements — from helium to iron — from primordial hydrogen. The peak in the $B/A$ curve around the iron–nickel region means that elements heavier than iron (Fe) cannot be made by ordinary fusion processes in stars. It is believed that many elements heavier than iron, and certainly all those heavier than lead, have been produced in supernovas. In these cataclysmic stellar explosions, processes exist that produce elements with $A > 56$, i.e. the right-hand end of Figure 3.14. We shall discuss fusion in stars further in Section 5.3. Fusion reactions are also the basis for fusion weapons, the so-called hydrogen bombs, and for the longstanding but as yet unfulfilled hopes for the controlled production of fusion energy on a commercial scale.

It is important to remember that the *total energy E* of a bound system is *minus* the binding energy, $E = -B$. Hence the shape of the *total energy per nucleon* curve is obtained by turning the binding energy per nucleon graph (Figure 3.14) upside down. You can then think of fission and fusion as the tendency to roll down the total energy curve towards its lowest most stable point, near iron ($^{56}$Fe).

## 3.4 The valley of stability

So far we have studied one aspect of nuclear stability: the stability of the most stable nuclei on or near the path of stability. This gave an understanding of why fission and α-decay of heavy nuclei and fusion of light nuclei release energy. We now turn to the stability of nuclei against β-decay.

Recall that β-decay does not change the value of $A$; only the values of $Z$ and $N$ change. Hence we wish to consider the relative stability of nuclei sharing the same value of $A$. Such nuclei are called **isobars**. In Figure 3.12, isobars lie along a straight line with $Z + N = A$, for $A$ constant. The line of isobars for $A = 60$ is shown in Figure 3.15, cutting the $N$- and $Z$-axes at 45°.

For small $A$, less than about 20, the lines of isobars are at right-angles to the path of stability. As we move in each direction along a line of isobars away from the path of stability, the values of $B/A$ decrease and the nuclei become unstable and subject to β-decay. Nuclei lying below the line of stability have too many neutrons for stability and tend to decay by β⁻-decay, while those above have too many protons and tend to decay by β⁺-decay or by electron capture.

It is useful to think of the *total energy per nucleon, E/A,* rather than the binding energy per nucleon; $E/A = -B/A$. As you move along a line of isobars in either direction away from the path of stability, the total energy per nucleon *increases* (as $B/A$ decreases). When we plot the *total energy per nucleon* for all known nuclei as points above the $Z$–$N$ plane, they all lie very close to a surface called the **valley of stability**, with the valley floor lying directly above the path of stability in the $Z$–$N$ plane, see Figure 3.16.

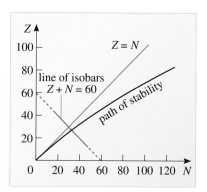

**Figure 3.15** A line of isobars, $Z + N = 60$, cutting across the path of stability.

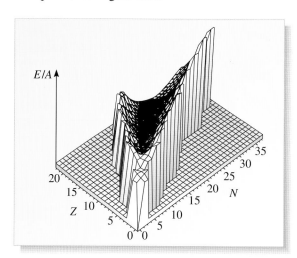

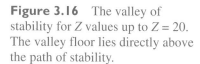

**Figure 3.16** The valley of stability for $Z$ values up to $Z = 20$. The valley floor lies directly above the path of stability.

Cutting across the valley, at right-angles to it for small $A$, are the lines of isobars. As one moves from the valley floor in either direction along a line of isobars, the total energies of the nuclei increase rapidly (the binding energies, of course, decrease) and the nuclei tend to have shorter β-decay lifetimes. You can now think of β-decay as a tumbling down the sides of the valley to the valley floor to find stable $N$–$Z$ combinations: β⁻-decay will reduce $N$ in neutron-rich isobars, and β⁺-decay, or more likely electron capture, will reduce $Z$ in proton-rich isobars. Figure 3.17 illustrates this by showing the relative ground-state energies of some of the $A = 137$ isobars and some of the β-decays leading down to the stable nucleus $^{137}_{56}\text{Ba}$.

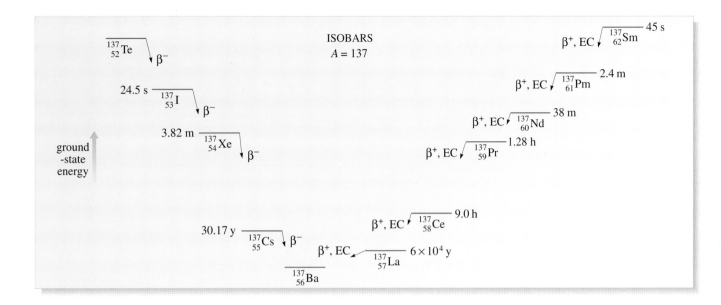

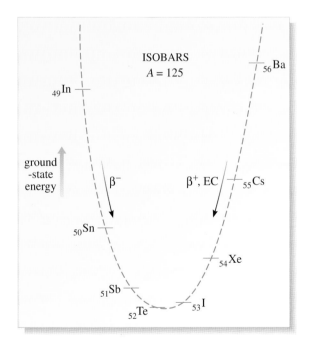

Figure 3.17 is essentially a cross-section of the valley for $A = 137$. Another cross-section, for $A = 125$, is shown in Figure 3.18. The cross-sections are roughly parabolic in shape.

**Figure 3.17** Ground state energies of some of the $A = 137$ isobars showing β-decay modes and lifetimes leading down to the stable $^{137}$Ba on the valley floor. (EC stands for electron capture, and the symbol β$^+$, EC indicates that both β$^+$-decay and electron capture occur as competing processes.)

**Figure 3.18** Ground state energies for some of the $A = 125$ isobars. The cross-section is roughly parabolic in shape.

The valley floor is not level. As you saw in Section 3.2 in connection with α decay, the total energy per nucleon has its lowest point near $^{56}$Fe. Thus the valley gives us a metaphorical picture of both β-decay and α-decay. The α-decays of heavy elements can be thought of as streams running from large $A$ to smaller $A$ down the valley floor, while β-decays are streams running down the valley walls. Remember that the lowest point is in the middle of the valley.

# 4 Nuclear structure

So far we have discussed the size and composition of atomic nuclei and described nuclear processes such as radioactivity, fission and fusion. These processes were thought of in terms of movement towards more stable configurations of nucleons, i.e. towards a lower total energy or greater binding energy. However, we have not yet explained *why* nuclei have the properties described; in particular, why the nuclear densities are more or less constant, and why the energies of nuclei change with $Z$ and $N$ in the way depicted in the binding energy per nucleon curve and the valley of stability. To answer these questions we must study the forces between nucleons and how they hold the nucleus together. Ultimately of course this is a problem in quantum mechanics. However, it is possible to understand the broad features of the binding energy per nucleon curve, Figure 3.14, using the so-called *semi-empirical model* of the nucleus based on empirical nuclear data and an analogy with a water droplet. This we do in Section 4.1. To go further, and to underpin the semi-empirical model, it is necessary to build a quantum-mechanical model of the nucleus using the Schrödinger equation and the Pauli exclusion principle. This leads to a *nuclear shell model*, described in Section 4.2, which is in some ways analogous to the electron shell structure of atoms.

## 4.1   The semi-empirical model of the nucleus

The **semi-empirical model** was developed by Hans Bethe and Carl von Weizsäcker in 1936 as an attempt to understand the overall trends of nuclear binding energies over the entire range of nuclear masses. The underlying idea is that the nucleus can be considered as a liquid drop, an idea first put forward by Gamow and developed into a model for describing such phenomena as nuclear fission by Bohr, Kalckar, Wheeler and others a few years later.

Subsequently, it has been found that there are many nuclear phenomena which fall outside the scope of the liquid drop model, but it endures as a source of understanding of the gross properties of nuclei and the global trends of nuclear sizes and nuclear binding energies. Before discussing the semi-empirical model we need to discuss the forces between nucleons.

**Figure 3.19**   George Gamow (1904–1968) and Niels Bohr (1885–1962).

There are two main types of forces acting between nucleons: the electrostatic (*Coulomb*) *force* of repulsion that acts between all pairs of protons, and the *strong nuclear force* that holds the nucleus together. Let's look at the main characteristics of these two forces.

The *repulsive Coulomb interaction* between any two protons separated by a distance $r$ is expressed by the potential energy function $V(r) = e^2/(4\pi\varepsilon_0 r)$, or equivalently by a repulsive radial force component

$$F_r(r) = -\frac{dV}{dr} = \frac{e^2}{4\pi\varepsilon_0 r^2}$$

the familiar inverse square law. The Coulomb potential energy function $V(r)$ is illustrated in Figure 3.20. The Coulomb interaction is said to be *long-range* because it falls off relatively slowly with distance, i.e. $V(r) \propto 1/r$ and $F(r) \propto 1/r^2$. This is in contrast to the strong nuclear force.

The **strong nuclear force** is extremely complicated and is not completely understood even today. However, we can outline some of its key properties.

- The strong nuclear force is a *very short-ranged attractive* force that acts between nucleons; it does not act on electrons and is almost independent of electric charge. There is a small **asymmetry effect**: the strong nuclear force is slightly stronger for unlike nucleons than for like nucleons, i.e., it is slightly stronger between a neutron and a proton than it is between two neutrons or between two protons.

- Although the strong nuclear force is very strongly attractive, its range is limited to distances of about 2 fm' beyond which it falls very rapidly to zero. This is illustrated by the potential energy function sketched in Figure 3.21. At very small separations, about 0.4 fm, the force is strongly repulsive. This is one reason why nuclear matter cannot easily be compressed.

- Remembering (Figure 3.7) that the nuclear radius varies from about 2 fm to 3 fm for the lighter nuclei and between about 6 fm to 8 fm for the heavier nuclei, you can see that the nucleons mainly interact, via the strong nuclear force, with their nearest neighbours. Thus in heavy nuclei, nucleons at a separation of approximately the nuclear radius will not be within range of one another.

There is in fact another type of nuclear force called the *weak interaction* which is responsible for β-decay. The weak interaction will be discussed in Chapter 4 of this book.

Gravitational forces are too weak to have any effect on nuclear structure.

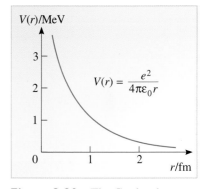

**Figure 3.20** The Coulomb potential energy $V(r)$ for two protons separated by $r$. The radial force component is given by minus the gradient, $-dV/dr$, and is positive everywhere, i.e. it is a repulsive force acting in the direction of increasing $r$.

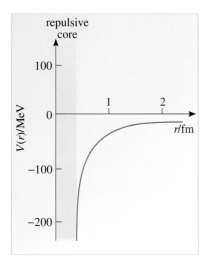

**Figure 3.21** A potential energy curve for the strong nuclear force between two nucleons. You can see that two nucleons within about 2 fm of one another are within a mutually attractive potential well.

The precise form of the strong interaction between two nucleons is in fact much more complicated than Figure 3.21 suggests; it depends on, among other things, the way the spins of the two nucleons are aligned and on their relative angular momentum.

• The short-range property leads to a fruitful analogy between an atomic nucleus and a water droplet. Water molecules are bound to one another in a droplet by van der Waals forces. These attractive forces are of very short range so that any water molecule is attracted only to its nearest neighbours. Thus molecules in the centre of a water droplet are more strongly bound than those on the surface which have only half the number of nearest neighbours. This accounts for so-called *surface tension* effects, such as the spherical shape of the droplet. By taking on the shape of a sphere the droplet minimises its surface area thereby reducing its total energy (i.e. increasing its total binding energy). Nuclei have a tendency to be spherical, partly because the short range of the strong nuclear force leads to similar surface effects.

To understand the binding energy of the nucleus we must consider the combined effects of the Coulomb force and the strong nuclear force. A key point is that although the Coulomb force is much weaker than the strong nuclear force when the nucleons are close, within about 2 fm, it has a much longer range and so dominates at larger distances. Thus the analogy with the water droplet requires the droplet to carry a net electric charge.

We begin by considering the effects of the strong nuclear force alone. Consider first, a heavy nucleus, i.e. one of large volume so that we can, to a good approximation, neglect *surface effects*. That is, we assume that each nucleon is completely surrounded by other nucleons. Since it interacts only with its nearest neighbours, each nucleon finds itself in the same environment as any other. Furthermore, if we also neglect all differences between neutrons and protons, such as the *asymmetry effects*, all nucleons see the same environment in all large nuclei regardless of different neutron–proton ratios. (This idea is supported by the fact that, as we have seen, nuclear densities near the nuclear centre are much the same.) Within these approximations we can regard all nuclei as being made of nuclear matter of constant density and having the same binding energy per nucleon, $B/A$. This constant value of $B/A$ can be calculated using detailed models of the nuclear force and is found to be just over 15 MeV per nucleon; it is shown in Figure 3.22 (top line) as the so-called **volume energy** contribution to the binding energy per nucleon curve. This is clearly much larger than the known values of between about 5 MeV and 9 MeV seen in

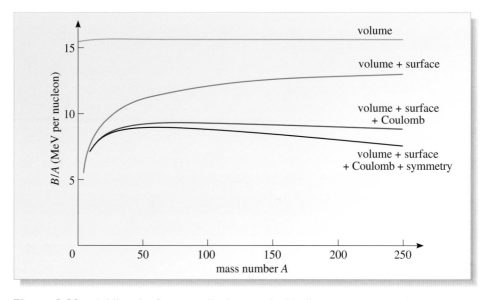

**Figure 3.22** Adding the four contributions to the binding energy per nucleon curve.

Figure 3.14. To obtain agreement we obviously need to bring in the effects we have neglected, especially surface effects and the Coulomb repulsion between protons.

We first bring in *surface effects*, which become progressively more important for the medium-sized and smaller nuclei. For the lightest nuclei it is clearly false that all nucleons are completely surrounded by other nucleons, since a large proportion of them lie on the surface. The surface nucleons have fewer neighbours and so are less strongly bound and the binding energy per nucleon is reduced. When this **surface energy** contribution to *B/A* is included we obtain a good description of the fall-off of the *B/A* curve for small *A*. This is shown by the *volume + surface* curve in Figure 3.22.

We now bring in the effects of Coulomb repulsion between protons. This has little effect for small nuclei where the strong nuclear force dominates, but the Coulomb repulsion is a *long-range* force and becomes increasingly important in large nuclei. Its effects are cumulative: each proton sees a repulsive Coulomb potential which has contributions from *all* other protons. The effect is progressively to reduce the binding energy per nucleon of heavy nuclei. Inclusion of the **Coulomb energy** contribution is shown in the third curve down in Figure 3.22.

Finally we include the **symmetry energy** contribution to the *B/A* curve. There are two contributions to the symmetry energy. One contribution is due to the fact that the strong nuclear force between unlike nucleons is slightly stronger than it is between like nucleons, as mentioned above. The other contribution is due to the Pauli exclusion principle which in this context lowers the total energy when $Z = N$. (The relevance of the Pauli exclusion principle here should become clearer in Section 4.2.) The result of both the Pauli and the unlike-pair effects is that the binding energy per nucleon due to the strong nuclear force is greatest if the number of protons and neutrons is equal. We have seen that as *Z* gets larger, *N* progressively exceeds *Z*. Thus *B/A* is progressively reduced by the symmetry energy as *A* increases.

You can see from Figure 3.23 that with all four contributions to *B/A* (the smooth curve) we obtain quite good overall agreement with the experimental values, although it has to be admitted that the parameters in the four contributions have been chosen to give a good fit. You can also see from Figure 3.23 that there are still some discrepancies in the small-scale structure, especially near small *A*.

Useful though the semi-empirical model is, there are many nuclear properties that it cannot describe and questions that it cannot answer. Why is there an excess of neutrons in heavy nuclei? How do we explain the valley of stability? Why are certain magic numbers of nucleons favoured? Why does each nucleus have its particular set of excited states? A quantum-mechanical approach is needed to answer these and many other questions. That's the topic of the next section.

**Question 3.13**   Explain in your own words why the various contributions, apart from the volume energy, reduce *B/A*.

**Question 3.14**   The binding energy per nucleon for the tin isotope $^{130}_{50}$Sn is less than that for $^{124}_{50}$Sn. Which contribution accounts for this? Give reasons for your answer. ■

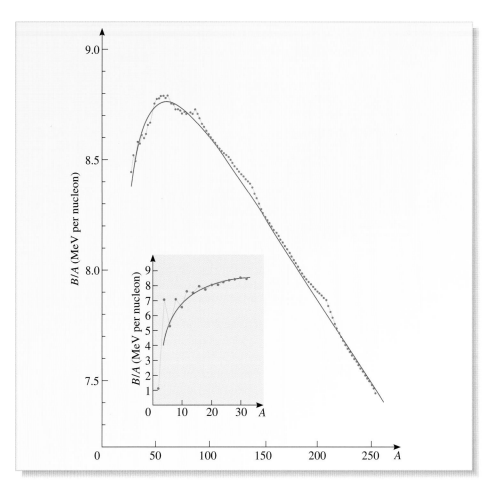

**Figure 3.23** The dots are the measured binding energies per nucleon. This is a version of Figure 3.14 with an expanded energy scale. The smooth curve is the total binding energy per nucleon curve, a version of the lowest curve in Figure 3.22 but with an expanded energy scale. The inset shows the values for $A$ less than 30.

## 4.2  The nuclear shell model

You have seen (*Quantum physics: an introduction*) that atomic structure is described in terms of electron wavefunctions and energy levels leading to the remarkable periodicities of the Periodic Table associated with the closure of electron shells. The picture of nuclei we have so far seems to be very different. Nuclear properties vary with $Z$ and $N$ in a relatively slow and regular fashion. We have seen that nuclei have more or less constant densities which, at the nuclear centre at least, are much the same for all but the lightest nuclei.

Yet, nuclei *do* have a shell structure, although shell effects in nuclei are much less noticeable than electron shell effects in atoms. We can see evidence for nuclear shell effects in discontinuities in the properties of nuclei as $N$ and $Z$ vary. For example, certain values of $N$ and $Z$ are strongly associated with naturally-abundant and stable nuclei, as was mentioned in Section 2.5 when we discussed Figure 3.11, pointing out that 2, 8, 20, 28, 50, 82 and 126 seemed to be favoured numbers. These are the so-called *magic numbers*. Nuclei with either $N$ or $Z$ magic are called **magic nuclei**. (Nuclei with both $N$ and $Z$ magic are said to be *doubly-magic*.) Binding energies per nucleon are greater for magic nuclei than for neighbouring nuclei. This is apparent in

Figures 3.14 and 3.23 where there are conspicuous departures from a smooth variation. There is much other evidence for shells in nuclei, and there is now a well-developed quantum-mechanical model of nuclei, the **nuclear shell model**, which reproduces very well the energy levels of nuclei and also many other properties that we cannot go into.

### Nucleon wavefunctions and energy levels

At the heart of the nuclear shell model is the idea that each nucleon moves in a potential energy well due to all the other nucleons. The nucleon wavefunctions and energy levels can be calculated using Schrödinger's equation once the potential energy functions describing the potential wells are found. The method is in some ways similar to the methods used to find electron wavefunctions and energy levels in heavy atoms (see *Quantum physics: an introduction*).

This approach leads to potential energy wells which more or less follow the nuclear density distributions described in Section 2. Essentially, this is because the strong nuclear force is short-ranged and so the potential energy of a nucleon at any point depends principally on the nucleon density in the immediate neighbourhood of that point. Potential energy wells for neutrons are shown in Figure 3.24b.

Potential energy wells for protons differ slightly but crucially from those of neutrons because of the Coulomb repulsion, as you will see in the next section.

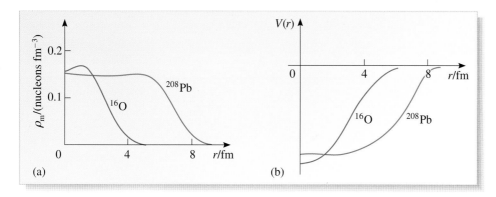

**Figure 3.24** (a) Nuclear matter densities $\rho_m(r)$ and (b) potential energy functions $V(r)$ for neutrons, in oxygen $^{16}$O and in lead $^{208}$Pb nuclei. The potential energy functions have approximately the same shape as the nuclear matter distributions, apart from a change of sign.

One of the key things left out of the story so far is the fact that while neutrons are distinguishable from protons, neutrons themselves are indistinguishable *fermions*, and so also are protons. This means that every neutron in a given nucleus lies in the same potential energy well as every other neutron in that nucleus, but no two neutrons can occupy the same quantum state in the well, and the same is true for the protons. This is an effect of the *Pauli exclusion principle*. As a result, the nucleon energy levels get filled up rather as electron energy levels do in atoms, thus leading to a shell-type structure. The magic numbers correspond to nucleon shell closures and arise in much the same way as the periods of elements arise in chemistry. The actual magic numbers are different from the numbers occurring in the Periodic Table because of the particular nature of the strong nuclear force.

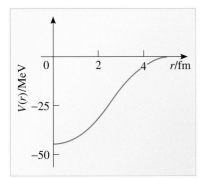

Coulomb energy is potential energy arising from the Coulomb (electrostatic) interaction.

**Figure 3.25** The potential energy well seen by a neutron in $^{17}_{8}O_9$.

### Explaining the neutron-proton ratio and β-decay

The potential energy wells seen by protons differ from those seen by neutrons in a very important way. The repulsive Coulomb interaction between protons is always present and, as you might guess, this is why there is an excess of neutrons in all but the lightest nuclei. To see how this comes about, consider first two particular isobars, $^{17}_{8}O_9$ and $^{17}_{9}F_8$. Note that a neutron in $^{17}_{8}O_9$ and a proton in $^{17}_{9}F_8$ are both bound to the same core nucleus $^{16}_{8}O_8$, so the difference in the energies of the two isobars is accounted for by the Coulomb energy of a proton (the neutron, being neutral, has no Coulomb energy).

Figure 3.25 shows the potential energy well seen by a neutron in $^{17}_{8}O_9$; it more or less follows the nuclear density. Now consider the potential energy well seen by a proton in $^{17}_{9}F_8$. This well is shown as the solid line in Figure 3.26 and is not the same as that of the neutron.

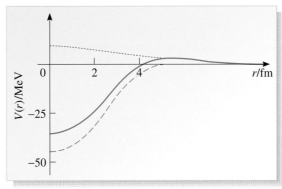

**Figure 3.26** The potential energy well seen by a proton in $^{17}_{9}F_8$ (solid blue line) is the sum of the neutron well of Figure 3.25 (red dashed line) and the Coulomb potential energy (green dotted line) due to the repulsion of the other eight protons.

It has a bump in the nuclear surface. The source of this bump can be seen by noting that the solid line is the sum of the dashed line and the dotted line. The dashed line is the same as the neutron potential energy in $^{17}_{8}O_9$ shown in Figure 3.25. The dotted curve in Figure 3.26 represents the repulsive interaction between the proton and the other eight protons. Note that besides the repulsive bump at the nuclear surface, the Coulomb repulsion results in the proton well being less deep overall than the neutron well.

Because the neutron well is deeper than the proton well the neutron energy levels fall below the corresponding proton energy levels. In other words, the neutrons are more tightly bound. To see some of the consequences of this, refer to Figure 3.27

**Figure 3.27** Potential energy well for a neutron (curve on the right) and for a proton (curve on the left with reflected $r$-axis) in $^{16}_{8}O_8$. The solid dots indicate nucleons in energy levels. The star shows the extra proton in $^{17}_{9}F_8$ and the unfilled circle indicates the extra neutron in $^{17}_{8}O_9$.

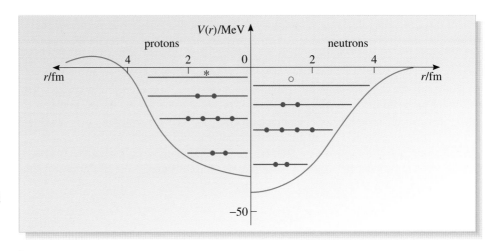

which depicts the 8 protons and 8 neutrons in the lowest filled levels, i.e. the nuclear ground state, for the stable oxygen nucleus $^{16}_{8}O_8$.

Now consider two possibilities: you could imagine adding a proton to the unfilled proton level to make the ground state of the fluorine nucleus $^{17}_{9}F_8$ (this extra proton is indicated by the star in Figure 3.27) or adding a neutron to the unfilled neutron level to make the ground state of $^{17}_{8}O_9$ (this extra neutron is indicated by the unfilled circle). Now the energy of the extra proton is distinctly higher than the energy of the extra neutron. The difference in this case is more than 1.3 MeV. Because of this a $^{17}_{9}F_8$ nucleus undergoes β-decay, specifically β⁺-decay with a half-life of 65 s, to form the stable $^{17}_{8}O_9$ nucleus. This kind of reasoning leads to the following conclusion.

> Nuclei that are stable against β-decay are those for which the proton and neutron potential wells are both filled to approximately the same energy.

If one well is filled much higher than the other, β-decay takes place until they are approximately equal. For heavy nuclei the cumulative effect of the Coulomb potential becomes very significant; the proton wells are so much shallower than the neutron wells that the β-stable nuclei have significantly more neutrons than protons. This is illustrated in Figure 3.28 which depicts the filled energy levels for three isobars with $A = 100$. The isobar on the valley floor is a stable isotope of ruthenium, $^{100}_{44}Ru_{56}$. The other two are an unstable isotope of tin, $^{100}_{50}Sn_{50}$, and an unstable isotope of strontium, $^{100}_{38}Sr_{62}$.

The important point to note is that is for $^{100}_{44}Ru_{56}$ the top-filled neutron and proton levels are at much the same energy. This is characteristic of a nucleus on the floor of the valley of stability. Look now at the case of $^{100}_{38}Sr_{62}$. Here you see that the top-filled neutron level is much higher than the top-filled proton level. This is a nucleus with too many neutrons for stability. It β⁻-decays with a short lifetime ($T_{1/2} = 202$ milliseconds) to an yttrium nucleus, $^{100}_{39}Y_{61}$, and this too β⁻-decays, and so on, all the way down towards the floor of the valley of stability. On the other side of the valley

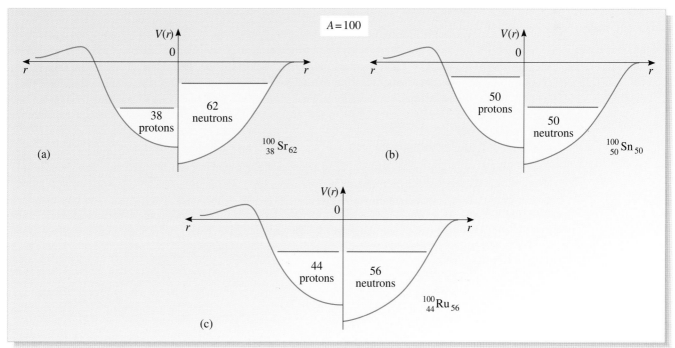

**Figure 3.28** The filled energy levels shown schematically for three isobars with $A = 100$.

floor, the nucleus $^{100}_{50}\text{Sn}_{50}$ ($T_{1/2} \approx 1$ s) has too many protons; it undergoes $\beta^+$-decay, the first of a chain, also leading to the floor of the valley. (The relatively long half-life of about 1 s is due to the fact that $^{100}_{50}\text{Sn}_{50}$ is *doubly magic*, i.e. both $N$ and $Z$ are magic.)

**Question 3.15** There is another doubly-magic isotope of tin. Write down its symbol in the form $^{A}_{Z}\text{Sn}_{N}$. Would you expect it to be stable? If not, what decay mode would you expect? (*Hint*: Refer to Figure 3.11 in Section 2.2.) ■

Most of these $\beta$-decays are accompanied by $\gamma$-ray emission. This is because the nucleus resulting from each decay is in an excited state which decays to the lowest unfilled energy level, the nuclear ground state, with the emission of one or more $\gamma$-ray photons.

### Symmetry energy and pairing energy

The version of the shell model we have outlined is greatly simplified and there are several other effects that influence the nucleon energy levels. We have already mentioned the *symmetry energy* in the semi-empirical model. If it were not for the Coulomb repulsion, the symmetry energy would keep $N = Z$ so as to maximise the relative number of unlike nucleon pairs (i.e. proton-neutron pairs). In fact you have seen that for heavy nuclei the Coulomb repulsion is strong enough to ensure that the neutron wells are significantly deeper than the proton wells and consequently $N > Z$ for stable nuclei. However, this neutron excess increases the symmetry energy (i.e. reduces the binding energy) for heavy nuclei by reducing slightly the depths of the neutron wells and thereby raising slightly the neutron energy levels. Thus the Coulomb repulsion and the symmetry effect are countervailing tendencies.

There are some other effects that are quite small but can nevertheless have huge significance. For example, the **pairing energy** increases the binding energy per nucleon for nuclei that have even numbers of protons and/or even numbers of neutrons, rather than odd numbers. This effect is not included in the semi-empirical curve in Figure 3.22 but can be seen as an odd-even zigzag in the measured $B/A$ values in Figure 3.23, most noticeably in the inset. It is of sufficient magnitude to make stable nuclei with odd values of both $Z$ and $N$ rare; the heaviest is $^{14}_{7}\text{N}_{7}$. Later, we shall see how the pairing energy makes all the difference between the fission properties of the uranium isotopes $^{235}\text{U}$ and $^{238}\text{U}$. The pairing energy in nuclei is a quantum-mechanical effect, essentially the same as electron pairing (in Cooper pairs) which leads to superconductivity in certain metals at very low temperatures.

### Deformed nuclei

You have seen that the total energy of the nucleus has contributions from a volume energy, a surface energy, Coulomb energy and symmetry energy, with an odd-even staggering from a subtle pairing effect, all within a shell structure. You can now begin to see how complex nuclear structure is. In some nuclei there is another small contribution of enormous significance.

You will recall that for a given mass number $A$, the surface energy is a minimum for a spherical shape. Many nuclei are in fact spherical for this reason. However, it turns out that when $N$ or $Z$ is far from the closed shell values, i.e. far from the magic numbers 2, 8, 20, 28, etc., there exist weaker but significant shell effects for other numbers of nucleons. These additional shell effects favour lower energy levels for nuclei that are deformed, in most cases rather like a rugby ball. For these **deformed nuclei** the advantage of having lower energy levels overcomes the surface effect.

Quantum mechanics requires that a spherical nucleus, which has nothing to distinguish one direction in space from another, cannot rotate. But a deformed nucleus can rotate, and this leaves its mark on the energy level structure. Deformed nuclei have discrete rotational energy levels that can be excited in high-energy scattering experiments. Roughly speaking, the more deformed the nucleus, the higher the probability of exciting the nucleus into states with rotational energy. Analysing such experiments allows the shapes to be deduced.

Nuclear deformation is quite common in certain regions of the valley of stability, in particular for nuclei in the region near $Z = 70$, $N = 100$ and near $Z = 90$. Deformation in the region near $Z = 90$ has important effects. The uranium isotopes $^{235}$U, $^{238}$U and the plutonium isotope $^{239}$Pu lie in this region. These nuclei readily undergo fission (on absorbing a neutron) because their ground state deformation gives the fission process a head start. Nuclear fission will be discussed further in Section 5.

## 4.3 Overview of the Z–N plane

We now take a broader view of the landscape in which the valley of stability has been such a dominant feature. Figure 3.29 shows the lie of the land emphasising somewhat different features compared with the history emphasized in Figure 3.11. The stable and known unstable nuclei are shown, and the closed shell values of $Z$ and $N$ are indicated. The regions where nuclei are deformed are shown, and it can be seen that in addition to the regions noted above, deformed-nuclei regions appear between about $Z = 40$ and $Z = 60$, above and below the path of stability. This suggests one of the reasons why there is so much interest in exploring regions as far from stability as possible — new phenomena occur in these regions.

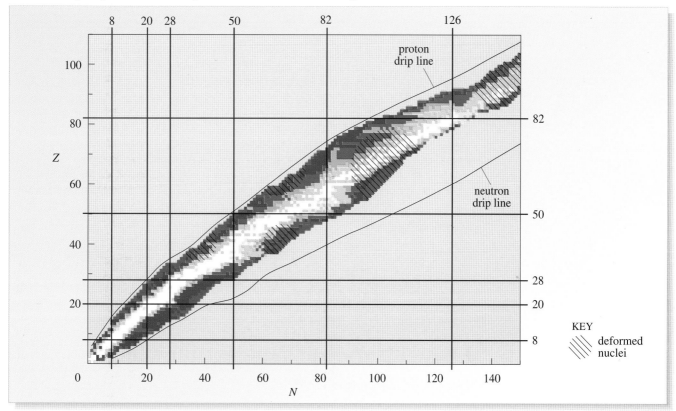

**Figure 3.29**   The Z–N plane showing stable and known unstable nuclei. The magic numbers are indicated (except 2), as are the regions of strong deformation. Also shown are the proton and neutron 'drip lines', the expected limits of particle-stable nuclei.

Figure 3.29 also shows the proton and neutron **drip lines**. These demarcate the region within which nuclei can exist. Beyond the *neutron drip line*, for example, additional neutrons are not bound. In general we are far from reaching the neutron drip line, although the *proton drip line* has been reached in places. As the proton drip line is approached, a variety of new phenomena appear, in particular *proton radioactivity* competes with electron capture as a decay mode, for example lutetium $^{150}$Lu can decay by proton emission

$$^{150}_{71}\text{Lu} \rightarrow \, ^{149}_{70}\text{Yb} + \text{p}.$$

The neutron drip line remains for the most part a distant goal. It is expected that new phenomena will be discovered there, for example nuclei with diffuse neutron skins, and the standard shell model could well break down or will at least have to be modified. Nuclear physicists hope that exploration of this area will answer many longstanding questions such as the precise role of relativistic effects in nuclei. Moreover, many of these short-lived nuclei play a role as intermediate steps in the production in supernova explosions of the heavy elements, those heavier than iron (Fe). A more complete knowledge of these nuclei is therefore essential for achieving a definitive account both of supernovas and of the production of the heavier elements we find about us. Generally, our theoretical models do a pretty good job of reproducing the basic properties of known nuclei, but the real test will only come when we get further out towards the drip lines!

**Question 3.16**   One way of investigating unstable nuclei well away from the floor of the valley of stability is to study fission products: Explain why the fission of a heavy nucleus into two nuclei of comparable mass numbers is bound to produce highly neutron-rich nuclei. Would you expect such fission products to be radioactive? If so, explain why.  ∎

Sustained efforts have also been made to artificially produce new **transuranic** elements with $Z > 92$ (uranium). These efforts started with the production of *neptunium* ($Z = 93$) by McMillan and Abelson in 1940. The most stable isotope of neptunium has a half-life $T_{1/2} = 2.1 \times 10^6$ years, but most transuranic elements have much shorter half-lives. As $Z$ increases these new elements become more and more unstable against $\alpha$-decay and spontaneous fission due to the increasing influence of Coulomb repulsion. For example, the most stable isotope of the element named *californium* ($Z = 98$) has $T_{1/2} = 800$ years, while that of *dubnium* ($Z = 105$) has $T_{1/2} = 34$ s. However, arguments have been put forward for a new proton closed shell at $Z = 114$ which could lead to an 'island of stability'. In fact as this chapter is being written claims have been made for the production of isotopes with $Z$ values of 114, 116 and 118.

Some of the most remarkable new nuclei have been discovered at the small $A$ end of the Periodic Table. These are the so-called *halo nuclei* which depart markedly from the general rule that the nuclear radius is about $(1.2 \text{ fm})^{1/3} A^{1/3}$. The nucleus $^{11}$Li is a fine example — its two last neutrons stretch out so far that it is effectively as large as a $^{40}$Ca nucleus.

# 5 Quantum-mechanical tunnelling in nuclear reactions

In the previous section we were concerned with the total energy of a nucleus as a measure of its stability against various processes, such as radioactive decay, fission and fusion. We have assumed that nuclei tend to undergo processes that would result in a lower total energy (or increased binding energy) but we have not enquired about the actual mechanisms by which these processes might occur. In fact a process will not necessarily occur just because it results in a lower total energy. An apple in a fruit bowl will not spontaneously jump down to the floor, even though its gravitational potential energy would be less there. It would first need to be given enough energy to climb up the side of the bowl before it could fall to the floor.

Now consider a proton inside a nucleus. A potential energy function for such a proton was shown in Figure 3.26; it is the sum of a strong nuclear potential energy and a Coulomb potential energy, and there is a bump or barrier near the nuclear edge. According to classical physics, a proton of total energy $E$ greater than zero but less than the energy at the top of the bump cannot escape from the nucleus, even though it could have a lower potential energy outside (see Figure 3.30); it would need some extra energy to get it to the top of the barrier before it could "roll down" the other side.

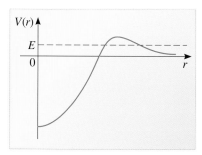

**Figure 3.30**  A proton of total energy $E$ cannot surmount the barrier.

According to quantum mechanics however, even though the proton does not have enough energy to surmount the barrier, it can tunnel its way through and emerge on the other side with total energy $E$. Spontaneous emission of a proton by this mechanism does actually occur as a very rare form of radioactivity. More commonly however, proton-rich heavy nuclei tend to emit $\alpha$-particles by the same tunnelling process. Tunnelling is also an essential mechanism for nuclear fission as we shall explain later in this section.

Another aspect of nuclear tunnelling is that it can also happen in reverse. A proton or an $\alpha$-particle approaching a nucleus from outside, without sufficient energy to climb the Coulomb barrier to the top of the bump, can actually *tunnel inwards*. Inward tunnelling is the key to important nuclear reactions that provide the energy for the Sun and other stars, and the creation of elements from hydrogen. Before discussing some of these reactions we first give the background to the discovery of tunnelling in 1928 as the mechanism by which $\alpha$-decay occurs.

## 5.1  Tunnelling and $\alpha$-decay

We have mentioned the extreme range of half-life values for $\alpha$-decay. In 1912, Geiger and Nuttall found a most dramatic relationship relating half-life to the kinetic energy $E$ of the emitted $\alpha$-particle. Roughly speaking, the higher the kinetic energy of the $\alpha$-particle, the shorter the half-life. A more precise statement reveals how dramatic the relationship is. It turns out that the half-life $T_{1/2}$ is given approximately by

$$T_{1/2} \propto \exp\left(\frac{Za}{\sqrt{E}}\right)$$

where $Z$ is the atomic number of the daughter nucleus, $a = \dfrac{e^2}{\varepsilon_0 \hbar}\sqrt{\dfrac{M}{2}} = 3.97\,\mathrm{MeV}^{1/2}$ and $M$ is the mass of the $\alpha$-particle.

The fact that doubling the energy of the emitted $\alpha$-particle can correspond to a factor of $10^{-16}$ in half-life is accommodated by this equation. The following exercise dramatises the huge range of half-lives corresponding to a modest range in energy.

**Question 3.17**  Consider the thorium isotopes $^{232}_{90}\mathrm{Th}$ and $^{218}_{90}\mathrm{Th}$. The former emits an $\alpha$-particle of energy 4.08 MeV and has $T_{1/2} = 1.405 \times 10^{10}$ years, while the latter emits an $\alpha$-particle of energy 9.85 MeV and has $T_{1/2} = 1.09 \times 10^{-7}$ s. Calculate the ratio of the half-lives and compare your answer with the prediction of the above equation. (Take 1 year to be $3.16 \times 10^7$ s.)  ■

In 1928 George Gamow, and also E. U. Condon and A. Gurney, explained the dramatic relationship between half-life and energy by modelling $\alpha$-decay as a

**quantum-mechanical tunnelling** process. They assumed that α-particles, being very tightly bound combinations of nucleons, exist as such inside heavy nuclei. The potential energy $V(r)$ of an α-particle is depicted in Figure 3.31a.

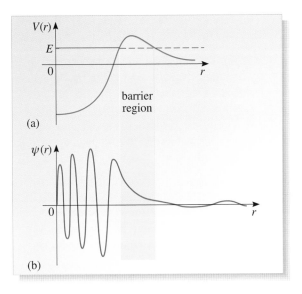

(a)

(b)

**Figure 3.31** (a) The potential energy well, $V(r)$, seen by an α-particle. (b) Schematic picture of the wavefunction $\psi(r)$ tunnelling into the classically-forbidden barrier region and leaking outside.

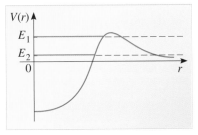

**Figure 3.32** The potential energy barrier for two different α-particle energies. The tunnelling probability depends *very* sensitively on the energy and is *very* much less for total energy $E_2$ than for $E_1$.

Outside the nucleus $V(r)$ is dominated by the Coulomb repulsion between the two protons in the α-particle and the $(Z - 2)$ protons in the residual nucleus, while inside the nucleus the strong attractive nuclear force is dominant. The result is the same sort of potential energy barrier that we have seen for protons. Figure 3.31b depicts a wavefunction $\psi(r)$ for an α-particle in an energy level $E$, obtained by solving the Schrödinger equation for an α-particle in the potential energy well $V(r)$. The wavefunction is of large amplitude inside the nucleus, indicating a large probability of being there, but it also penetrates into the classically-forbidden barrier region, falling off rapidly with distance and feeding a tiny wave leaking out. This leak represents a very small probability of finding the α-particle outside the nucleus with total energy $E$ after tunnelling through the barrier. It is found that the fall-off inside the barrier depends *very strongly* on the width and height of the barrier above $E$. Thus the lower $E$, the greater the width and height of the barrier, as shown in Figure 3.32, and the smaller the probability of the α-particle leaking out. This, essentially, is why the lifetime increases so dramatically with decreasing energy of the α-particle.

The explanation of α-decay as a tunnelling process in 1928 was to revolutionize nuclear physics. But before looking at its consequences, let's go back a decade.

## 5.2  Tunnelling and the first nuclear accelerator

In 1919 Rutherford reported an experiment in which a high-energy α-particle from a natural radioactive source collided with a nitrogen nucleus in air and knocked out a proton. This was the first artificial nuclear reaction. However, it was inconvenient to have to work with the restrictions of intensity and energy imposed by the use of natural radioactive sources. Obviously it would be better to build accelerators that could produce well-controlled beams of high-energy particles to explore nuclear structure. But in the 1920s, straightforward calculations using classical physics made it clear that the energies to which beams of protons or α-particles could be accelerated using the then-existing technology, a few hundred keV or so, were not enough to surmount the Coulomb barriers of several MeV around nuclei.

## Example 3.2

Estimate the height (in MeV) of the Coulomb barrier for a proton approaching a lithium $^{7}_{3}\text{Li}_4$ nucleus.

## Solution

We can assume that the height of the Coulomb barrier is the Coulomb potential energy of the proton and lithium nuclei when they just 'touch', i.e. when their separation $R$ is equal to the sum of their radii. Thus, estimating the radii from Section 2.2, $r \approx (1.2\,\text{fm})A^{1/3}$, we find $R \approx 3.5\,\text{fm}$, and the potential energy is

$$V(3.5\,\text{fm}) = \frac{e \times 3e}{4\pi\varepsilon_0 R} = \frac{9 \times 10^9 \times 3(1.6 \times 10^{-19})^2}{3.5 \times 10^{-15}}\,\text{J}$$
$$= 2.0 \times 10^{-13}\,\text{J} = 1.2\,\text{MeV}.$$

In fact, a beam of protons incident on stationary lithium nuclei would need more energy than the 1.2 MeV calculated in Example 3.2. This is because momentum is conserved in a collision, and so the product particles will carry away some kinetic energy. This kinetic energy has to be provided by the incident protons.

The existence of tunnelling demonstrated by Gamow and others in 1928 was electrifying news, since it showed that achievable accelerators could make beams of particles which had enough energy to *tunnel inwards*. The first such accelerator was built by John Cockcroft and Ernest Walton (Figure 3.33) in Cambridge, and in 1932 they employed it in the first nuclear reaction studied with an artificially-accelerated beam. The beam was of protons and the target was lithium (as considered in Example 3.2 above). The result was the disintegration of the lithium nucleus and the creation of two α-particles, which we can write as

$$\text{p} + {}^{7}\text{Li} \rightarrow 2\alpha$$

or $\qquad {}^{1}\text{H} + {}^{7}\text{Li} \rightarrow {}^{4}\text{He} + {}^{4}\text{He}.$

Because of tunnelling, this reaction can occur with protons of energy as low as 120 keV (although Cockcroft and Walton's machine could provide 700 keV protons), very much less than the Coulomb barrier of more than 1.2 MeV. This experiment was certainly a turning point in our understanding of nuclei. Soon afterwards, Ernest Lawrence at Berkeley, California, invented a more powerful accelerator, the *cyclotron* (discussed in Chapter 4), thus initiating a line of development that continues to this day.

**Figure 3.33**
J. D. Cockcroft (1897–1967) and
E. T. S. Walton (1903– ).

## 5.3 Tunnelling before there were stars

Some of the most exciting applications of nuclear physics are in the cosmology of the expanding Universe, the evolution of stars and the synthesis of the elements. The evolution of the Universe following the Big Bang is not yet completely understood and remains an area of active research, but it is generally believed that, following the Big Bang, protons and neutrons 'condensed' out of a *quark–gluon plasma*. Once this had happened, the possibility of deuterium formation arose:

$$p + n \rightarrow {}^2H + \gamma.$$

This exothermic reaction releases the 2.22 MeV binding energy of the deuteron in the form of the γ-ray photon. In fact, the Universe had to cool down before this reaction created significant deuterium. This is because when the universe had a temperature equivalent to 2.22 MeV, about $2 \times 10^{10}$ K, this equation is more properly written as

$$p + n \leftrightarrow {}^2H + \gamma.$$

In such an equilibrium reaction, ${}^2H$ is destroyed by γ-ray disintegration (the reverse reaction) as fast as it is made.

**Question 3.18**   Use the ideal gas model (*Classical physics of matter*) to confirm the statement above that a temperature of about $2 \times 10^{10}$ K is equivalent to an energy of about 2.22 MeV.   ■

Once the Universe had cooled enough for deuterium to be made, it was rapidly absorbed by combining with protons or neutrons in strongly-exothermic reactions leading to ${}^3H$ (tritium) and ${}^3He$. There is also a reaction involving two deuterons leading to the production of ${}^4He$, which is highly exothermic because of the very large binding energy of ${}^4He$.

**Question 3.19**   Given that the binding energy of ${}^4He$ is 28.30 MeV and that of a deuteron ${}^2H$ is 2.22 MeV, calculate the energy released when two deuterons fuse to form ${}^4He$.   ■

The processes which took place during this cool period are complicated but have been modelled using nuclear physics and thermodynamics, with tunnelling as an essential mechanism for reactions between positively-charged particles. Such models lead to a satisfactory prediction of the proportions of the protons and ${}^4He$, with traces of ${}^2H$, ${}^3He$ and ${}^7Li$, that made up the tenuous medium out of which the first generation of stars condensed. Our Sun is a later generation star, incorporating heavy elements created in and expelled by explosive events (such as supernovas) in the history of earlier generations of stars.

For many purposes, you can regard the Sun as starting out essentially as a ball of mainly hydrogen ${}^1H$ which is being converted to helium ${}^4He$ in a manner we now discuss.

## 5.4 Tunnelling in the Sun and other stars

At about the same time that the tunnelling explanation of α-decay stimulated the work of Cockcroft and Walton, it suggested to many physicists a solution to an old problem. Where did the energy in the Sun come from? How could it last so long without running out of fuel? It was apparent that a lot of energy could be released in nuclear reactions, but in the days before tunnelling was known about, the Coulomb barrier seemed to present a problem. The sort of temperatures which were envisaged

Quarks and gluons will be discussed in Chapter 4 of this book.

in stars corresponded to energies that were far too low for two protons ever to approach each other closely enough to overcome the Coulomb barrier and fall within the range of the strong nuclear force. According to our present understanding, the temperature at the centre of a star like the Sun is about $T = 1.5 \times 10^7$ K, and the corresponding kinetic energy $E$ (using $E \approx 3kT/2$) is only about 1 keV. Such an energy is much less than the height of the Coulomb barrier between two separated protons. Without tunnelling, the fusion of two protons

$$p + p \rightarrow {}^2H + e^+ + \nu_e$$

could not happen at all at the temperatures of the Sun's core.

We now know that two protons can tunnel through the Coulomb barrier between them, and that this reaction is the first step in a series of reactions that is both the main source of energy for the Sun and also the first step towards creating all elements heavier than hydrogen. In fact, given the possibility of tunnelling, it is the very slow rate of the β-process that determines the rate of the above fusion reaction. It takes, on average, about 5 billion years for a pair of protons to become a deuteron ($^2H$).

Once deuterons are formed in the Sun by the above reaction, they are rather quickly destroyed by the proton capture reaction you met in Section 3.3,

$$p + {}^2H \rightarrow {}^3He + \gamma$$

which releases 5.5 MeV. This reaction does not involve a β-process, and deuterons last only about a second thanks to tunnelling.

The next most important process is the formation of an α-particle:

$$^3He + {}^3He \rightarrow {}^4He + p + p.$$

This chain of nuclear reactions is referred to as the **proton–proton chain**, or pp chain, the net effect of which is that protons (hydrogen nuclei) fuse to become helium:

$$4p \rightarrow {}^4He + 2e^+ + 2\nu_e + 2\gamma.$$

The energy released in the pp chain in the central region of the Sun is the main source of the Sun's energy.

But where do heavy elements like mercury and lead originate? The production of heavier elements takes place largely in the later stages of a star's evolution. We cannot go into the physics of stars here, but the key point is that nuclei with large $Z$ are made by the fusion of nuclei which already have quite high $Z$. This requires tunnelling through higher barriers than for the pp chain. Remember that the repulsive coulomb potential energy function for two nuclei with proton numbers $Z_1$ and $Z_2$, has a factor $Z_1 Z_2$ in the numerator. You will recall also the extremely sensitive relationship between total energy and barrier height for α-particles tunnelling outwards in α-decay; exactly the same relationship applies to the inwards tunnelling when nuclei fuse in stars. For high-$Z$ nuclei to have the energy to tunnel through these high barriers, they must be in a very hot environment where $kT$ and hence the kinetic energies of colliding nuclei are large — hotter than the centre of stars like the Sun. It is in such hot places that the lead on your roof was made. Where are such places?

In brief outline, as the pp chain process which powers stars like the Sun becomes exhausted, the outgoing radiation, which holds the star up, is reduced. The star then collapses and the resulting release of gravitational energy heats it up. When the star is hot enough, the higher barriers associated with higher $Z$ can be tunnelled through, producing heavier elements. The star goes through various further stages in which successively heavier elements are produced, but only nuclei of $A$ up to about 60 (near iron and nickel) can be made in this way. This is because the $B/A$ curve, Figure 3.11, curves downwards after $A \approx 60$, so that fusion reactions that produce heavier nuclei are endothermic. To make heavier elements Nature resorts to stellar cataclysms known as supernovas. These release vast quantities of energy and, crucially, an intense flux of neutrons. So intense is the neutron flux that many neutrons can be absorbed by a nucleus before even very rapid β-decay processes have a chance to take place. This leads to the production of very unstable nuclei out toward the neutron drip line; these eventually β-decay to the heavy elements we find on Earth. It turns out that detailed models involving both stellar theory and nuclear theory, give a rather good explanation of why the elements exist in the proportions that are found.

## 5.5  Nuclear fission — a case study

We have already seen certain aspects of nuclear fission, most particularly the large amount of energy released when a heavy nucleus undergoes fission. It is this, of course, in the context of nuclear weapons and nuclear reactors, which has brought fission into the public arena. Here we first give some background facts about fission and then attempt to explain some of them in terms of the processes, including tunnelling, that have been described in this chapter.

### Background to fission

Uranium ($Z = 92$) forms about 2.7 parts per million in the Earth's crust, which means that your back garden, down to a metre or so, may contain up to a kilogram of uranium. Naturally occurring uranium is a mixture of isotopes: 99.275% $^{238}$U, 0.720% $^{235}$U and about 0.0056% of $^{234}$U.

Most uranium nuclei decay by α-emission ($T_{1/2} = 4.51 \times 10^9$ years for $^{238}$U and $7.04 \times 10^8$ years for $^{235}$U) but **spontaneous fission** does occur as an extremely rare process. If α-decay were somehow suppressed, $^{235}$U would have a half-life of order $10^{16}$ years. On the other hand, the fission process discovered in the late 1930s and exploited in nuclear reactors is *induced by neutrons* that have been slowed down to *thermal energies*. These so-called **thermal neutrons** have average kinetic energies of $3kT/2 \approx 0.08$ eV, corresponding to the temperatures inside the reactors, a negligible amount of energy on the nuclear scale. Nuclei such as $^{235}$U that have a significant probability of undergoing **induced fission** after absorbing a thermal neutron, are called **fissile**. The isotope $^{235}$U is fissile but the more common $^{238}$U is not. Hence the interest in separating uranium isotopes so as to enhance the proportion of the fissile $^{235}$U for use as nuclear fuel.

$^{238}$U can fission by the absorption of a fast neutron of energy above about 1 MeV.

When a $^{235}$U nucleus undergoes fission, it releases a number of neutrons (from 1 to 4 with an average of 2.47) with energies of a few MeV, as well as the two large fission fragments each with kinetic energies in the 100 MeV range. For example, one such induced fission reaction is

$$^{235}\text{U} + \text{n} \rightarrow {}^{236}\text{U} \rightarrow {}^{93}\text{Rb} + {}^{141}\text{Cs} + 2\text{n}.$$

The neutrons, when slowed down to thermal energies, can lead to further fissions, thus giving rise to a 'chain reaction' in a large mass of $^{235}$U.

In a reactor the neutrons are slowed down by elastic collisions with *light* nuclei in a material called a **moderator**. The light moderator nuclei absorb neutron kinetic energy by recoiling during collisions until the neutrons are thermalized, i.e. slowed down to thermal equilibrium with the moderator. Historically, the important moderators have been $^{12}$C in the form of graphite and $^{2}$H in the form of 'heavy water'; ordinary water is also used in some reactors but $^{1}$H absorbs neutrons.

The non-fissile isotope $^{238}$U can absorb a thermal neutron to produce the unstable $^{239}$U which β-decays to $^{239}$Np (neptunium) which itself β-decays to a fissile plutonium isotope $^{239}$Pu:

$$^{238}\text{U} + \text{n} \rightarrow {}^{239}\text{U} \rightarrow {}^{239}\text{Np} + \text{e}^- + \overline{\nu}_e$$

$$^{239}\text{Np} \rightarrow {}^{239}\text{Pu} + \text{e}^- + \overline{\nu}_e.$$

This is how fissile plutonium $^{239}$Pu is made in reactors. Its significance lies in the fact that the chemical extraction of $^{239}$Pu is far easier than the isotopic separation of the fissile $^{235}$U from natural uranium.

**Question 3.20**  (a) Assuming that nuclei have radii given by $1.2 \times A^{1/3}$ fm, determine how much energy is required to bring together the fission product nuclei $^{93}$Rb and $^{141}$Cs to the point where they just touch. (Consult Figure 3.2 for the Z values.) Explain why this is of the same order as the energy released per fission. (b) Assuming that about 200 MeV is released per fission of a $^{235}$U nucleus estimate how long a 1000 MW nuclear power station could be kept running by the fission of 1 kg of $^{235}$U. Assume an overall efficiency of 20%. ■

*The fission process*

You have seen (Section 4.2) that the uranium nucleus in its lowest energy state has a deformed rugby-ball shape as a result of a balance between weak shell-effects that give a lower energy for a deformed shape, and surface effects which tend to make the nucleus spherical. You can picture the fission process as a sequence of larger and larger deformations until a neck appears and finally the nucleus splits into two parts. This is illustrated in Figure 3.34 which shows a sequence of nuclear shapes and associates these with a deformation parameter $s$.

You can think of $s$ as the distance between the two parts as indicated in Figure 3.34. The value $s = 0$ corresponds to a spherical nucleus. A small positive value of $s$ corresponds to a rugby-ball shape of the nucleus, and increasing $s$ further corresponds to the increasingly-deformed shapes shown in Figure 3.34. After the neck has developed, higher values of $s$ track the nucleus as it is pushed on its path to fission by the Coulomb repulsion of the two separating parts.

Now suppose we could somehow freeze the nucleus at any deformation $s$ and ask what ground-state energy $E(s)$ it would have for each value of $s$. We can then plot the $E(s)$ values as a function of $s$ and this has been done in Figure 3.35.

The fall-off at very large $s$ by about 200 MeV corresponds to the much lower nuclear binding energies of the two lighter fission products, which is just what we expect from the $B/A$ curve, Figure 3.14. The ground-state energy of the nucleus is not actually at $s = 0$ because, as stated above, the uranium nucleus is deformed in its lowest energy state, with a deformation parameter corresponding to the local minimum of $E(s)$.

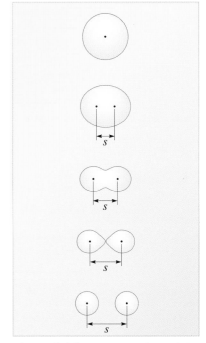

**Figure 3.34**  A series of possible distorted nuclear shapes leading to fission. The distortion parameter $s$ is shown as the distance between the centres of the two parts.

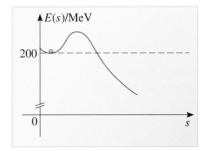

**Figure 3.35**  The ground-state energy function $E(s)$.

The energy function $E(s)$ in Figure 3.35 is the nuclear ground-state energy and not a potential energy as in Figure 3.31, but it can nevertheless be treated in the same way as a potential energy in the tunnelling theory.

According to classical physics, i.e. without tunnelling, the fission process would require the nucleus to move from the lowest energy, at the bottom of the well in Figure 3.35, climb up the hill and then slide down the far side. The problem with this classical picture is that a nucleus at the bottom of the well cannot climb the hill unless an energy, equal to the height of the hill above its resting point, is given to it.

However, nuclei are described by quantum mechanics and so the possibility of tunnelling through the barrier, known as the **fission barrier**, arises. We have seen that the tunnelling probability is extremely sensitive to the energy relative to the top of the barrier. Spontaneous fission of uranium by tunnelling through the fission barrier does occur, and can be observed despite the fact that it is a highly improbable process with a half-life of some $10^{16}$ years, and the competing process of $\alpha$-decay with a much shorter lifetime is likely to occur instead.

But consider what happens when a thermal neutron is absorbed by a uranium nucleus. The neutron goes from being free to being bound in a nucleus, and so an amount of binding energy approximately equal to the binding energy per nucleon of the resulting nucleus is released. This released energy puts the resulting nucleus in an excited state. When a $^{235}_{92}\text{U}$ nucleus absorbs a thermal neutron, the resulting nucleus is $^{236}_{92}\text{U}$ which has both $Z$ and $N$ even and therefore has an relatively large $B/A$ due to the *pairing energy*, discussed in Section 4.2 As a result, the binding energy released when the neutron is absorbed is enough to put the excited state of the resulting $^{236}_{92}\text{U}$ nucleus *above* the top of its fission barrier, and so the $^{236}_{92}\text{U}$ nucleus is unstable and immediately undergoes fission. This is why $^{235}_{92}\text{U}$ is fissile. On the other hand, when $^{238}_{92}\text{U}$ absorbs a neutron, the resulting nucleus is $^{239}_{92}\text{U}$ which has odd $N$ and therefore a relatively low $B/A$ due to the pairing energy. In this case the amount of binding energy released when the neutron is absorbed is only enough to excite the $^{239}_{92}\text{U}$ nucleus to about 1 MeV *below* the top of its barrier. This energy gap is enough to make the tunnelling probability extremely small and hence fission extremely unlikely. There are many competing decay processes for the $^{239}_{92}\text{U}$ nucleus, including $\gamma$-decay to a lower energy state, and the production of the fissile plutonium isotope $^{239}_{94}\text{Pu}$ through the two-step $\beta$-decay process given above. Thus although $^{238}_{92}\text{U}$ itself is not fissile, the absorption of thermal neutrons does lead to the production of fissile plutonium $^{239}_{94}\text{Pu}$ .

**Question 3.21**   Which of the two nuclei, $^{233}\text{U}$ and $^{240}\text{Pu}$, might you expect to be fissile? Explain why.   ∎

We have now seen briefly how many of the physical processes we have discussed in this chapter play a role in the fission process which, however uncomfortable it seems, is an inescapable presence in our lives. Currently (1999) about 30% of our electricity (about 80% in France) comes from fission nuclei rolling down the hill in Figure 3.35.

> Open University students may leave the text at this point and do the optional S103 multimedia package *Nucleons in nuclei*. When you have completed this you should return to the text. This activity will occupy about one hour.

# 6   Closing items

## 6.1   Chapter summary

1   An atomic nucleus is composed of $Z$ protons and $N$ neutrons held together by the strong nuclear force in a region of radius

$$r \approx (1.2 \text{ fm})A^{1/3}$$

where the mass number $A = Z + N$. The atomic number $Z$ specifies the particular chemical element and is also equal to the number of electrons in the neutral atom.

2   A particular nuclear species is specified by a symbol $^{A}_{Z}X_{N}$, or in the non-redundant form $^{A}X$, where X is denotes the chemical symbol of the element.

3   Most nuclei are unstable, i.e. radioactive. In $\alpha$-decay the nucleus emits a high-energy $\alpha$-particle ($^{4}_{2}He_{2}$). In $\beta$-decay, the nucleus emits a high-energy electron ($\beta^{-}$-decay) or positron ($\beta^{+}$-decay), or captures an atomic electron (electron capture). Neutrinos (or antineutrinos) are also emitted in $\beta$-decay. These particles are almost undetectable but carry away some of the energy, all of it in the case of electron capture. Both $\alpha$-decay and $\beta$-decay are usually accompanied by the emission of $\gamma$-rays, high-energy photons like X-rays. Some heavy nuclei undergo spontaneous nuclear fission, and some nuclei exhibit proton emission.

4   Radioactive decay is governed by the exponential decay law

$$N(t) = N_0 \exp(-0.6931t/T_{1/2}) \qquad (3.5)$$

where $0.6931 = \log_e(2)$ to 4 decimal places, $N_0$ is the initial number of radioactive nuclei and $N(t)$ is the average number left at time $t$. The half-life, $T_{1/2}$, characterises the nucleus and decay mode. The number of undecayed nuclei is halved during any period equal to the half-life.

5   An overview of known nuclei can be seen by plotting them on the $Z$–$N$ plane (Figure 3.11). Nuclei that are naturally abundant on Earth lie along the path of stability, a straight line $Z = N$ for light nuclei ($Z$ less than about 20), curving towards the $N$-axis for medium and heavy elements. The further out a nucleus is from the path of stability the more recently it is likely to have been discovered and the more rapidly it is likely to decay.

6   The binding energy $B$ of a nucleus is the minimum energy required to disassemble the nucleus into its constituent nucleons; it is also equal to the energy released when the nucleus is formed from its constituents. The binding energy per nucleon, $B/A$, is a measure of the relative stability of nuclei.

7   $B/A$ for most nuclei is within 10% or so of 8 MeV per nucleon, with a maximum value of about 8.8 MeV per nucleon near $A = 56$ (iron and nickel), then falling off slowly for heavier nuclei and falling off quite rapidly for the light nuclei. This means that energy can be released when heavy nuclei undergo fission or when light nuclei undergo fusion.

8   When the values of the total energy per nucleon ($-B/A$) for all known nuclei are plotted as points above the $Z$–$N$ plane they lie near a surface having the shape of a valley called the valley of stability, with the valley floor lying directly above the path of stability. Then the $\beta^{-}$-decay of neutron-rich isobars and the $\beta^{+}$-decay (or electron capture) of proton-rich isobars can be thought of as streams running down the valley walls, while $\alpha$-decay of heavy nuclei is like a stream running down the valley floor.

9    The semi-empirical model of the nucleus incorporates the short-range strong nuclear force and the repulsive Coulomb force and is based on an analogy between a nucleus and a charged liquid drop. It identifies four contributions to $B/A$: a volume energy, a surface energy, a Coulomb energy and a symmetry energy. Using empirical parameters, the model fits the overall trend of measured $B/A$ values quite well but there are discrepancies in the small-scale structure suggestive of shell effects.

10   In the nuclear shell model each proton and neutron is represented by a wavefunction and occupies an energy level in a potential energy well produced by the other nucleons. The potential energy wells for neutrons follow the nuclear matter densities, while those for protons have contributions from Coulomb repulsion, resulting in a bump or barrier at the nuclear edge, and wells that are less deep than the corresponding neutron wells. Nuclei with magic numbers of nucleons, corresponding to closed shells, are particularly stable. Because of pairing, nuclei with even numbers of protons and neutrons tend to be more stable than those with odd numbers.

11   The proton and neutron energy levels are filled starting with the lowest energies and subject to the Pauli exclusion principle. Nuclear stability favours the filling of the neutron wells and the shallower proton wells to the same energy. This explains the neutron excess in medium and heavy nuclei and the tendency for isobars far from the valley floor to exhibit β-decay.

12   Many nuclei are spherically symmetric due to surface effects, but weak additional shell effects favour distorted rugby-ball shapes for some nuclei.

13   Quantum-mechanical tunnelling is an essential mechanism for many nuclear reactions involving charged particles. α-decay occurs when α-particles inside nuclei tunnel to the outside through the Coulomb barrier. The tunnelling probability depends very sensitively on the energy of the α-particle relative to the top of the barrier. This explains the huge range of α-decay lifetimes corresponding to a small range of α-particle energies.

14   Many nuclear reactions between charged particles occur by the particles tunnelling inwards through their mutual Coulomb barrier so that they can interact by the strong nuclear force. Fusion reactions in the Sun can only occur at the prevailing temperatures by tunnelling.

15   Nuclear fission of uranium and some other heavy nuclei can be modelled by progressive distortions of a charged liquid drop aided by the initial distortion of the nucleus due to weak shell effects. Fission occurs by tunnelling through the fission barrier.

16   Spontaneous fission is very rare, but some isotopes are fissile, i.e. they undergo fission on absorbing thermal neutrons. This induced fission releases energy as well as additional neutrons which can lead to a chain reaction. The $^{235}$U isotope is fissile, partly because of the pairing energy of the resulting $^{236}$U nucleus, but the more abundant isotope $^{238}$U is not.

## 6.2  Achievements

Now that you have completed this chapter, you should be able to:

A1   Explain the meanings of all the newly defined (emboldened) terms introduced in this chapter.

A2   Outline experiments from which information on nuclear charge and matter distributions are obtained, and estimate the radii of specified nuclei.

A3  Describe the main modes of radioactive decay: α-decay, β-decay and γ-decay. Distinguish between the various modes of β-decay.

A4  Write an equation describing a given decay process using the $_Z^A X_N$ symbols (or the non-redundant forms of these symbols) for the nuclei.

A5  Use the exponential decay law to relate the numbers of undecayed nuclei in a sample at different times.

A6  Use the concepts of binding energy and binding energy per nucleon to discuss the relative stability of the naturally abundant nuclei, and to calculate energy changes occurring in fission and fusion.

A7  Describe the main features of the valley of stability.

A8  Describe the semi-empirical model of the nucleus.

A9  Describe the nuclear shell model and use it to discuss the neutron–proton ratios of the naturally abundant nuclei and the β-instability of nuclei far from the valley floor.

A10  Discuss the role of quantum-mechanical tunnelling in α-decay and in fission and fusion reactions.

## 6.3  End-of-chapter questions

**Question 3.22**  (a) Write equations describing the α-decay of the thorium nucleus $_{90}^{224}Th$, the β⁻-decay of the platinum nucleus $^{197}Pt$ and the electron capture of the mercury nucleus $^{197}Hg$. (Consult Figure 3.2 for the $Z$ values.) (b) State which one of the following $A = 197$ isobars you might expect to be stable, giving your reasons: $^{197}Au$, $^{197}Tl$ and $^{197}Pb$.

**Question 3.23**  The oxygen isotope $^{15}O$ is a positron emitter with a half-life of 124s and is used in positron emission tomography. A sample is inhaled by a patient at 9.00 a.m. Determine what fraction of the $^{15}O$ isotope is left in the patient's body at 9.05 a.m. the same day. Explain why the isotope has to be produced on-site at the hospital (using a cyclotron).

**Question 3.24**  Figure 3.36 depicts a β-decay mode of the iodine nucleus $^{126}I$. Some excited energy levels of the daughter xenon nucleus $^{126}Xe$ are shown with their energies above the $^{126}Xe$ ground state. Accompanying γ-ray emissions are indicated by the vertical arrows. Determine the wavelength of the highest energy γ-ray emitted.

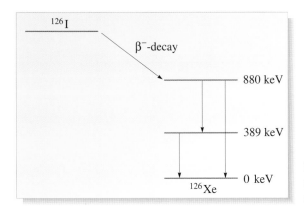

**Figure 3.36**  β-decay of $^{126}I$ and some energy levels of daughter $^{126}Xe$.

**Question 3.25** Consider the nuclear reaction $^6\text{Li} + \text{n} \rightarrow {}^3\text{H} + {}^4\text{He}$. Determine whether the reaction is exothermic or endothermic and state how much energy is released or absorbed by the reaction. Take the binding energies per nucleon to be: 5.33 MeV per nucleon for $^6\text{Li}$, 2.83 MeV per nucleon for the triton and 7.07 MeV per nucleon for the $\alpha$-particle.

**Question 3.26** (a) The solar constant (the solar power per square metre at the Earth's surface) is $1.4\,\text{kW}\,\text{m}^{-2}$. Determine the total power output from the Sun. Take the radius of the Earth's orbit to be $1.5 \times 10^{11}\,\text{m}$. (b) The main source of solar power is the fusion of hydrogen into helium as described by the proton–proton chain, the net result of which is

$$4{}^1_1\text{H} \rightarrow {}^4_2\text{He} + 2\text{e}^+ + 2\nu_\text{e} + 2\gamma$$

with the release of 26.7 MeV. The mass of the Sun is about $2 \times 10^{30}\,\text{kg}$. Assuming this mass is almost entirely hydrogen and about one-third of this will undergo fusion, estimate how many such fusion chains occur per second, and how long it will take for all the Sun's hydrogen to be consumed. (Ignore the energy carried off by neutrinos.) (c) Estimate the total solar power output per kilogram of solar mass. How does this compare with the power output per kilogram of the human body; assume a 50 kg person generates about 70 W. ∎

# Chapter 4  Particle physics

## 1  The challenge of particle physics

One of the main aims of physics is to understand the fundamental constituents of matter and the forces between them. At one time, it was thought that the basic building blocks of matter were atoms. However, you now know that an atom consists of a nucleus and one or more electrons. The nucleus in turn is made up of a number of protons and neutrons. You have also seen that other subatomic particles exist: neutrinos and muons (the heavy cousin of the electron), as well as particles of antimatter such as the positron (the antielectron) and the antineutrino. This picture prompts obvious questions. Do other subatomic particles exist? Are the proton, neutron and electron truly fundamental particles, or are they made up of even more fundamental constituents?

In the third quarter of the twentieth century, collision  studies using cosmic rays and particle accelerators have revealed a whole plethora of subatomic particles. By the early 1960s, there were 13 subatomic particles known for certain, plus dozens of what seemed to be extremely short-lived particles called *resonances.* By the end of the 1970s there were literally hundreds of known particles, most of them unstable. Attempts were made to classify the growing list of particles and to describe them in terms of a smaller number of truly fundamental entities. The milestone here is the *quark hypothesis* put forward in 1964 by Murray Gell-Mann and George Zweig, independently of one another. They proposed that most subatomic particles, including protons and neutrons as well as scores of others, are combinations of a very much smaller number of unobservable fundamental constituents which Gell-Mann called *quarks* (sometimes pronounced 'quorks'). This hypothesis has since developed into the so-called *standard model*, the currently accepted theory of fundamental particles and their interactions. The main aim of this chapter is to describe the discovery of subatomic particles and the development of our modern understanding of them.

Section 2 is an historical account of the methods used to study the subatomic world and the early discoveries of subatomic particles including the electron, proton, neutron, and muon, etc. Section 3 describes the families into which these particles have been grouped and introduces some of the basic terminology of the field. When you reach the end of this section, you may well experience some of the bewilderment felt by physicists in the 1960s and 1970s when the list of newly discovered particles seemed to grow almost daily. Obviously, you are not expected to memorize lists of particles and their properties. The important physical ideas are the patterns and unifying themes, especially the *conservation laws*, which bring order to particle interactions by restricting the possible outcomes.

Section 4 brings simplicity to the world of subatomic particles by introducing the quark hypothesis. You will see how hundreds of the known subatomic particles were interpreted as combinations of quarks held together by other particles called *gluons* that are ultimately responsible for the short-range *strong interaction*. Thus, the sequence from molecule to atom to nucleus takes another step to a deeper level of structure (Figure 4.1).

Section 5 sketches modern developments. The standard model has been enormously successful, yet many physicists believe it to be incomplete and efforts continue towards finding a deeper and more unified understanding of particles and their interactions.

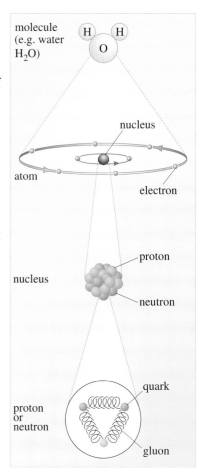

**Figure 4.1**   A deeper level of structure. Protons, neutrons and a host of other subatomic particles are thought to be composed of quarks held together by gluons mediating the short-range strong interaction force. (The gluons are represented schematically by spring symbols.)

In this introduction we give a brief preview of material, discussed in more detail in Section 4, describing what are now believed to be the truly fundamental particles of which all matter is composed. This will give you a relatively simple and coherent picture to hang onto, while grappling with the twists and turns of the experimental discoveries and theoretical advances that led to the standard model.

## 1.1  Leptons and quarks — a brief preview

Fundamental particles are sometimes called *elementary particles*.

Today, it is believed that there are two families of **fundamental particles**, called *leptons* and *quarks*. By *fundamental* we mean that there is no evidence that these particles are composed of smaller or simpler constituents. There are just six leptons and six quarks, together with an equal number of their antiparticles. All the familiar forms of matter are ultimately composed of these particles.

Do not take flavours too literally — you cannot taste leptons!

The six leptons, along with their electric charges, are depicted in Figure 4.2a. The six different types are often referred to as different *flavours* of lepton, and the three pairs are said to represent the three *generations* of leptons. The first generation consists of the familiar electron ($e^-$) and its β-decay partner, the *electron neutrino* ($\nu_e$). The second pair of leptons consists of the *muon* ($\mu^-$) and another type of neutrino called a *muon neutrino* ($\nu_\mu$). The muon is similar to the electron except that it is about 200 times heavier and unstable with a fairly long lifetime of a few microseconds. (We referred to muons in Chapter 3 in the context of *muonic atoms*.) The third generation of leptons consists of a particle called a *tauon* ($\tau^-$) and a third type of neutrino called a *tauon neutrino* ($\nu_\tau$). The tauon is similar to and even heavier than the muon and has a much shorter lifetime. These two heavier leptons, being unstable, are not normally constituents of matter, but are created in high-energy collisions between other subatomic particles.

Associated with these six leptons are the six *antileptons*, particles of *antimatter*. These include the positron ($e^+$) which is the antiparticle of the electron and the antineutrino ($\overline{\nu}_e$). We shall return to the leptons in Section 3.

|  | 1st generation | 2nd generation | 3rd generation |
|---|---|---|---|
| leptons with charge $-e$ | $e^-$ | $\mu^-$ | $\tau^-$ |
| leptons with charge $0$ | $\nu_e$ | $\nu_\mu$ | $\nu_\tau$ |

(a)

|  | 1st generation | 2nd generation | 3rd generation |
|---|---|---|---|
| quarks with charge $+\frac{2}{3}e$ | u | c | t |
| quarks with charge $-\frac{1}{3}e$ | d | s | b |

(b)

**Figure 4.2**  (a) The three generations of leptons and their electric charges. (b) The three generations of quarks and their electric charges.

The pattern of Figure 4.2a is repeated for the *quarks*. Figure 4.2b shows the six types (or flavours) of quark labelled (for historical reasons) by the letters u, d, c, s, t and b, which stand for up, down, charm, strange, top and bottom. (You will find that vivid terminology like this is common throughout particle physics.) Like the leptons, the quarks are paired off in three generations on the basis of their mass. To each quark, there corresponds an *antiquark*, with the opposite electric charge and the same mass. The antiquarks are denoted by $\bar{u}$, $\bar{d}$, $\bar{c}$, $\bar{s}$, $\bar{t}$, and $\bar{b}$.

Unlike leptons, the quarks and antiquarks have *never* been observed in isolation. They only seem to occur bound together in combinations held together by gluons. For example, the familiar proton is a combination of two up quarks and a down quark, which we can write as uud. This is illustrated in Figure 4.3a. Note that each up quark carries an electric charge of $2e/3$ and a down quark carries a charge of $-e/3$, so the combination uud does indeed give a net electric charge equal to the charge $e$ on a proton. Similarly, a neutron is the combination udd which has a net electric charge of zero. This is illustrated in Figure 4.3b.

Observable particles consisting of combinations of quarks are collectively called *hadrons* and there are literally hundreds of them, the proton and neutron being the most familiar.

There are three recipes for building hadrons from quarks:

A hadron can consist of

- three quarks (in which case it is called a *baryon*);
- three antiquarks (in which case it is called an *antibaryon*);
- one quark and one antiquark (in which case it is called a *meson*).

This is illustrated in Figure 4.4.

This tally of six leptons and six quarks, each with its own antiparticle, may seem like a huge number of fundamental particles. However, it is a very small number compared with the huge number of hadrons that are known to exist. Moreover, everything around us is made up of merely the first generation of each type, namely electrons, up quarks and down quarks, with electron neutrinos and antineutrinos being created in β-decay processes. The second and third generations of leptons and quarks have similar properties to their first generation counterparts except that they are more massive.

You may have noticed that the photon is not included in the above lists of fundamental particles. The photon is the particle associated with electromagnetic radiation — the particle of light. In the standard model, the photon concept is rather broader and more subtle than this — the photon can act as a kind of link between electrically charged particles, mediating the electromagnetic interaction between them. Two forces acting at the subatomic level, called the *strong interaction* and the *weak interaction*, also have mediating particles associated with them. Collectively, these mediating particles are called *exchange particles* and they are discussed in Section 5, but we shall not discuss them further at this stage.

**Figure 4.4**  The three recipes. Quarks and antiquarks with a charge of $\pm 2e/3$ are shown in purple, those with a charge of $\pm e/3$ are shown in orange. The symbol q represents a quark and these are shown with black borders. In contrast, $\bar{q}$ represents an antiquark, and these are shown with white borders. Gluons are not shown. (a) Four possible combinations making a baryon — two particular examples are shown in Figure 4.3. (b) Four possible combinations making an antibaryon. (c) Four possible combinations making a meson.

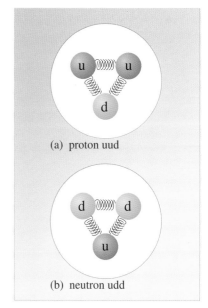

(a) proton uud

(b) neutron udd

**Figure 4.3**  (a) A proton is a combination of three quarks uud. (b) A neutron is the combination udd.

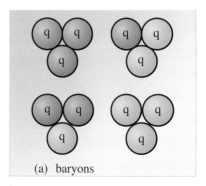

(a) baryons

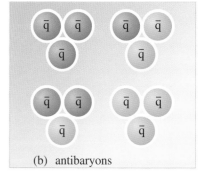

(b) antibaryons

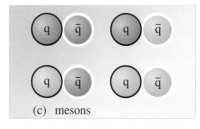

(c) mesons

# 2 Introduction to particle physics and its techniques

This section gives a largely historical account of some of the important discoveries and techniques of particle physics, bringing together topics covered in earlier parts of *The Physical World* and concentrating on experimental techniques for creating and detecting subatomic particles.

## 2.1 Early discoveries

The story of elementary particle physics really begins with the discovery of the *electron* by the Cambridge physicist, J. J. Thomson, in 1897. On the basis of a series of brilliant experiments, Thomson concluded that all the familiar forms of matter contained a common constituent — a light, negatively charged particle that he initially referred to as a 'corpuscle' but which soon came to be called the electron. Further work by Thomson and others, particularly the American Robert Millikan, led to precise determinations of the mass and charge of the electron.

Investigations into the behaviour of electrons in atoms established that electrons have another characteristic property called *spin*. Spin is an inherently quantum property, so it would be wrong to think of the electron as though it were a classical object. The electron is not like a miniature planet, spinning on its axis. Indeed, there is no convincing evidence that the electron is anything other than a point-like particle. Nonetheless, the electron's spin does mean that there is a certain fixed amount of *angular momentum* intrinsically associated with each electron, just as there is a certain amount of angular momentum associated with the Earth's rotation about its axis. In the case of the electron, the magnitude of the intrinsic angular momentum is indicated by saying that the electron has *spin angular momentum quantum number $s$* = 1/2. This is often abbreviated to the statement that:

> the electron has spin 1/2.

Not all subatomic particles have spin 1/2. You will see that there are particles with spin 0, spin 1, spin 3/2, and so on.

Another important discovery — that every atom possesses a *nucleus* — eventually led to the discovery of another two subatomic particles, the *proton* and the *neutron*. In a crucial experiment, two research students of Ernest Rutherford (1871–1937), Hans Geiger and Ernest Marsden, investigated the behaviour of energetic α-particles (helium nuclei) passing through a thin metal foil. It was expected that the α-particles would lose some of their energy and that many of the particles would be slightly deflected as a result of travelling through the foil. However, the observed behaviour was much more surprising. Some of the α-particles suffered very large deflections and effectively bounced back from the foil as though they had collided with a dense concentration of matter within the foil. To explain this discovery, Rutherford proposed that most of the mass of every atom was concentrated in a tiny central nucleus of enormous density. According to Rutherford, this tiny central nucleus was positively charged and was consequently able to hold a number of negatively charged electrons in orbit, thus accounting for the previously established facts that atoms existed and contained electrons.

The discovery of the atomic nucleus is also discussed in Chapter 1 of *QPI* and Chapter 3 of this book.

In 1920, after further research, Rutherford suggested that all atomic nuclei contained positively charged particles, which he named *protons*. He suggested that the simplest hydrogen nucleus consisted of a single proton but heavier nuclei contained more protons. The existence of the proton is now well established. It is known to have spin 1/2, one unit of positive charge and a mass of about 1840 electron masses.

Note that the charge of the proton is opposite in sign, but equal in magnitude, to that of the electron. All the subatomic particles that have so far been observed, have been found to carry simple multiples (0, 1, 2, etc.) of the charge on the electron or the proton. (The fractional charges carried by quarks (Figure 4.2) have never been directly observed — quarks, remember, are not observed in isolation.) This quantization of charge, in units of $e$, currently remains one of the fundamental mysteries of particle physics.

If protons were the sole constituents of nuclei, the charge of a nucleus would be roughly proportional to its mass. As you have seen this is not correct, except for the very lightest elements (recall plots of $Z$ against $N$ in Chapter 3). The other constituents of nuclei were discovered in the 1930s when the work of James Chadwick (Figure 4.5) and others revealed that nuclei contained electrically neutral particles. Chadwick made his discovery while investigating a highly penetrating form of radiation that was given off by the light element beryllium when bombarded by energetic electrons. By studying the way in which atoms recoiled when struck by the beryllium radiation, he was able to show that it actually consisted of neutral particles with a mass similar to that of the proton. Chadwick called the new particles *neutrons*, although it took some time for him to become convinced that they were really particles in their own right and not merely the result of binding together a proton and an electron.

It was the accurate determination of the neutron's mass that finally convinced Chadwick. As you know, it is quite common for particles to bind together to form composite structures, but when they do so, it is always the case that the resulting structure has a total mass that is less than that of the sum of its parts when they are widely separated. This mass deficit is equivalent to the *binding energy* of the composite structure. It was found that the neutron's mass was slightly greater than the sum of the proton and electron masses. This made it seem likely that the neutron was an elementary particle in its own right. The neutron is now known to have spin 1/2, which would be impossible if it were composed of an electron and a proton.

The discovery of the neutron completed the first phase in the development of elementary particle physics. It finally settled the nature of the α-particle, which had long been known to be related to the nucleus of the helium atom, but which could now be seen as a bound combination of two protons and two neutrons. It also paved the way for the development of nuclear physics.

**Figure 4.5**   Sir James Chadwick (1891–1974).

## 2.2  High-energy collisions

Much of our knowledge of subatomic particles has been learned from observing them collide at high energy. Collisions allow us to probe the internal structure of particles. Moreover, collisions of known particles sometimes produce entirely new types of particles. The energies of the particles involved in these collisions are usually expressed in terms of MeV ($10^6$ eV) or GeV ($10^9$ eV), although the energies involved in very high-energy collisions (by current standards) might involve several TeV ($10^{12}$ eV). Particles with these energies, or higher, occur naturally in the form of *cosmic rays* and can be supplied on demand by artificial devices called *particle accelerators*. Cosmic rays and particle accelerators will be discussed in more detail a little later.

We are using SI prefixes: mega (M) = $10^6$; giga (G) = $10^9$; tera (T) = $10^{12}$.

Although a typical collision is usually initiated by just two particles, it is often the case that more than two particles will emerge from the collision. When the energy is high enough for this to happen, the particles that emerge may be quite different from those that initiated the event. The possibility of creating particles in this way is a consequence of Einstein's special theory of relativity, which tells us that the *mass energy*, or *rest energy*, for a particle of mass $m$ is given by

$$E_{\text{mass}} = mc^2.$$

*(PM 3.21)*

This equation implies that the creation of a particle of mass $m$ requires the availability of an amount of energy $mc^2$, usually in the form of kinetic energy.

The energies involved in particle physics are very high and the particle speeds are extremely close to that of light. Consequently, it is usually necessary to use the relativistic expressions for energy and momentum, rather than the Newtonian ones, when analysing a collision process. You know from *Dynamic fields and waves* and *Predicting motion* that, according to special relativity, a particle of mass $m$ and velocity $\boldsymbol{v}$ has *relativistic momentum* $\boldsymbol{p}$ and (total) *relativistic energy* $E_{\text{tot}}$ given by

$$\boldsymbol{p} = \frac{m\boldsymbol{v}}{\sqrt{1 - v^2 / c^2}} \qquad\qquad (PM\ 3.20)$$

and

$$E_{\text{tot}} = \frac{mc^2}{\sqrt{1 - v^2 / c^2}}. \qquad\qquad (PM\ 3.22)$$

Since the relativistic energy $E_{\text{tot}}$ of a moving particle can be regarded as consisting of a *rest energy* contribution, $E_{\text{mass}} = mc^2$, as well as a *relativistic (translational) kinetic energy*, $T$, it follows that

$$T = E_{\text{tot}} - E_{\text{mass}}.$$

In a high-energy collision of elementary particles, total relativistic energy and momentum are always *conserved* (that is, they have the same values before and after the collision), but the total mass and the relativistic kinetic energy are not separately conserved. For example, suppose two identical particles, each of mass $m$, and with kinetic energies $T_1$ and $T_2$, collide (as indicated in Figure 4.6), and that the collision produces four particles of mass $m$ with kinetic energies $T_3$, $T_4$, $T_5$ and $T_6$. Then conservation of total relativistic energy requires that

$$T_1 + T_2 + 2mc^2 = T_3 + T_4 + T_5 + T_6 + 4mc^2.$$

More generally, we can write

$$\begin{pmatrix} \text{kinetic energy} \\ \text{of incident} \\ \text{particles} \end{pmatrix} + \begin{pmatrix} \text{rest energy} \\ \text{of incident} \\ \text{particles} \end{pmatrix} = \begin{pmatrix} \text{kinetic energy} \\ \text{of emergent} \\ \text{particles} \end{pmatrix} + \begin{pmatrix} \text{rest energy} \\ \text{of emergent} \\ \text{particles .} \end{pmatrix}$$

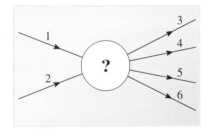

**Figure 4.6** A schematic collision in which two particles enter but four emerge. Understanding the detailed processes that take place during such a collision (represented by the question mark) is the main challenge that confronts theories of elementary particles.

In addition to the conservation of total (relativistic) energy, similar conservation laws apply to each of the components of relativistic momentum, to angular momentum (including the spin contributions) and to electric charge.

**Question 4.1**    From the relation $E_{\text{mass}} = mc^2$, it is clear that mass can be measured in the same units as $E/c^2$ and that a suitable unit of mass is therefore the MeV/$c^2$. Express the value of $1\,\text{MeV}/c^2$ in terms of kilograms, and hence express the mass of the electron, the proton and the neutron in terms of MeV/$c^2$.

**Question 4.2**    Give an example of a conservation law that would prevent two colliding electrons from producing a final state that consisted of just three electrons.    ■

You will see later in this chapter that there are other physical properties that are conserved under certain circumstances, such as *lepton number*, *baryon number*, *strangeness* and *charm*. The *conservation laws* that apply to the various physical properties of subatomic particles provide us with one of the most powerful theoretical tools for analysing collisions and bringing order to the world of particle physics.

## 2.3 Experimental techniques

### The cosmic ray era

Most of the discoveries concerning subatomic particles that were made in the 1930s and 1940s involved the observation of cosmic rays. **Cosmic rays** are energetic particles that reach the Earth from outer space. Most of them are protons, but there are also some heavier nuclei and some neutral particles. The existence of cosmic rays was first convincingly demonstrated in 1911–12 by the Austrian physicist Victor Hess (Figure 4.7). During a series of high-altitude balloon ascents, Hess showed that the rate at which charged particles were detected in the Earth's atmosphere increased with height, indicating that there must be an extraterrestrial source of such particles. The origin of cosmic rays is still something of a mystery, although it is generally accepted that the majority come from exploding stars and other sites of energetic astronomical activity.

Particles coming directly from space, sometimes called **primary cosmic rays**, arrive at the top of the Earth's atmosphere with a range of energies. The most energetic protons have energies in excess of $10^{18}$ eV, and are quite rare, but less energetic particles are common. As they pass through the atmosphere and approach the ground, a number of these primary cosmic rays collide with atomic nuclei in the atmosphere. The outcome of such high-energy collisions may well be a shower of secondary particles, as indicated in Figure 4.8. The investigation of such showers, and the study of cosmic rays in general, revealed the existence of several other elementary particles apart from the electron, proton and neutron. The first of these discoveries was that of the *positron* by Carl Anderson in 1932 (Figure 4.9).

**Figure 4.7** Victor Hess (1883–1964).

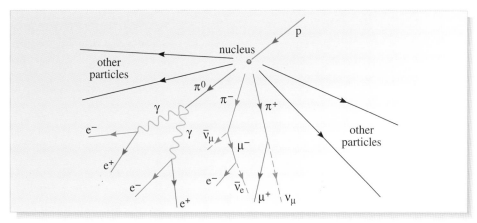

**Figure 4.8** A primary cosmic ray collides with a nucleus in the Earth's atmosphere and produces a shower of secondary particles including electrons, positrons, pions, muons and neutrinos. (These particles are discussed in later sections.)

The positron was the first **antiparticle** to be discovered, and the first example of an elementary piece of antimatter. The positron is the antiparticle of the electron. It has the same mass and spin as the electron, but its charge is $e$ rather than $-e$. The existence of the positron had been predicted by the British physicist Paul Dirac (Figure 4.10), as a consequence of his pioneering theoretical work combining quantum physics with Einstein's special theory of relativity. It is now believed that the observation of any kind of subatomic particle implies the existence of a corresponding kind of antiparticle; electrons imply positrons, protons imply antiprotons, neutrons imply antineutrons and so on. Consequently, when we introduce new types of particles, or count the different kinds of particle, we shall often not bother to mention or include antiparticles — their existence can be taken for granted. However, you should note that some electrically neutral particles are their own antiparticles; the photon is a case in point, although the neutron is not. So there is no distinct antiphoton, but there is an antineutron. By convention, a bar over

**Figure 4.9** Carl Anderson (1905–1991).

**Figure 4.10** Paul Dirac (1902–1984).

In particle physics it is common to refer to the *mean lifetime* (or *lifetime*) of an unstable particle rather than its half-life $T_{1/2}$. The mean lifetime is in fact $T_{1/2}/0.6931$.

the symbol for the particle usually denotes an antiparticle. For example, the proton and antiproton are denoted by p and $\bar{p}$ respectively. However, this is not a rigorous rule; although e⁻ denotes the electron, the positron is usually denoted by e⁺.

The positrons observed by Anderson were detected by means of a device known as a **cloud chamber**. This kind of detector allows small but visible water droplets to form around atoms that have been ionized when charged particles pass through air containing supersaturated water vapour. Cloud chambers provided the main means of detecting particles in many of the early cosmic ray experiments and resulted in many important discoveries. One such finding occurred in 1937 when Carl Anderson and fellow American Seth Neddermeyer discovered the **muon**. Muons occur with either positive or negative charge, the positive muon ($\mu^+$) being the antiparticle of the negative muon ($\mu^-$). A muon has a mass of about 207 electrons and generally behaves like a heavy electron although, unlike the electron, it is unstable and typically decays into other particles within a few millionths of a second of its creation.

### Lifetimes of moving particles

*Dynamic fields and waves* discussed a relativistic effect known as *time dilation* which allows unstable particles observed to be travelling at speeds close to that of light to live much longer than identical particles that are observed at rest. For example, the *mean lifetime* of a muon in its rest frame is $2.2 \times 10^{-6}$ s, while a muon moving through the Earth's atmosphere at a speed of $0.98c$ would have a mean lifetime of $1.1 \times 10^{-5}$ s as measured by an observer on Earth. The mean lifetimes quoted throughout this chapter are those that would be measured by an observer relative to whom the particles are at rest, i.e. they refer to the mean lifetime measured in the particle's rest frame.

The continued investigation of cosmic rays, aided by the development of new methods for detecting and tracking charged particles, resulted in the discovery of even more kinds of subatomic particles. The direct recording of particle tracks in packages of highly sensitive **photographic emulsion**, flown in balloons or exposed at high altitude mountain sites, was pioneered by Cecil Powell (1903–1969) at the University of Bristol, and led to the discovery of positively and negatively charged **pions** ($\pi^+$ and $\pi^-$) in 1947. These particles were about 30% heavier than muons, but had mean lifetimes that were about a hundred times shorter. Within a few years, a combination of cosmic ray and accelerator techniques confirmed the existence of a third, uncharged, pion ($\pi^0$) which turned out to have a mean lifetime that was less than a billionth of the muon's. The extreme rapidity of this decay, which usually resulted in the production of two high-energy photons, indicated that the underlying decay mechanism was quite different from that responsible for the much slower decay of muons or charged pions. The decay of the muons and charged pions is actually a manifestation of what is now known as the *weak interaction*. This is the interaction involved in β-decay. The rapid $\pi^0$-decay was caused by the more powerful *electromagnetic interaction* that is generally associated with the emission or absorption of photons.

It was eventually established that all three kinds of pion had spin 0, and that the $\pi^0$ is its own antiparticle. (In Section 4, you will see that pions are examples of *mesons* — combinations of a quark and an antiquark. In contrast, the muon, like the electron, belongs to the family of leptons, so it is not composed of quarks.)

Another important discovery of the cosmic ray era was that of the so-called *strange particles*. This will be described in Section 3.2 together with some other widely used terminology regarding families of particles.

**Question 4.3** What are the values of the mass (in MeV/$c^2$), spin and electric charge of the positron? How much energy is required to create an electron–positron pair? ■

### The accelerator era

Cosmic rays are a useful source of high-energy particles, but they are mainly a high altitude phenomenon and inherently uncontrollable. The systematic investigation of subatomic particles really requires the use of laboratory-based facilities where artificially created beams of identical particles with fixed energy can be used to create collisions more-or-less on demand. Such beams are produced using a device called a **particle accelerator**.

There are essentially two types of particle accelerator:

1  **Linear accelerators** (or **linacs**) in which charged particles are accelerated in a straight line.

2  **Cyclic accelerators** in which charged particles follow closed or nearly closed paths so that they can be accelerated repeatedly by the same parts of the machine.

In both types of machine, the particles are accelerated in a vacuum and the acceleration (together with any guidance or focusing that may be necessary) is achieved by applying carefully controlled electric and magnetic fields. That is why the particles must be charged; the fields would have no effect on neutral particles. Beams of neutral particles may however be produced by forming beams of unstable charged particles that decay in flight to yield the required neutral particles.

As described in Chapter 3, the first linear accelerator was constructed in 1932 by John Cockcroft and Ernest Walton who succeeded in accelerating protons to energies of 0.7 MeV and used them to produce the first artificial nuclear reactions. The world's largest linear accelerator is currently the 3 km linac at the Stanford Linear Accelerator Center (SLAC) in the USA. Electrons or positrons can be accelerated to energies of 50 GeV (i.e. $50 \times 10^9$ eV) using this accelerator. The SLAC linac (Figure 4.11) employs the electric fields of travelling electromagnetic waves contained within conducting cavities to accelerate electrons and positrons. The particles gain energy from these waves in a manner analogous to that of a surfer picking up energy from a water wave.

**Figure 4.11** The Stanford Linear Accelerator Centre (SLAC).

The cyclotron is also discussed in *Static fields and potentials*.

The first of the cyclic accelerators to be developed was the **cyclotron** (Figure 4.12), invented by Ernest Lawrence (Figure 4.13) in which a vertical magnetic field was used to bend the path of particles that circulated horizontally within two 'dee'-shaped metal cavities. The dees were separated by a small gap and were maintained at opposite voltages so that a horizontal electric field existed between them. The idea was that the electric field would accelerate the particles each time they crossed the gap. However, in order to ensure that the electric field would always cause the particles to speed up rather than slow down, it was necessary to change the direction of the field during the time that it took a particle to complete half a circuit. Fortunately, this was not difficult thanks to the fact that as a particle was accelerated the magnetic field was less effective at bending its path and the radius increased in consequence. Lawrence realized that at relatively low energy, this combination of increased speed and increased radius had the effect of keeping the time between gap crossings constant. Consequently, the necessary field reversals could be achieved by applying an alternating voltage to the dees. Lawrence's original cyclotron, built in 1930, was just 13 cm across and accelerated protons to 80 keV, but it was the forerunner of the modern cyclic accelerator.

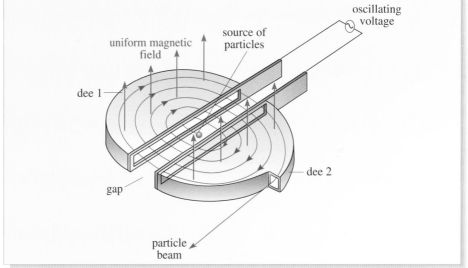

**Figure 4.12** A cyclotron. A charged particle moving in a horizontal plane and subject to a vertical uniform magnetic field is repeatedly accelerated by the oscillating electric field between two hollow metal dees. As a result of its increasing energy, the particle moves in a spiral of gradually increasing radius at increasing speed.

**Figure 4.13** Ernest Lawrence (1901–1958).

*CERN* is the European Organization for Nuclear Research and straddles the Swiss–French border near Geneva. *Fermilab* is 50 km west of Chicago in the USA.

A consequence of the laws of electromagnetism is that an accelerating charged particle emits electromagnetic radiation and therefore loses some energy. This poses an extra problem for cyclic accelerators since in addition to the speed of the particle increasing (as in a linac) the particle is also undergoing centripetal acceleration towards the centre of its circular track. Radiation energy losses are therefore more serious for cyclic accelerators than for linacs.

Cyclotrons are still used in nuclear physics and nuclear medicine. However, the fact that the time for a full circuit is independent of the particle energy holds true only at low energies, where speeds are much less than the speed of light and relativistic effects can be ignored. Consequently, in modern particle physics laboratories, such as CERN or Fermilab, the main cyclic accelerator is a **synchrotron** (Figure 4.14). In this kind of machine, particles are accelerated in bunches and made to move in closed paths of large fixed radius (typically several kilometres) by a number of separate bending magnets (Figure 4.15). Since the beam consists of distinct bunches of particles, rather than a continuous beam, it is possible to adjust the frequency of the applied accelerating fields to the speed of the particles.

(a)

(b)

(c)

**Figure 4.14** (a) Outside view of the Super Proton Synchrotron at CERN. (b) Inside the Super Proton Synchrotron. This old machine now serves a much bigger one, extending underground for many kilometres. (c) DESY laboratories in Hamburg. The old PETRA ring now serves the much bigger HERA ring..

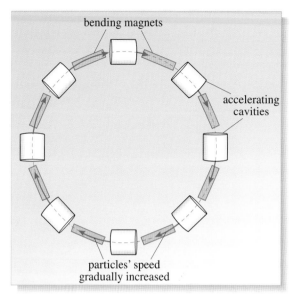

**Figure 4.15** A synchrotron (schematic). Particles are accelerated in bunches; bending and focusing are achieved by appropriate magnetic fields and the bunches of particles are accelerated repeatedly, at increasing frequency, by carefully controlled electric fields.

Beam focusing and acceleration take place between the bending magnets, with the acceleration being achieved by electric fields. Due to the fixed radius, the frequency with which bunches of particles pass through the accelerating region increases with time, implying that the frequency of the accelerating voltage must also increase. These voltage oscillations must be carefully synchronized with the increasing energy of the particles — hence the name synchrotron. Relativistic mechanics is thoroughly confirmed by the success of these machines and by the pattern of the emitted electromagnetic radiation, called *synchrotron radiation*.

The statistics of modern synchrotrons are awe-inspiring. The main accelerator at Fermilab can accelerate protons or antiprotons to energies of about 1 TeV ($10^{12}$ eV). At this energy, the particles are travelling at 99.999 95% of the speed of light. The ring that contains these particles is 2 km in diameter and contains 2000 electromagnets, many of which use superconducting technology. In order to attain their maximum energy, the protons must cover a total distance of about $3 \times 10^6$ km (about 8 times the distance to the Moon) and this requires about 20 seconds of acceleration. The accelerator can deliver a bunch of $2 \times 10^{13}$ particles every 60 s. A more powerful accelerator with a 27 km circumference, the Large Hadron Collider (LHC), is under construction at CERN.

**Question 4.4** In all of these accelerator devices, the accelerating process involves an electric field, even if a magnetic field is also used to produce a curved trajectory. Why cannot a magnetic field alone be used to increase the kinetic energy of a charged particle? ■

The purpose of accelerating particles to high energy is, of course, to provide a controlled source of high-energy collisions for detailed study. One way of producing collisions is to direct the beam at one or more **fixed targets** surrounded by appropriate detectors. Another technique is to direct the beam at another beam of particles moving more-or-less directly opposite to the original beam (see Figures 4.16 and 4.17). Although such **colliding beams** are difficult to arrange and tend to produce a much lower collision rate than fixed targets, they are of such significance that all of the world's large particle accelerators now provide them.

The great advantage of using colliding beams rather than fixed targets concerns the energy available for the creation of new particles. We have already noted that,

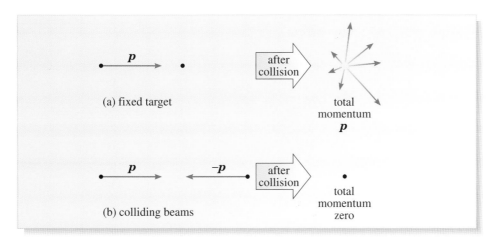

**Figure 4.16** (a) In a fixed target experiment, the final momentum of the projected particle plus the momentum of the particles knocked out of the target is equal to the initial momentum of the projected particle. Hence, not all the initial kinetic energy can go into creating new particles. (b) A colliding beam experiment makes more efficient use of the kinetic energy. If two particles having the same mass and speed collide head-on, their total momentum is zero. In such circumstances, all of their kinetic energy can go into the creation of new particles.

**Figure 4.17**   The former intersecting storage rings at CERN. Protons circulating in the rings collided nearly head-on in the region where the rings cross. The first collisions were observed using these rings on 27 January 1971.

according to the principle of conservation of relativistic energy, the kinetic energy of colliding particles may be used to create new particles in accordance with the equation $E = mc^2$. However, when particles collide, linear momentum must also be conserved. In the case of a fixed-target experiment, this means that the outgoing particles must have the same total linear momentum as the incident particle. This implies that the outgoing particles must also carry some kinetic energy. Thus, some of the initial kinetic energy must go into the kinetic energy of the outgoing particles. In contrast, if the collision involves two identical particles with the same kinetic energy meeting head on, then the total linear momentum prior to the collision is zero. Consequently, there is no requirement that any of the initial kinetic energy should be expended in ensuring momentum conservation — all the energy can go into the creation of new particles.

The current beam types and maximum particle energies at the major laboratories are listed in Table 4.1.

**Table 4.1**  Beam types and maximum particle energies at the major accelerator laboratories. Note that beam energies are expressed in GeV, where $1\,\text{GeV} = 10^9\,\text{eV}$.

| Laboratory | Accelerator | Beam 1 | Beam 2 |
| --- | --- | --- | --- |
| SLAC (USA) | SLC | $e^-$ 50 GeV | $e^+$ 50 GeV |
| Fermilab (USA) | Tevatron | p 1000 GeV | $\bar{\text{p}}$ 800 GeV |
| CERN (Europe) | LEP II | $e^-$ 100 GeV | $e^+$ 100 GeV |
|  | LHC (for 2005) | p 7000 GeV | $\bar{\text{p}}$ 7000 GeV |
| DESY (Germany) | HERA | p 820 GeV | $e^-$ or $e^+$ 27 GeV |

The development of particle accelerators has been paralleled by a similar development in particle detectors. For almost 30 years, from the early 1950s to the early 1980s, the standard 'workhorse' detector in most fixed-target experiments was a device called a **bubble chamber** (Figure 4.18), the operation of which is somewhat similar to that of the cloud chamber. A bubble chamber contains a 'cryogenic liquid', such as liquid hydrogen, which is held at its boiling temperature (about 20 K for hydrogen). Prior to a measurement, boiling is prevented by holding the liquid under pressure. However, when the pressure is released, there is often a delay before boiling commences. During this time, the liquid is said to be *superheated*, and while in this condition tiny bubbles of gas will form along the track of any fast-moving charged particles that enter the liquid. This provides a means of detecting charged particles. The sort of event that might be seen in a bubble chamber is shown schematically in Figure 4.19. Notice that applying a uniform magnetic field within the chamber assists the identification of particles. The magnetic field causes the track of any charged particle to curve in a way that is determined by the particle's mass, charge and speed, and the strength of the magnetic field.

**Figure 4.18**  The 3.7 m European Bubble Chamber. Large as it may appear, its dimensions are dwarfed by modern multi-purpose detectors and their supporting electronics.

In the past 20 years or so, developments in detector technology and computer analysis, together with the need to analyse the large number of events produced by colliding beams, led to the replacement of bubble chambers by solid-state detectors and other solid-state devices that can surround the site of a particle collision and send their findings directly to a computer. We shall not go into the details of these (though solid-state detectors were mentioned in Chapter 2); suffice it to say that they provide the modern electronic equivalent of a bubble chamber.

The combination of very powerful particle accelerators and detectors has led to many discoveries in particle physics. Before describing some of these discoveries, we briefly consider the nature of particle collisions in general, and some of the quantities that might be measured with the aid of a particle accelerator and an appropriate detector.

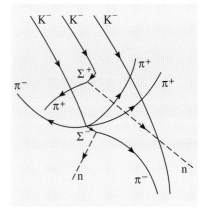

**Figure 4.19** A schematic illustration of a bubble chamber event in which a K⁻ particle encounters target protons. Note that neutral particles do not leave tracks in a bubble chamber. Their presence can be inferred from the behaviour of the charged particles that are observed. Dashed lines indicate the tracks of the unseen neutral particles. The symbols $K^-$, $\Sigma^+$ and $\Sigma^-$ represent particles to be introduced in the next section. $\pi^-$ and $\pi^+$ are charged pions.

### Particle scattering cross-sections

When a beam of particles encounters a target, some of the beam particles will be *scattered*, i.e. deflected from their original course due to their interaction with the target. If the energy of the beam particles is sufficiently high, the interaction may lead to changes in the nature of the beam or target particles or even to the production of additional particles.

Encounters in which the beam and target particles do not change their nature (although they may exchange energy and momentum) are said to be examples of **elastic scattering**. An example is the scattering of α-particles from a nucleus as in the experiments by Geiger and Marsden discussed in Section 2.1. No new particles are created, and the α-particle and nucleus do not break up; they simply exchange kinetic energy and momentum.

Encounters in which the nature or number of the particles is changed are said to be examples of **inelastic scattering**. An example is the encounter of a very high energy cosmic ray proton with a nucleus as indicated in Figure 4.8 where the nucleus is smashed up and a large number of secondary particles is created. Inelastic scattering generally involves the conversion of kinetic energy into mass.

The scattering of elementary particles is an inherently quantum-mechanical process. Consequently, it is usually not possible to say that any particular beam particle will *definitely* be scattered when it encounters a given type of target particle at some specified energy. The most that can be done is to predict the *probability* that a certain kind of scattering will occur at a specified energy.

The physical quantity that provides this probabilistic information is called a **cross-section** and is often denoted by the symbol $\sigma$ (Greek 'sigma'). Cross-sections generally depend on the kinetic energy of the colliding particles. You might, for example, ask 'what is the cross-section for the elastic scattering of a positive pion ($\pi^+$) by a target proton (p) at an energy of 10 GeV?' The larger the cross-section, the greater the probability that the pion will be scattered by the proton. A great deal of experimental effort has gone into the measurement of various cross-sections across a large range of energies and a great deal of theoretical ingenuity has gone into predicting and explaining the measured cross-sections.

As the name 'cross-section' suggests, $\sigma$ is measured in the same units as area. It could be expressed in m² (metres squared) but the cross-sections of interest to particle physicists are so small that it is more conventional to use a non-SI unit called the **barn** (represented by the symbol b), which is defined as 1 barn = $10^{-28}$ m². Very roughly, the cross-sections of processes involving the various interactions at typical experimental energies are:

- 10 millibarns (10 mb) for the strong interaction;
- 100 microbarns (100 μb) for the electromagnetic interaction;
- 1 picobarn (i.e. $10^{-15}$ b) for the weak interaction.

You can see that, although small by everyday standards, the barn is a relatively large unit of cross-section. Note that 1 barn = $(10\,\text{fm})^2$ so the barn is approximately equal to the effective cross-sectional area associated with a heavy nucleus such as that of lead (Pb), as indicated by the nuclear radii shown in Figures 3.6 and 3.7. The name 'barn' originated in a humorous reference to the expression 'as big as a barn door'.

A cross-section may be thought of as the effective area that each target particle presents to the incident beam. However, it is vital to note the presence of the word 'effective' in the preceding sentence. Elementary particles are described by quantum physics rather than classical physics; they do not behave like billiard balls and it would be naive and misleading to think of a scattering cross-section as a measurement of the 'true' cross-sectional area of the particles concerned.

When particles interact with one another there are many possible outcomes; they may undergo elastic scattering, or inelastic scattering with the conversion of kinetic energy into different kinds of particles. The relative probabilities of these possible outcomes can in principle be predicted by quantum mechanics and described in terms of appropriate cross-sections. A given cross-section can be measured for a particular outcome, or for all possible outcomes. In the latter case, it is called a **total cross-section**. Figure 4.20 shows some examples of total cross-sections.

**Figure 4.20** Some total cross-sections as functions of energy for a variety of processes. The middle curve, labelled $\pi^+$p, shows the total cross-section for positive pions scattering from protons. This includes elastic scattering as well as inelastic events in which new particles are created.

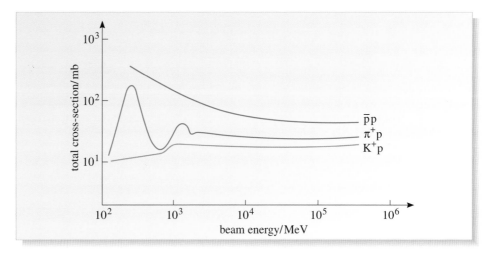

# 3 Particles and their interactions

Before introducing any more of the subatomic particles that have been discovered, it is worth saying a few words about the ways in which particles have been found to interact with one another. This is only a first glimpse of a topic to which we shall return later.

## 3.1 The fundamental forces

From the results of an enormous number of experiments, we know that elementary particles interact in essentially four different ways. These different kinds of interaction are often referred to as the **fundamental forces** since, among other things, they allow the interacting particles to exchange energy and momentum.

Two of the four forces are familiar from everyday life. The **gravitational interaction** operates between all particles but is normally too weak to have any observable effect at the subatomic level. The **electromagnetic interaction** operates between charged particles, and its various manifestations (electric and magnetic forces) were discussed at length in *Static fields and potentials* and *Dynamic fields and waves*. These two interactions are said to have *infinite range*, meaning that the forces fall off relatively slowly, as $1/r^2$, with distance $r$ from a point source, which is why we are so familiar with them.

The other two forces fall off very much more rapidly than this and have ranges that are limited to about $10^{-15}$ m or less, which is the size of a typical nucleus. The **strong interaction** is responsible for binding quarks into hadrons and for holding the nucleons together in atomic nuclei. Although this force is stronger than the electromagnetic force, its short range means that it goes essentially unnoticed in the everyday world, even though our existence depends upon it. (Without the strong nuclear force, there could be no nuclei, no nuclear fusion to power the Sun and no life on Earth.) Finally, there is the **weak interaction**, the best-known manifestation of which is in the process of $\beta^-$-decay where a neutron decays to produce a proton, an electron and an antineutrino. It too is of vital consequence in the Sun.

In the context of nuclear physics, the strong interaction and the weak interaction are referred to, respectively, as the *strong nuclear force* and the *weak nuclear force*.

Table 4.2 lists the four fundamental forces in order of strength. It should be noted that the relative strengths depend on the energy at which they are measured. At higher energies, particles can come closer together in a collision and so the short-range forces become more important. The strengths quoted here are those relevant to short-range collisions and to collision energies of a few GeV.

**Table 4.2**  The four fundamental forces.

| Force (or interaction) | Strength (relative to the strong force) | Range/m |
|---|:---:|:---:|
| strong | 1 | $10^{-15}$ |
| electromagnetic | $10^{-2}$ | infinity |
| weak | $10^{-5}$ | $10^{-17}$ |
| gravitational | $10^{-38}$ | infinity |

The relative strengths of the four fundamental forces depend on the energies at which they are measured, and so the numbers in Table 4.2 may differ from those quoted elsewhere.

These fundamental forces underpin all of the pushes and pulls that are normally called 'forces' in everyday life. For instance, the reason that the downward pull of gravity doesn't make you crash through the floor and fall towards the centre of the Earth is that the electrons on the soles of your shoes are repelled by the topmost electrons on the surface of the floor. The floor doesn't collapse because electromagnetic forces also hold it up, as with the ground beneath it, and so on all the way to the centre of the Earth.

**Question 4.5**   Using Newton's law of gravitation and Coulomb's law of electrostatics, compare the magnitudes of the electrical and gravitational forces between two protons that are separated by a distance of $10^{-14}$ m.  ∎

## 3.2  Families of particles

From the discoveries made using cosmic rays in the late 1940s and early 1950s, it became increasingly clear that there was a world of subatomic, or even subnuclear, particles. It was also clear that the various types of particle belonged to larger families; there were, for example, clear similarities between electrons and muons, which led to them both being regarded as members of a family, called the *leptons*. When this term was first introduced in 1946, it simply indicated that the electron and muon were particles of relatively low mass. The pions were subsequently assigned to the family of *mesons* (middleweight particles) while the proton and the neutron were both *baryons* (heavyweight particles). As the number of known particles grew, these family names were destined to persist, but the simple link with particle masses was not. The three families can now be defined as follows:

### Families of particles and some conservation laws

**Leptons** are spin 1/2 particles. They can interact by the weak interaction and, if charged, by the electromagnetic interaction. They never interact by the strong interaction. (Electrons and muons are examples of leptons.)

**Mesons** are particles with zero spin or integer spin, i.e. spin 0, 1, 2, etc. They can interact by the strong interaction, the weak interaction and, if charged, by the electromagnetic interaction. (Pions are examples of spin 0 mesons.)

**Baryons** are particles with half odd-integer spin, i.e. spin 1/2, 3/2, etc. They can interact by the strong interaction, the weak interaction and, if charged, by the electromagnetic interaction. (Protons and neutrons are examples of spin 1/2 baryons)

One of the predictions of particle physics is that all particles with half odd-integral spin are fermions, while particles with integral or zero spin are bosons. Thus, all leptons and baryons are fermions but mesons are bosons.

As far as is known at the time of writing, the leptons are a family of truly fundamental particles. In contrast, the mesons and baryons, which are collectively referred to as **hadrons** — meaning strongly interacting — are now known to be composite particles made up from quarks. All hadrons can interact strongly, i.e. they can feel the strong force, but hadrons may also interact weakly and, if charged, electromagnetically. In fact, even the neutron has electromagnetic interactions, via its internal electric charge distribution (which sums to give an overall electric charge of zero) and via its very small magnetic moment.

The significance of these families becomes particularly evident when particles decay. A muon (a lepton), for example, can decay in a number of ways, but the decay products always include an electron (another lepton), ensuring that the leptonic character of the muon is preserved, even though the muon itself is not. This suggests that there is a conserved quantity, called **lepton number** $L$. Similarly, when a neutron decays, the decay products always include a proton (i.e. another baryon) implying the existence of a conserved quantity called **baryon number** $B$, that is common to both particles.

By convention, the lepton number of the electron is $L = 1$, and the baryon number of the proton is $B = 1$. The baryon numbers and lepton numbers of other particles can be determined by observing their behaviour in decays or collisions. (If you look ahead to Table 4.3, you will see values of $B$ and $L$ listed for some 'common' particles. The allocation of baryon and lepton numbers for these particles may seem rather arbitrary, but in fact they form a consistent set of values, assigned after observations of numerous particle decays and collisions.

It turns out that all mesons have $B = 0$, all baryons have $B = 1$ and all antibaryons have $B = -1$. All leptons have $L = 1$ and all antileptons have $L = -1$.

Another important quantity associated with any particle is the charge number $Q$. (The **charge number** is defined so that if we multiply $Q$ by $e$, we get the electric charge of the particle. The charge number of an atomic nucleus for example is the atomic number $Z$.) The results of numerous experiments enable us to make the following statement:

> **Some conservation laws**
>
> Baryon number $B$, lepton number $L$ and charge number $Q$ are conserved in all known processes.

Conservation of baryon number, conservation of lepton number and conservation of charge number are three of the conservation laws used to bring order to the complexity of particle decay and collision processes — we shall introduce more conservation laws later in this chapter.

**Question 4.6**  The electron cannot decay because there is no lighter charged particle of any kind to which it can pass its charge. However, the positron is lighter than a proton and both have the same positive charge. What accounts for the stability of the proton?  ■

## 3.3  More hadrons — strangeness and resonances

A major and unexpected discovery of the cosmic-ray era was the existence of so-called *strange particles*. These particles help to exemplify the family structures we have just been discussing and also provide evidence of another conservation law. The first examples of **strange particles** to be discovered were spin 0 mesons that are now called *kaons*. There are four kinds of kaon ($K^+$, $K^-$, $K^0$, $\overline{K}^0$), each of which has approximately half the mass of a proton. ($\overline{K}^0$ is pronounced 'kay zero bar'.)

Other strange particles, that are heavier than the proton or neutron, were discovered throughout the 1950s. These included the electrically neutral *lambda particle* ($\Lambda^0$), the *sigma particles* ($\Sigma^+$, $\Sigma^-$, $\Sigma^0$) and the *xi particles* ($\Xi^-$, $\Xi^0$). Xi particles are sometimes called *cascade particles*, for reasons that you will see shortly. All of these heavier particles are baryons with spin 1/2.

The Greek letter $\Lambda$ is the upper case version of $\lambda$ and $\Lambda^0$ is pronounced 'lambda zero'.

The Greek letter $\xi$, written in upper case as $\Xi$, is pronounced 'ksi'.

It eventually became clear that the distinctive feature of all these particles, whether they were baryons or mesons, was their possession of a new property. This new property was called **strangeness**. (Again, we see the use of vivid terminology which you should not take literally.) In some respects, strangeness is similar to charge or baryon number, but it is very different in other respects. The strangeness of a particle is usually indicated by the letter $S$ and takes simple whole number values. The $K^+$ and $K^0$ mesons have $S = +1$ while the $K^-$ and $\overline{K}^0$ have $S = -1$. The $\Lambda^0$ and the three sigma particles ($\Sigma^+$, $\Sigma^-$ and $\Sigma^0$) also have $S = -1$, but the two xi particles $\Xi^-$ and $\Xi^0$ have $S = -2$.

Unlike charge and baryon number, strangeness is not always conserved, although it is conserved in some circumstances. It is this fact that accounts for some of the distinctive features of the strange particles. As a consequence of numerous experiments, we can make the following statement:

Conservation of strangeness (not in weak interactions)

Strangeness, $S$, is conserved in processes that are caused by the strong or electromagnetic interaction, but not in weak interactions.

The presence of strangeness prevents strange baryons (such as xi particles) from decaying to lighter non-strange baryons (such as protons or neutrons) by strong or electromagnetic interactions. Restricting such decays to weak interactions implies that the strange baryons live for the relatively long mean lifetime of about $10^{-10}$ seconds (see Table 4.3) and this makes them relatively easy to detect. (You may not have thought of $10^{-10}$ s as being a 'relatively long time'! In this context, a 'short' time is the time taken by light to travel one nuclear diameter. This is about $10^{-23}$ s, so $10^{-10}$ s is a very long time indeed in particle physics terms.)

Moreover, it is observed that when decays that do not conserve strangeness do take place, the strangeness only increases or decreases by 1. For example, for an $S = -2$ baryon such as a xi minus ($\Xi^-$) to pass its baryon number to a proton with $S = 0$, there must be at least two decays. The $\Xi^-$ first decays weakly (i.e. by the weak interaction) to produce a baryon with $S = -1$, and then this baryon decays weakly to produce the proton. In practice, the intermediate ($S = -1$) baryon is always a lambda zero ($\Lambda^0$). The fact that the $\Xi^-$ has to decay in such a sequence of steps explains why xi particles are also known as *cascade* particles.

The properties of the particles that we have mentioned so far in this chapter are given in Table 4.3. This table is for reference — you are not expected to remember the various particles listed or their properties.

### Antibaryons

One group of particles that was subjected to early accelerator-based investigation was the heavy antiparticles, particularly the **antibaryons**. The antiparticles of low mass particles such as the electron, the muon, the pion and the (charged) kaon were already known from cosmic ray experiments. However, it was the development of accelerators that made possible the creation and systematic investigation of the various antibaryons, starting with the antiproton ($\bar{p}$) in 1955. The antineutron ($\bar{n}$), antilambda ($\bar{\Lambda}$), antisigma ($\bar{\Sigma}$) and antixi ($\bar{\Xi}$) followed in due course. In each case, the antiparticle was found to have the same mass as the particle to which it corresponded.

Accelerators also permitted the investigation of the antiparticle of the $K^0$ meson. In contrast to the $\pi^0$, which is its own antiparticle, the anti-$K^0$ is a distinct particle, denoted by $\bar{K}^0$. The $K^0$ has strangeness $S = 1$, and is thereby distinguished from its antiparticle, the $\bar{K}^0$, which has $S = -1$. The physics of the $K^0$ and $\bar{K}^0$ exhibits a number of interesting and important features, but is also rather complicated so we will not consider it here.

### Hadron resonances

Another set of discoveries initiated in the 1950s revealed the existence of a variety of very short-lived mesons and baryons collectively referred to as **hadron resonances**. Most of the unstable particles we have met so far have mean lifetimes of about $10^{-10}$ s. These mean lifetimes are sufficiently long for a particle moving at a speed close to that of light to travel an observable distance in a bubble chamber or in a photographic emulsion. The resonances, in contrast, have mean lifetimes of about $10^{-24}$ s — such a short time that they do not leave measurable tracks in detectors. The reason for these

**Table 4.3**  Some early discoveries in particle physics. The mean lifetime of the $K^0$ has been omitted due to subtleties in $K^0$ physics. The charge number $Q$ indicates the electric charge in units of $e$, while $L$, $B$ and $S$ indicate lepton number, baryon number and strangeness respectively. The signs of $Q$, $L$, $B$ and $S$ are all reversed (if applicable) for the corresponding antiparticles. N/A indicates 'not applicable' which, for the purposes of calculation, may be interpreted as 0.

| Particle | Symbol | Mass (MeV/$c^2$) | Spin | Mean lifetime/s | $Q$ | $L$ | $B$ | $S$ |
|---|---|---|---|---|---|---|---|---|
| photon | $\gamma$ | 0 | 1 | stable | 0 | N/A | N/A | N/A |
| electron | $e^-$ | 0.511 | 1/2 | stable | −1 | 1 | N/A | N/A |
| positron | $e^+$ | 0.511 | 1/2 | stable | 1 | −1 | N/A | N/A |
| muon | $\mu^-$ | 105.6 | 1/2 | $2 \times 10^{-6}$ | −1 | 1 | N/A | N/A |
| antimuon | $\mu^+$ | 105.6 | 1/2 | $2 \times 10^{-6}$ | 1 | −1 | N/A | N/A |
| proton | p | 938.3 | 1/2 | stable | 1 | N/A | 1 | 0 |
| neutron | n | 939.6 | 1/2 | 917 | 0 | N/A | 1 | 0 |
| pion | $\pi^+$ | 140 | 0 | $2 \times 10^{-8}$ | 1 | N/A | 0 | 0 |
| pion | $\pi^-$ | 140 | 0 | $2 \times 10^{-8}$ | −1 | N/A | 0 | 0 |
| pion | $\pi^0$ | 135 | 0 | $0.8 \times 10^{-16}$ | 0 | N/A | 0 | 0 |
| kaon | $K^-$ | 494 | 0 | $1.2 \times 10^{-8}$ | −1 | N/A | 0 | −1 |
| kaon | $K^+$ | 494 | 0 | $1.2 \times 10^{-8}$ | 1 | N/A | 0 | 1 |
| kaon | $K^0$ | 498 | 0 | | 0 | N/A | 0 | 1 |
| lambda | $\Lambda^0$ | 1115 | 1/2 | $2 \times 10^{-10}$ | 0 | N/A | 1 | −1 |
| sigma | $\Sigma^+$ | 1189 | 1/2 | $0.8 \times 10^{-10}$ | 1 | N/A | 1 | −1 |
| sigma | $\Sigma^-$ | 1197 | 1/2 | $1.5 \times 10^{-10}$ | −1 | N/A | 1 | −1 |
| sigma | $\Sigma^0$ | 1192 | 1/2 | $6 \times 10^{-20}$ | 0 | N/A | 1 | −1 |
| xi | $\Xi^0$ | 1315 | 1/2 | $2.9 \times 10^{-10}$ | 0 | N/A | 1 | −2 |
| xi | $\Xi^-$ | 1321 | 1/2 | $1.6 \times 10^{-10}$ | −1 | N/A | 1 | −2 |

very short mean lifetimes is easy to explain. The decay of resonances proceeds via the strong interaction, but all the other decays we have discussed so far involve either the weak interaction, or in a few cases the electromagnetic interaction. However, the involvement of the strong interaction makes the decay process much more rapid.

In view of the absence of detectable tracks, you may wonder how the hadron resonances were detected at all. The answer, which also partly explains their peculiar name, is that they were first observed as enhancements in certain cross-sections. For instance, the total cross-section for $\pi^+$p scattering (the collision of a positively charged pion, $\pi^+$, with a proton) shows a high broad 'bump' or peak centred at a beam energy of around 200 MeV (see Figure 4.21). The shape of this peak was similar to that seen in other branches of physics when composite particles collided and was taken to indicate that the pion and the proton were somehow able to combine to form a short-lived conglomerate that rapidly decayed again. Careful analysis is needed to separate what might be contributions from several such short lived resonances, but once this had been done, the lifetime of each resonance could be gauged by using the energy–time form of Heisenberg's uncertainty principle:

$$\Delta E \Delta t \geq \hbar / 2. \qquad (QPI\ 2.8)$$

In the context of resonances, we may interpret $\Delta t$ as the mean lifetime $\tau$ of the resonance particle, and $\Delta E$ as the **decay width** $\Gamma$ of the resonance. The decay width $\Gamma$ can be taken to be the width of the resonance bump at the point where the scattering cross-section has half of its maximum value. Thus, for minimal uncertainty, $\Gamma\tau \approx \hbar/2$. The decay widths $\Gamma$ of the so-called $\Delta$ resonances seen in $\pi^+$p scattering (Figure 4.21) are typically a few hundred MeV, so the uncertainty principle implies that the mean lifetimes of those particles are about $10^{-24}$ s, as claimed. The properties of a small selection of meson and baryon resonances are listed in Table 4.4. This table is for reference only; you are not expected to memorize its contents.

**Figure 4.21** The total cross-section for $\pi^+$p scattering plotted against beam energy, showing the effect of the $\Delta^{++}$(1232) resonance. The decay width $\Gamma$ is a few hundred MeV.

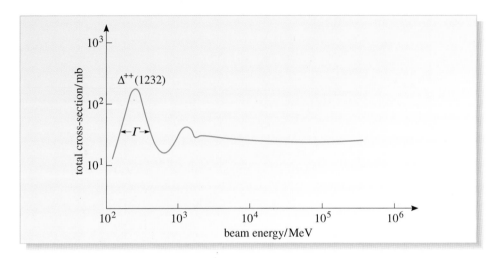

**Question 4.7** Estimate the lifetime of the $\Delta^{++}$(1232) resonance (refer to Table 4.4). ■

**Table 4.4** Some mesonic ($B = 0$) and baryonic ($B = 1$) resonances. The signs of $Q$, $B$ and $S$ are reversed for the corresponding antiparticles. Since a single symbol (e.g. N or $\Delta$) is generally used for many different resonances of the same type, it is customary to include the mass (in MeV/$c^2$) with the symbol when referring to a particular resonance. For example, the first resonance in the table is usually referred to as the N$^+$(1470).

| Resonance type | Symbol | Mass (MeV/$c^2$) | Spin | Width $\Gamma$/MeV | $Q$ | $B$ | $S$ | Main decay mode |
|---|---|---|---|---|---|---|---|---|
| nucleon | N$^+$ | 1470 | 1/2 | $\approx 200$ | 1 | 1 | 0 | p$\pi^0$ |
| nucleon | N$^-$ | 1470 | 1/2 | $\approx 200$ | $-1$ | 1 | 0 | n$\pi^-$ |
| nucleon | N$^+$ | 1520 | 3/2 | $\approx 125$ | 1 | 1 | 0 | p$\pi^0$, n$\pi^+$ |
| nucleon | N$^-$ | 1520 | 3/2 | $\approx 125$ | $-1$ | 1 | 0 | n$\pi^-$ |
| delta | $\Delta^{++}$ | 1232 | 3/2 | $\approx 115$ | 2 | 1 | 0 | $\pi^+$p |
| delta | $\Delta^+$ | 1650 | 1/2 | $\approx 140$ | 1 | 1 | 0 | p$\pi^+\pi^-$, p$\pi^0\pi^0$, n$\pi^+\pi^0$ |
| lambda | $\Lambda$ | 1815 | 5/2 | $\approx 80$ | 0 | 1 | $-1$ | n$\overline{\text{K}}^0$, p$\overline{\text{K}}^-$ |
| rho | $\rho$ | 770 | 1 | $\approx 160$ | 0 | 0 | 0 | $\pi^+\pi^-$ |
| omega | $\omega$ | 783 | 1 | $\approx 10$ | 0 | 0 | 0 | $\pi^+\pi^-\pi^0$ |

## 3.4  More leptons — neutrinos and the tauon

The development of accelerators not only revealed the existence of many new hadrons (mesons and baryons); it also added extra members to the much smaller family of leptons. The possibility of leptons other than the electron and the muon gained wide acceptance long before there was any direct evidence that such particles existed. As early as the 1930s, theoretical studies of the $\beta^-$-decay of nuclei led to the suggestion by Pauli that the underlying process was one in which a neutron decayed via the weak interaction to produce a proton and an electron, together with an unobserved neutral particle of very low mass. The Italian physicist Enrico Fermi (1901–1954) named this unseen neutral particle the **neutrino**, meaning 'little neutral one.'

A near massless, neutral lepton is inevitably hard to detect since it interacts neither strongly nor electromagnetically. Because its most powerful interaction is the feeble weak interaction, a neutrino can travel through enormous amounts of ordinary matter with very little chance of being stopped or even deflected. (About a million million neutrinos, coming from the nuclear reactions that power the Sun, are thought to flood through your head every second of your life without doing you any harm.) The fact that neutrinos really are common participants in weak interactions is indicated by the fact that energy and momentum would not be conserved in processes such as $\beta^-$-decay if neutrinos (antineutrinos actually) were not also emitted. Nonetheless, direct evidence of the neutrino was not obtained until the mid-1950s, when a Nobel Prize-winning experiment by Clyde Cowan and Frederick Reines used the neutrinos flooding out of a nuclear reactor to trigger detectable nuclear reactions.

In 1962, in what was to become another Nobel Prize-winning discovery, a team working with the accelerator at Brookhaven National Laboratory in the USA found clear evidence that there were actually two different kinds of neutrino: one associated with the electron, the other with the muon. These two kinds of neutrino are now represented by the symbols $\nu_e$ and $\nu_\mu$, and are referred to as the **electron neutrino** and the **muon neutrino** respectively.

In 1975, a team working at SLAC's colliding beam facility discovered a third charged lepton called the **tauon** ($\tau$), with a very much shorter mean lifetime than the muon. This new particle has the same charge as the electron and the muon but a mass of about $1777\ \mathrm{MeV}/c^2$, almost twice the mass of a proton. The existence of this particle implied the existence of an antitauon, but it also strongly suggested that there should be a **tauon neutrino** and a corresponding antineutrino. Evidence of the new neutrino was eventually obtained in another collider experiment, completing the third generation of leptons and bringing the total number of leptons to six, along with six antileptons. The six leptons are shown in the top part of Figure 4.2.

The six types, or six **flavours**, of leptons are summarized in Table 4.5 where they are shown grouped in three pairs or **generations**, each generation consisting of a charged particle and its neutrino.

The association of a different neutrino with each type of charged lepton justifies

**Table 4.5** The six flavours of leptons, arranged in three generations of increasing mass. The charge number $Q$ and lepton numbers $L_e$, $L_\mu$ and $L_\tau$ have the opposite signs for the corresponding antiparticles. The masses of the neutrinos have not yet been determined but there is some evidence that the mass of the electron neutrino, although small, is not zero.

| Particle | Symbol | Mass (MeV/$c^2$) | Spin | Mean lifetime /s | $Q$ | $L_e$ | $L_\mu$ | $L_\tau$ |
|---|---|---|---|---|---|---|---|---|
| electron | $e^-$ | 0.511 | 1/2 | stable | −1 | 1 | 0 | 0 |
| electron neutrino | $\nu_e$ | <0.000 015 | 1/2 | stable | 0 | 1 | 0 | 0 |
| muon | $\mu^-$ | 105.6 | 1/2 | $2 \times 10^{-6}$ | −1 | 0 | 1 | 0 |
| muon neutrino | $\nu_\mu$ | <0.17 | 1/2 | stable | 0 | 0 | 1 | 0 |
| tauon | $\tau^-$ | 1777 | 1/2 | $3 \times 10^{-13}$ | −1 | 0 | 0 | 1 |
| tauon neutrino | $\nu_\tau$ | <18 | 1/2 | stable | 0 | 0 | 0 | 1 |

refining the concept of lepton number and its conservation. Rather than using a lepton number $L$, it is possible to identify three separately conserved lepton numbers $L_e$, $L_\mu$ and $L_\tau$ called **electron number**, **muon number** and **tauon number**, respectively. The values of $L_e$, $L_\mu$ and $L_\tau$ assigned to each lepton are indicated in Table 4.5. (The signs should be reversed for the corresponding antileptons.) Taken together, these three conserved quantities have the same effect as the conservation of lepton number $L$, but separately they provide even more constraints on possible processes. For example, one of the ways in which a tauon decays is via the process:

$$\tau^- \rightarrow \mu^- + \overline{\nu}_\mu + \nu_\tau$$
$$L_\tau : 1 \rightarrow 0 + 0 + 1$$
$$L_\mu : 0 \rightarrow 1 + (-1) + 0.$$

As indicated, this is consistent with the conservation of tauon number and muon number, as well as the less restrictive conservation of lepton number. However, the following decay has never been observed.

$$\tau^- \rightarrow \mu^- + \overline{\nu}_\tau + \nu_\tau$$
$$L_\tau : 1 \rightarrow 0 + (-1) + 1$$
$$L_\mu : 0 \rightarrow 1 + 0 + 0.$$

This is because, although overall lepton number would be conserved, separate conservation of tauon number and muon number forbids the decay.

**Question 4.8**  (a) Which one of the following decays or reactions is forbidden by the conservation of baryon number? (b) Which one is forbidden by conservation of one (or more) of the three lepton numbers? (Refer to Tables 4.3–4.5.)

$\mu^- \rightarrow e^- + \nu_\mu + \overline{\nu}_e$　　　　$n \rightarrow p + e^- + \overline{\nu}_e$

$\pi^+ \rightarrow \mu^+ + \nu_\mu$　　　　　　　$\mu^- \rightarrow e^- + \nu_\mu + \overline{\nu}_e$

$\pi^+ + p \rightarrow \Delta^{++}$　　　　　　　　$p + \overline{p} \rightarrow 2\pi^+ + \pi^0 + 2\pi^-$

$n \rightarrow e^+ + e^- + \nu_e + \overline{\nu}_e$　　　$p \rightarrow n + e^+ + \overline{\nu}_e.$

# 4 The quark model for hadrons

You have seen that there are only six flavours of leptons (Table 4.5) plus, of course, the six antiparticles, and these twelve particles and antiparticles are believed to be truly fundamental, i.e. they are not composite structures. The same is *not* true of the hadrons. By the early 1960s, a large number of hadrons were known, including 80 or so resonances. Although the number of known particles was large, it was clear that many of them were related to one another. Consequently, a classification scheme was needed that would order and relate the particles in a way that reflected their underlying (and not then understood) physics. There was also a feeling in some quarters that, with so many different hadrons being observed, it could not be the case that all were equally fundamental. Some, it was felt, might be composite particles.

As it turned out, these two issues were related. In 1961, a classification scheme was proposed by Murray Gell-Mann (who had been one of those to propose the introduction of strangeness some years earlier) and, independently by Yuval Ne'eman (1925–). This scheme, known as the *eightfold way*, brought order to the world of subatomic particles just as the Periodic Table had brought order to the elements, and was successful in predicting the properties of several particles before they were detected. A few years later it was realized that this classification scheme was consistent with the idea that the observed hadrons were combinations of a small number of unobserved fundamental constituents that Gell-Mann called **quarks**.

## 4.1 The six quarks

The original version of the quark model, proposed in 1964 by Murray Gell-Mann and independently by George Zweig (1937–), assumed the existence of just three different flavours of quark: **up** (u), **down** (d) and **strange** (s), together with an equal number of antiquarks ($\bar{u}$, $\bar{d}$, $\bar{s}$).

Subsequent discoveries (described below) led to the introduction of three more flavours of quark: **charm** (c), **top** (t) and **bottom** (b), and three more antiquarks ($\bar{c}$, $\bar{t}$, $\bar{b}$). Each of these three new quarks is endowed with its own new and distinctive property, similar to the strangeness of the strange quark. These three new properties are called **charm**, **top** (or topness) and **bottom** (or bottomness). Like strangeness, they are conserved in strong and electromagnetic interactions, but not in weak interactions.

Like the leptons, the quarks are grouped into three generations of increasing mass, the first generation consisting of the up and down quarks (and their antiparticles). The properties attributed to the six quarks and six antiquarks are listed in Table 4.6. Note that all quarks have fractional baryon number $B$ and fractional charge number $Q$; this distinguishes them from all the other particles that have been discussed so far.

**Figure 4.22** Murray Gell-Mann (1929–). In the late 1960s, Gell-Mann was widely regarded as the world-leading theoretical physicist, his only serious rival being Richard Feynman. Both men at that time worked at Cal Tech (the Californian Institute of Technology) which was, and still remains, a major centre for theoretical particle physics. At the time of writing, Gell-Mann is associated with the Santa Fe Institute, where he works on various aspects of chaos and complexity.

**Table 4.6** The six flavours of quarks (and the six antiquarks) grouped in three generations of increasing mass. Quarks are not observed as isolated particles, so their masses must be deduced by indirect observations. There is still some controversy about the correct way of doing this and hence about the mass values.

| Quark | Symbol | Mass (MeV/$c^2$) | Spin | $Q$ | $B$ | Charm | Strangeness | Top | Bottom |
|---|---|---|---|---|---|---|---|---|---|
| up | u | 5 | 1/2 | 2/3 | 1/3 | 0 | 0 | 0 | 0 |
| down | d | 9 | 1/2 | −1/3 | 1/3 | 0 | 0 | 0 | 0 |
| charm | c | 1400 | 1/2 | 2/3 | 1/3 | 1 | 0 | 0 | 0 |
| strange | s | 170 | 1/2 | −1/3 | 1/3 | 0 | −1 | 0 | 0 |
| top | t | 175 000 | 1/2 | 2/3 | 1/3 | 0 | 0 | 1 | 0 |
| bottom | b | 4400 | 1/2 | −1/3 | 1/3 | 0 | 0 | 0 | −1 |
| antiup | $\bar{u}$ | 5 | 1/2 | −2/3 | −1/3 | 0 | 0 | 0 | 0 |
| antidown | $\bar{d}$ | 9 | 1/2 | 1/3 | −1/3 | 0 | 0 | 0 | 0 |
| anticharm | $\bar{c}$ | 1400 | 1/2 | −2/3 | −1/3 | −1 | 0 | 0 | 0 |
| antistrange | $\bar{s}$ | 170 | 1/2 | 1/3 | −1/3 | 0 | 1 | 0 | 0 |
| antitop | $\bar{t}$ | 175 000 | 1/2 | −2/3 | −1/3 | 0 | 0 | −1 | 0 |
| antibottom | $\bar{b}$ | 4400 | 1/2 | 1/3 | −1/3 | 0 | 0 | 0 | 1 |

(You may be surprised that Table 4.6 does not include columns indicating values of 'upness' and 'downness' similar to charm or strangeness. The reason for this is that the u and d quarks are deemed to possess a more subtle property called *isospin*. Discussion of isospin is beyond the scope of this book.)

## 4.2  Using the quark model

In its simplest form, the basic idea of the quark model may be stated as follows:

- All baryons are composed of three quarks.
- All antibaryons are composed of three antiquarks.
- All mesons are composed of a quark and an antiquark.

Many of the properties of the baryons, antibaryons and mesons can be understood as simple sums of the corresponding properties of the quarks that they contain. This is true, for example, of the charge, baryon number, strangeness and charm. Other properties, such as the spin and particularly the mass, are more difficult to work out, but they too are supposed ultimately to be determined by the composite nature of the hadrons.

To see how the simple quark model works in practice, let's examine the way in which quarks are assigned to a few of the hadrons you have already met. The proton, for example, is assumed to acquire its properties from the quark combination uud. We will indicate this by writing p = uud. Writing the values of baryon number $B$ and charge number $Q$ below the corresponding particles, we see that the simple quark model does indeed account for the charge and the baryon number of the proton:

$$p = u \quad u \quad d$$
$$B : 1 \quad = \tfrac{1}{3} + \tfrac{1}{3} + \tfrac{1}{3}$$
$$Q : 1 \quad = \tfrac{2}{3} + \tfrac{2}{3} - \tfrac{1}{3}.$$

Furthermore, the spin of the proton (1/2) can be explained by assuming that two of the quarks are in a 'spin-up' state while the third is 'spin-down', giving the uud combination the required net spin of 1/2.

A similar argument applies to the neutron:

$$n = u \quad d \quad d$$
$$B : 1 = \tfrac{1}{3} + \tfrac{1}{3} + \tfrac{1}{3}$$
$$Q : 0 = \tfrac{2}{3} - \tfrac{1}{3} - \tfrac{1}{3}$$

and to the antineutron, although in this case all the quarks must be replaced by the corresponding antiquarks:

$$\bar{n} = \bar{u} \quad \bar{d} \quad \bar{d}$$
$$B : -1 = -\tfrac{1}{3} - \tfrac{1}{3} - \tfrac{1}{3}$$
$$Q : 0 = -\tfrac{2}{3} + \tfrac{1}{3} + \tfrac{1}{3}.$$

In the case of the mesons, the combination of a quark and an antiquark (with baryon numbers 1/3 and −1/3, respectively) automatically guarantees that the overall baryon number will be $B = 0$, as it must be for a meson. The quark model also explains how some neutral mesons (those consisting of quarks and antiquarks of the same type, such as $u\bar{u}$ or $d\bar{d}$) manage to be their own antiparticles, while others, such as the $K^0$ (a $d\bar{s}$ combination), are different from their antiparticles ($\overline{K}^0 = s\bar{d}$).

More quark assignments are listed in Table 4.7, which is for reference only. Note that some combinations of quarks appear more than once, even in this short list. For instance, the neutron and the $\Delta^0$ are both represented by the combination udd. In such cases they are distinguished by their mass, and possibly spin.

Table 4.7   Some quark assignments.

| Baryons | Antibaryons | Mesons |
|---|---|---|
| p = uud | $\bar{p} = \bar{u}\,\bar{u}\,\bar{d}$ | $\pi^+ = \bar{u}\bar{d}$ |
| n = udd | $\bar{n} = \bar{u}\,\bar{d}\,\bar{d}$ | $\pi^- = d\bar{u}$ |
| $\Lambda^0$ = uds | $\overline{\Lambda}^0 = \bar{u}\,\bar{d}\,\bar{s}$ | $\pi^0 = u\bar{u}$ and $d\bar{d}$ |
| $\Sigma^-$ = dds | | $K^+ = u\bar{s}$ |
| $\Sigma^0$ = uds | $\overline{\Sigma}^0 = \bar{u}\,\bar{d}\,\bar{s}$ | $K^- = s\bar{u}$ |
| $\Sigma^+$ = uus | | $K^0 = d\bar{s}$ |
| $\Xi^-$ = dss | | $\overline{K}^0 = s\bar{d}$ |
| $\Xi^0$ = uss | $\overline{\Xi}^0 = \bar{u}\,\bar{s}\,\bar{s}$ | $\phi = s\bar{s}$ |
| $\Delta^{++}$ = uuu | | $\psi = c\bar{c}$ |
| $\Delta^+$ = uud | | $\psi' = c\bar{c}$ |
| $\Delta^0$ = udd | $\overline{\Delta}^0 = \bar{u}\,\bar{d}\,\bar{d}$ | $D^+ = c\bar{d}$ |
| $\Delta^0$ = ddd | | $D^0 = c\bar{u}$ |
| $\Lambda_c^+$ = udc | | $\Upsilon = b\bar{b}$ |

**Question 4.9**   (a) Use the quark assignments in Table 4.7 and the data in Table 4.6. to determine the charge number, baryon number, strangeness and charm of the following particles: $\Sigma^0$, $\Delta^{++}$, $\Lambda_c^+$, $\overline{K}^0$, and $\psi$.

(b) What would be the expected charge, baryon number, strangeness and charm of the antiparticle of $\Lambda_c^+$?

(c) Some neutral mesons are their own antiparticles. Explain how this can be. Is it possible for any baryon to be its own antiparticle? (Your answer should be a general one, not restricted to the particles in Table 4.7.)   ∎

## 4.3  Evidence for the existence of quarks

The ability of the simple quark model to summarize the properties of the 200 or so known hadrons in a few lines is impressive, but in itself it is not enough to convince anyone that quarks really exist. How then have the quarks acquired a status similar to that of the leptons as a set of (as far as we know) truly fundamental particles?

One of the earliest successes of the quark model, actually a success for the classification scheme that preceded the quark model, concerned a group of spin 3/2 baryons that included the $\Delta^{++}(1232)$ resonance. In 1962, there were nine known members of this group, but Gell-Mann's scheme for classifying particles implied that the quark content of the particles could be displayed as in Figure 4.23a. Gell-Mann predicted that there should be a tenth member of the group, represented by the question mark. This should be an sss baryon with negative electric charge, strangeness −3 and a mass of about 1685 MeV/$c^2$. Gell-Mann pointed out that this particle, which he called the *omega minus* ($\Omega^-$), would decay far more slowly than its nine partners (in about $10^{-10}$ s, rather than $10^{-23}$ s), implying that it should leave a track that could be observed in a bubble chamber. The experimental discovery of the $\Omega^-$ was duly announced in February 1964 by a team using the 80-inch hydrogen bubble chamber at Brookhaven National Laboratory (BNL). Their initial value for the mass was $(1686 \pm 12)$ MeV/$c^2$ and all ten members of the group are shown in Figure 4.23b. The relatively long mean lifetime of the $\Omega^-$ is a consequence of the fact that all combinations of hadrons with strangeness −3 (such as $\overline{\Lambda}^0$, $\overline{K}^0$ and $\overline{K}^-$) have higher mass than the $\Omega^-$, so the $\Omega^-$ is unable to decay strongly.

| | | | | | | | | | |
|---|---|---|---|---|---|---|---|---|---|
| $\Delta^-$ ddd | $\Delta^0$ udd | $\Delta^+$ uud | $\Delta^{++}$ uuu | (1232 MeV/$c^2$) | $\Delta^-$ ddd | $\Delta^0$ udd | $\Delta^+$ uud | $\Delta^{++}$ uuu | (1232 MeV/$c^2$) |
| | $\Sigma^-$ dds | $\Sigma^0$ uds | $\Sigma^+$ uus | (1384 MeV/$c^2$) | | $\Sigma^-$ dds | $\Sigma^0$ uds | $\Sigma^+$ uus | (1384 MeV/$c^2$) |
| | | $\Xi^-$ dss | $\Xi^0$ uss | (1534 MeV/$c^2$) | | | $\Xi^-$ dss | $\Xi^0$ uss | (1534 MeV/$c^2$) |
| | | **?** | | | | | $\Omega^-$ sss | | (1686 MeV/$c^2$) |
| (a) | | | | | (b) | | | | |

**Figure 4.23**   (a) The group of nine spin 3/2 baryons including the $\Delta^{++}(1232)$ resonance known in 1962. (b) The complete group of ten spin 3/2 baryons.

More direct evidence relating to the reality of quarks was discovered in 1969. A team working at the Stanford Linear Accelerator Centre (SLAC) found that electrons colliding with protons at very high energies behaved as though the electrons were striking point-like targets. This was indicated by the high number of electrons that underwent relatively large deflections, just as, many years before, large deflections of α-particles had convinced Rutherford that atoms contained nuclei. Other experiments had already shown that the proton was not a point-like object, so the clear implication was that the point-like objects being struck were the quarks within the proton. Further experiments of this general kind, known as **deep inelastic scattering** experiments, eventually provided good evidence that quarks not only existed, but that they also had just the properties predicted for them. The accelerator known as HERA (Hadron Electron Ring Accelerator) at the DESY laboratory in Germany (see Table 4.1) continues this kind of work today, but at much higher energies.

The SLAC experiments in 1969 convinced many particle physicists of the reality of quarks, but theoretical arguments indicated that the original suggestion by Gell-Mann and Zweig, that there were just three quark types (u, d and s), could not be correct. According to these arguments, there had to exist a symmetry between quarks and leptons that would enable them to be gathered together into generations of four particles. The first generation of leptons and quarks, consisting of the up and down quarks together with the electron and the electron neutrino, was complete. However, in the early 1970s, when this quark–lepton symmetry was proposed, the second generation containing the muon and its neutrino, together with the strange quark, had an embarrassing gap that could only be filled by a fourth quark with charge number $Q = 2/3$. This led to the prediction of the charm quark (c) and to the predicted existence of a whole class of new hadrons containing charm quarks. Figure 4.24 shows the first observation of a charmed baryon.

The event that really confirmed the existence of quarks was the discovery of a relatively long-lived meson (seen as a very narrow resonance), with all the properties expected of a $c\bar{c}$ composite. The psi ($\psi$), as it is called, was discovered independently by experimental groups working at BNL and SLAC in November 1974 — an event still sometimes referred to as the 'November revolution'. Many other charmed hadrons have been observed since.

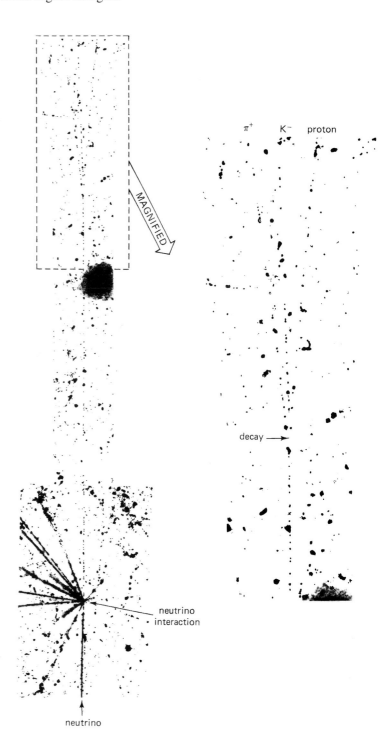

**Figure 4.24** First observation of a charmed baryon. The baryon $\Lambda_c^+$ (= cud) is produced by high-energy neutrinos colliding with an atomic nucleus in a photographic emulsion. This is shown by the tracks at the bottom of the figure. Tracks showing the subsequent decay of the $\Lambda_c^+$ ($\Lambda_c^+ \rightarrow p + K^- + \pi^+$) are seen at the top of the figure, and magnified on the right.

187

History repeated itself to some extent following the 1975 discovery of the tauon, the heavy partner of the electron and the muon. The existence of a third charged lepton implied the existence of a third generation of fundamental particles that should include a tauon neutrino and two more quarks — now called the top quark (t) and the bottom quark (b). Evidence of the bottom quark was found in 1977 when an experimental team using a large new proton synchrotron at Fermilab discovered the *upsilon particle* ($\Upsilon$), a $b\bar{b}$ combination. After much further searching, another Fermilab experiment provided evidence of the top quark, although this experiment required so much energy, because of the large mass of the top quark, that the discovery was not made until 1995. (By this time, a second accelerator ring had been built at Fermilab and the facility had become the only one in the world capable of producing proton–antiproton collisions at energies of 2 TeV.) The evidence indicated that the mass of the top quark was 175 GeV/$c^2$, making it as heavy as a gold atom. The three known generations of fundamental particles are shown in Figure 4.25.

| | 1st generation | | 2nd generation | | 3rd generation | |
|---|---|---|---|---|---|---|
| quarks | u | d | c | s | t | b |
| leptons | $e^-$ | $\nu_e$ | $\mu^-$ | $\nu_\mu$ | $\tau^-$ | $\nu_\tau$ |

**Figure 4.25** The known quarks and leptons arranged in three generations.

Murray Gell-Mann was awarded the Nobel Prize for physics in 1969. Nobel Prizes were also awarded to those mainly responsible for the 1969 SLAC deep inelastic scattering experiments, and to the discoverers of the $\psi$ (psi), $\tau^-$ (tauon) and $\Upsilon$ (upsilon) particles.

Despite these successes of the quark model, there has never been a direct observation of an isolated quark. The failure to observe 'free' quarks makes it difficult to determine quark masses, but this is not seen as conflicting with the reality of quarks. Rather, the challenge has been to construct credible theories that predict the **quark confinement** within the observed baryons and mesons, so that the quarks cannot be observed as free particles.

# 5 Quantum physics of fundamental particles

The modern theory of particle physics recognizes the three generations of leptons and quarks (and their antiparticles) as the fundamental particles of matter. Although the fundamental particles are a crucially important part of the theory, what really lies at its heart is our modern understanding of three of the four fundamental forces: the strong, electromagnetic, and weak interactions. Modern theories of these interactions combine special relativity and quantum physics and are known as quantum field theories. We introduced the fundamental forces in Section 3.1 as the means by which elementary particles exchange energy and momentum. There we represented such interactions by mysterious 'circles' of the kind shown in Figure 4.26; the quantum field theories tell us what goes on inside such circles.

# 5.1 Quantum field theories — a brief overview

The basic idea of a **quantum field theory** is that the fundamental forces are 'exchange forces' in which particles can exchange momentum and energy by emitting and absorbing yet another kind of particle called an **exchange particle**. All known exchange particles are spin 1 bosons and are sometimes called *exchange bosons*. Each of the fundamental forces is associated with a particular kind of exchange particle or with a family of exchange particles.

The first fundamental force to be understood as an exchange force was the electromagnetic interaction. In the late 1940s, after almost two decades of effort, Richard Feynman and others formulated a highly successful quantum field theory of the electromagnetic interaction. This theory is known as **quantum electrodynamics** or **QED**. Its aim is to predict the probability or likelihood of a prescribed process such as one in which two electrons, with specified energies and momenta, collide and scatter elastically. Feynman's approach to calculating the probability of such a process involves drawing a series of diagrams, now called **Feynman diagrams**, that represent all possible subprocesses that contribute to the process. Three such diagrams are shown in Figure 4.27. Each diagram is drawn in accordance with certain rules and is actually a pictorial representation of a mathematical expression that can be explicitly written down and (in principle) evaluated. The contributions from all relevant Feynman diagrams are added together, and then finally squared to find the probability of the process. The fact that the probabilities are added *before* squaring leads to the possibility of interference between different contributions.

If you look at the Feynman diagrams in Figure 4.27, you will see the role of the exchange particle (represented by the wavy line) in carrying energy and momentum between the incoming and outgoing electrons. The exchange particle in QED is the photon; it has spin 1 and is therefore a boson. The photons exchanged in Feynman diagrams are called **virtual photons**. There is a difference between these virtual photons and the real photons entering your eye. Virtual photons are not required to obey the condition $E = pc$ that links the energy and momentum magnitudes of real photons, and are not directly observable. They are referred to as *virtual* photons in order to emphasize this fact.

*Electroweak theory*

Quantum electrodynamics is a very successful theory. Some of the most accurate predictions made in any branch of science have been achieved using QED. It is therefore not surprising that physicists have tried to find an analogous theory for the weak and strong interactions. In the case of the weak interaction, the initial efforts to formulate an exchange theory failed. However, it was eventually realized that a key ingredient in the success of QED was the fact that the theory had a mathematical property known as *gauge invariance*, a generalization of the basic idea in quantum mechanics that only the square of the amplitude of a wavefunction is measurable. In the case of QED, one of the consequences of gauge invariance is that real photons should be massless, as they are observed to be.

The search for a similarly gauge-invariant theory of the weak interaction eventually led to a theory that required the introduction of three new spin 1 exchange particles known as the **W⁺**, **W⁻** and **Z⁰ bosons** (see Figure 4.28). However, far from being

The search for a similarly gauge-invariant theory of the weak interaction eventually led to a theory that required the introduction of three new spin 1 exchange particles known as the **W⁺**, **W⁻** and **Z⁰ bosons** (see Figure 4.28). However, far from being

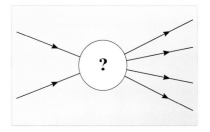

**Figure 4.26** A schematic interaction of particles. The circle with a question mark indicates our need for a theoretical understanding of the action of the fundamental forces in such a process.

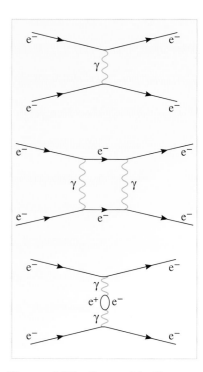

**Figure 4.27** Some of the Feynman diagrams that contribute to a calculation of the cross-section for electron–electron elastic scattering. You read the diagrams from left to right — the direction of increasing time. Thus, the first diagram shows two electrons approaching one another, then interacting by exchange of a virtual photon (the vertical wavy line labelled γ) and finally moving off with changed momenta and energy. The virtual photon carries the energy and momentum that is exchanged between the two electrons.

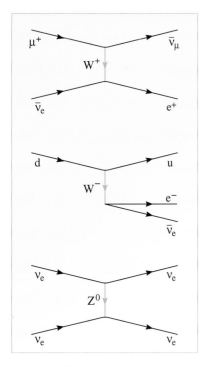

**Figure 4.28** Feynman diagrams showing the exchange of W⁺, W⁻ and Z⁰ particles. Reading from left to right (increasing time), the first diagram shows the collision of a positive muon and an electron antineutrino. The vertical line indicates the exchange of a W⁺ particle which carries electric charge, energy, momentum and other properties from one particle to the other. The result is the emergence of a muon neutrino and a positron.

In 1999 Gerardus 't Hooft and Martinus Veltman received a Nobel Prize for developing electroweak theory to the point where it could be stringently tested by the high-precision experiments of the 1990s.

massless, the theory implied that these exchange particles would have quite large masses (more than 80 times the proton's mass). Moreover, the neutral Z⁰ particle was expected to combine with the photon in various situations to produce effects that could only be attributed to a unification of the electromagnetic and weak interactions. For this reason, the new theory is usually referred to as the **electroweak theory**. However, the fact that the weak and electromagnetic interactions are actually different aspects of a single unified force was only expected to become manifest at very high energies.

Steven Weinberg and Abdus Salam proposed the electroweak theory in the late 1960s. The predictions were subjected to stringent experimental tests throughout the 1970s, which culminated in the early 1980s with the production (in p$\bar{\text{p}}$ collisions at CERN) of the W⁺, W⁻ and Z⁰ bosons, with masses that fell within the predicted range. By the time this landmark discovery was made, Sheldon Glashow, Weinberg and Salam had already been awarded the Nobel Prize for physics, but the experimental discoverers of the W⁺, W⁻ and Z⁰ bosons were also awarded the Nobel Prize in recognition of their achievement.

**Figure 4.29** Steven Weinberg (1933– ) left and Abdus Salam (1926–1996) right.

## Quantum chromodynamics (QCD)

With the realization that the fundamental particles involved in the strong interaction were quarks (and antiquarks), rather than mesons and baryons, a number of physicists more or less simultaneously formulated a gauge-invariant theory of the strong interaction. A property of quarks called **colour charge**, or simply **colour**, plays an important role in the theory, similar to that played by electric charge in QED. (Quark colour has no connection with the ordinary sense in which we use the word colour. This is just another example of the vivid terminology often used in particle physics.) One reason for introducing this new quark property arises from the fact that quarks, having spin 1/2, are fermions and therefore obey the Pauli exclusion principle. However, it is known that some baryons consist of three quarks of the same flavour in, apparently, the same quantum state. An example is the $\Delta^{++}$ particle, which is a composite of three up quarks (see Table 4.7). This combination can only be possible if the quantum states of the three up quarks in $\Delta^{++}$ are different in some way. Colour was introduced initially as the new property that would distinguish the otherwise identical quarks in hadrons such as the $\Delta^{++}$.

You can think of quark colour as just another physical property of quarks, akin to electric charge, but whereas there are only positive or negative electric charges, there are three colours, referred to as red, blue and green. In addition, there are the three anticolours: antired, antiblue and antigreen. Quarks carry a colour charge while antiquarks carry anticolour. Another way in which colour charge differs from electric charge is that no observed particles are coloured. The three quarks in a hadron always carry the three different colours, red, blue and green. This makes the hadron itself colour neutral, or colourless. Similarly, the two quarks in a meson consist of a colour–anticolour pair, so that the meson itself is colourless.

Because the gauge theory of the strong interaction involves colour, the theory is called **quantum chromodynamics**, or **QCD** for short. In QCD, the exchange particles are called **gluons**, and there are eight of them. Gluons, like quarks, carry colour and anticolour charge. (This is in contrast to photons, which do not carry electric charge.) Gluons may be exchanged only between particles with colour or anticolour, i.e. between quarks and antiquarks or between other gluons. Some Feynman diagrams representing processes in which gluons are exchanged are shown in Figure 4.30.

The fact that gluons can be exchanged between other gluons as well as between quarks or antiquarks is thought to be responsible for the confinement of quarks, i.e. the inability of individual quarks or gluons to escape from hadrons in the way that electrons, say, can escape from atoms.

We have no precise picture of how confinement works, but there is a useful analogy, not to be taken too literally, illustrated in Figure 4.31. You know that the north and south poles of a magnet cannot be separated by breaking a magnet in half. All that happens is that a new south pole appears at one broken end and a new north pole at the other, so each broken piece is itself a magnet with a north and a south pole. Now consider a quark–antiquark pair held together by a set of gluons in a meson. (Some mesons and their quark–antiquark components are shown in Table 4.7.) Suppose you were to supply energy to the meson hoping to prise the quark–antiquark pair apart. You can imagine the sets of field lines representing the gluons 'stretching' and eventually 'breaking'. However, this doesn't happen; the energy you supply goes into the creation of a new antiquark attached to one set of broken ends and a new quark attached to the other. Thus, you end up with two quark–antiquark pairs, i.e. two mesons.

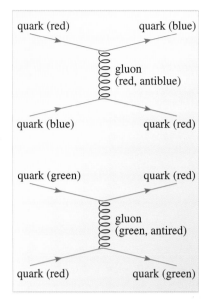

**Figure 4.30** Feynman diagrams showing the exchange of gluons. The colours carried by the quarks and gluons are indicated. The first diagram shows the exchange of colour mediated by a red–antiblue gluon.

**Figure 4.31** Analogy with a broken magnet — a highly simplified picture of confinement. The sets of gluon field lines holding the quark–antiquark pair in a meson are indicated by a horizontal line.

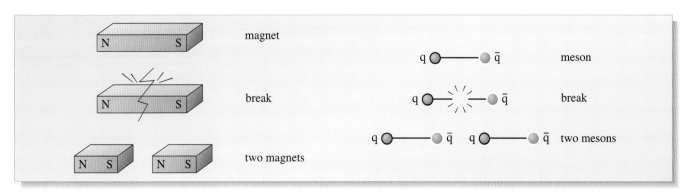

Although detailed mathematics is required to work out the predictions of QCD, the theory has stood up well to a number of experimental tests. One area in which there has been success concerns the production of jets of hadrons. These originate from the primary interactions of quarks, via gluon exchange, and gluon radiation. QCD predicts that in colliding beam experiments a pair of quarks may sometimes undergo large deflections, by exchanging a gluon that transfers a large amount of momentum. The result registers in the detectors as a pair of jets of hadrons, carrying the transferred momentum. Such an outcome is indicated in Figure 4.32, where the jets are detected at large angles to the other products of the collision. Different outcomes are possible. For example one of the quarks may radiate a gluon, which then produces a third jet, not shown in this case.

**Figure 4.32**  Jet production — an observable consequence of QCD.

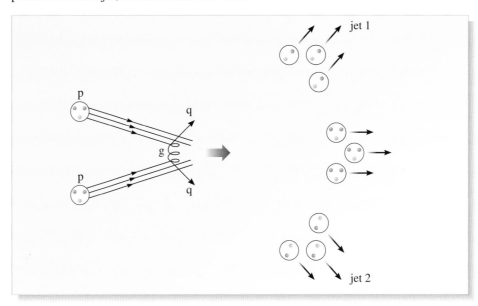

In principle, it is possible to use QCD to predict many properties of hadrons, including their masses. However, the calculations involved are enormously complicated and even these days are seriously limited by the amount of computer power available to theorists. QCD is certainly the best theory of the strong interaction we possess, but the need for more calculations and further experimental tests continues.

Table 4.8 summarizes the properties of the exchange particles that underpin the fundamental forces. The gravitational interaction has been included along with a hypothetical spin 2 exchange boson called the *graviton* that is sometimes associated with it. However, you should realize that attempts to formulate a consistent graviton exchange theory have not yet met with success, so the status of the graviton is very different from that of the other exchange particles in the table.

**Question 4.10**   In terms of QCD, why are electrons unable to 'feel' the strong interaction?  ■

## 5.2  The standard model and beyond

At the present time, the most widely accepted theory of particle physics is that provided by what is known as the **standard model**. This is a quantum field theory that incorporates fields representing the 12 fundamental quarks and leptons listed in Tables 4.5 and 4.6 and the exchange particles of the strong, electromagnetic and weak interactions listed in Table 4.8. (Fields representing the antiquarks and

**Table 4.8** The fundamental forces and their associated exchange particles. Note that the weak and electromagnetic forces are actually combined into a single electroweak force that becomes manifest at sufficiently high energy.

| Force | Exchange bosons: | | Mass | Charge |
| | name | symbol | (in units of GeV/$c^2$) | (in units of $e$) |
| --- | --- | --- | --- | --- |
| electromagnetic | virtual photon | $\gamma$ | 0 | 0 |
| weak | W-plus | $W^+$ | 82.22 | 1 |
| | W-minus | $W^-$ | 82.22 | $-1$ |
| | Z-zero | $Z^0$ | 91.17 | 0 |
| strong | gluons (8) | $g_1, g_2, \ldots g_8$ | 0 (but confined) | 0 |
| gravitational | graviton (not part of the standard model; no generally accepted quantum theory yet exists in this case) | | | |

antileptons are also included.) As far as is known, this theory is not in conflict with any of the experimentally established features of particle physics, and its construction is regarded as a major triumph in the battle to understand the physics of elementary particles. The standard model is constantly being probed and tested by experimentalists, but one of its most interesting tests concerns the predicted existence of a class of massive spin 0 particles called **Higgs bosons**.

Higgs bosons are required by the standard model in order to account for some of the observed features of the electroweak interaction and also play a crucial role in determining particle masses. They are thought to be so massive (perhaps in excess of 200 GeV/$c^2$ — equivalent to about 200 proton masses) that producing them in collisions is beyond the capacity of any existing particle accelerator. However, it is hoped that the Large Hadron Collider (LHC) under construction at CERN will lead to their discovery by the year 2010. If so, it would be a stunning success for the standard model. To fail to find the Higgs boson would, if anything, be even more exciting, since it is some time since experiment has provided thought-provoking surprises, rather than confirmation of the standard model.

Despite its successes, it is widely felt that the standard model is far from being the last word in particle theory. Experimental work may well expose the need for minor modifications or extensions, but we also look forward to a more fundamental revision. One very unsatisfactory feature of the standard model is that it contains a large number (about 35) of *experimental parameters*, i.e. parameters that we cannot predict from the theory but have to be obtained by experimental measurements. These parameters include the so-called *coupling constants* that fix the strengths of the interactions, and the 30 or so other parameters including the masses of the quarks and leptons.

One of the reasons why this number of experimental parameters is so high is that the standard model effectively *combines* two older theories (QCD and the electroweak theory) rather than truly *unifying* them in a comprehensive theoretical structure.

## GUTs and strings

Attempts have been made to construct so-called **grand unified theories (GUTs)** that would truly unify the strong and electroweak interactions. Such attempts are encouraged by the fact that the three effective coupling constants that enter into the standard model have values that actually depend on energy, and which show signs of naturally coming together at a very high energy (about $10^{19}$ GeV) usually known as the grand unification scale (Figure 4.33).

**Figure 4.33** Grand unification of the strong, electromagnetic and weak interactions. At very high energies, the strong interaction appears to get weaker and the weak and electromagnetic interactions get stronger, so that they should have the same strength at a certain very high energy known as the grand unification scale. Superunification refers to the unification of all the fundamental forces, including gravitation. The point marked 'superunification (?)' needs something we lack: a viable quantum theory of gravity. This is important at energies even higher than the proposed scale of grand unification.

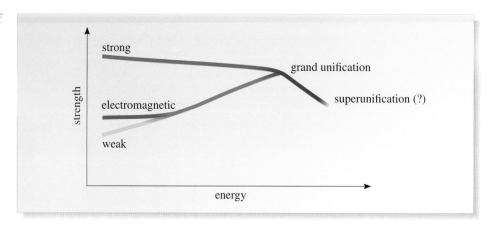

An even more ambitious attempt to go beyond the standard model is being made by those who would like to construct a theory that also includes the gravitational interaction. Theories of this kind, often referred to as **super unified theories** and also known, rather optimistically, as 'theories of everything', have been under active discussion for a number of years. Several have been based on an assumed new symmetry of nature called **supersymmetry** which, if it really exists, implies that the known particles should each have a 'superpartner' of different spin. No evidence to support this contention has yet been found, but the superpartners might be very heavy, putting them beyond the range of current experiments, so the search continues. At one time, it was hoped that a quantum theory involving supersymmetry but based on extended objects called **strings** might provide a realistic superunified theory. The initial enthusiasm for such theories waned when it became clear that there were a range of apparently self-consistent string theories, rather than one unique theory. However, interest revived when it was discovered that the different string theories might be viewed as different aspects of a single theory known as *M-theory*. At the time of writing, this remains one of the most active frontiers of research in mathematical physics.

## 5.3 Optional multimedia activity

Open University students may leave the text at this point and do the optional multimedia package *Quarks*. If you have done this package as part of S103, then you may skip it here. If you choose to do it now, you should return to the text on completion. The activity takes about one hour.

# 6 Closing items

## 6.1 Chapter summary

1  Particle physics concerns the fundamental constituents of matter and the interactions between them.

2  The subatomic particles, such as the electron, proton and neutron, can be characterized by their values of certain basic properties such as mass, spin and charge.

3  The high-energy particles that are used for collision studies in experimental investigations of particle physics may be obtained naturally from cosmic rays or produced artificially in linear accelerators (linacs) or cyclic accelerators such as the cyclotron and synchrotron.

4  When particles collide, certain quantities such as total electric charge, total (relativistic) momentum and total (relativistic) energy are always conserved. The particles may undergo elastic collisions where the colliding particles simply exchange

energy and momentum, or inelastic collisions where (relativistic) kinetic energy may be converted into rest energy and so the nature and number of particles may change.

5   The likelihood of a particle being scattered in a given process at a specified energy is described by a measurable quantity called a cross-section. Cross-sections are measured in barns (1 barn = $10^{-28}$ m²).

6   There are four fundamental forces or interactions through which particles can interact: the strong, electromagnetic, weak and gravitational interactions. These are summarized in Table 4.2.

7   Experiments indicate that all fundamental particles of matter that have been observed can be assigned to one of three families:

(i) *leptons*: a family of six particles that includes the electron, muon and tauon, together with the corresponding types of neutrino. Leptons are spin 1/2 particles and have lepton number $L = 1$. There is a corresponding family of 6 antileptons with $L = -1$. Rather than using a lepton number $L$, it is possible to identify three separately conserved lepton numbers $L_e$, $L_\mu$ and $L_\tau$ called electron number, muon number and tauon number, respectively. Leptons feel the weak interaction and the charged ones also feel the electromagnetic interaction, but leptons do not feel the strong interaction.

(ii) *baryons*: an extensive family of particles that have half odd-integer spin (1/2, 3/2, etc.) and baryon number $B = 1$. This family includes the proton and the neutron. There is a corresponding family of antibaryons with $B = -1$.

(iii) *mesons*: an extensive family of particles that have zero or integer spin (0, 1, 2, etc.) and baryon number $B = 0$. This family includes the pion and kaon.

The baryons and mesons are collectively referred to as hadrons. They can interact through the strong interaction as well as the weak interaction and, if charged, the electromagnetic interaction.

In addition, there are exchange particles associated with the fundamental forces: the photon, the $W^+$, $W^-$ and $Z^0$ particles, and the eight kinds of gluon. The exchange of these spin 1 particles is thought to account for the operation of the electromagnetic, weak and strong interactions.

8   The leptons are, as far as is currently known, truly fundamental particles that have no more basic constituents. There are six flavours of leptons grouped into three generations of increasing mass (Figure 4.2a and Table 4.5). The baryons and mesons, i.e. the hadrons, are not thought to be fundamental. Rather, they are generally supposed to be composed of strongly interacting particles called quarks, that are bound together by the exchange of gluons. According to the simplest version of the quark model, each baryon is composed of three quarks, and each meson is composed of a quark and an antiquark.

9   There are six flavours of quark (up, down, charm, strange, top and bottom) and six corresponding kinds of antiquark. Like the leptons, the quarks are grouped into three generations of increasing mass (Figure 4.2b and Table 4.6) All quarks have $B = 1/3$ and all antiquarks have $B = -1/3$. Some other properties are summarized in Table 4.6.

10  The strong, electromagnetic and weak interactions of the fundamental particles (and their antiparticles) may be described theoretically by quantum field theories in which the forces are mediated by exchange particles. These quantum field theories have been developed into the standard model of particle physics. This is not thought to be the 'final' theory, but its construction is regarded as a major triumph since, by using Feynman diagrams and other techniques, it permits the evaluation of measurable quantities such as cross-sections and mean lifetimes. Going beyond the standard model might require the formulation of a grand unified theory, or a string theory, and might involve supersymmetry.

## 6.2 Achievements

Now that you have completed this chapter, you should be able to:

A1 Explain the meaning of all the newly defined (emboldened) terms introduced in this chapter.

A2 Outline the means by which high-energy particles may be produced and detected in experiments. Also, briefly describe the specific circumstances that led to the discovery of a number of elementary particles.

A3 List the fundamental forces in order of strength, describe the basic features of each and briefly outline the exchange theory of fundamental forces.

A4 Explain the importance of high-energy collisions in the investigation of particle physics.

A5 Recall the conservation laws which apply in particle collisions and decays and, given appropriate lists of particle properties, identify those processes that are forbidden by specific conservation laws.

A6 Given appropriate information about specific elementary particles, identify those particles as leptons, mesons, baryons or exchange particles.

A7 List those particles (leptons and quarks) that are currently believed to be truly fundamental, and explain their relationship to the elementary particles that are directly observed in experiments.

## 6.3 End-of-chapter questions

**Question 4.11** What is the main difference between a linac and a synchrotron? What advantage is there in using a cyclic accelerator rather than a linear one? What are the disadvantages?

**Question 4.12** List the four fundamental forces and write down the basic properties of each.

**Question 4.13** Which of the following strong interaction processes are forbidden by conservation of charge, baryon number or strangeness? Refer to Table 4.3.

(a) $\pi^- + p \rightarrow n + p$

(b) $\Sigma^- + p \rightarrow K^0 + n$

(c) $K^- + p \rightarrow K^0 + n$.

**Question 4.14** Identify each of the following particles as either an exchange particle, a lepton, a meson or a baryon:

(a) muon; (b) gluon; (c) pion; (d) photon; (e) $Z^0$.

**Question 4.15** The $D^+$ meson has spin 0, charge number $Q = 1$ and charm $C = 1$. Which of the following represents a plausible quark model of the $D^+$? Explain your answer by referring to Table 4.6.

(i) uuc; (ii) udc; (iii) $s\bar{s}c$; (iv) $c\bar{d}$; (v) $c\bar{u}$; (vi) $d\bar{c}$; (vii) $s\bar{c}$.

(c)

# Chapter 5   Consolidation and skills development

## 1   Introduction

This final chapter has two aims. First, it will review some concepts that have appeared in this book. We will not attempt to give a comprehensive summary — for that, you can turn to the summaries that accompany each chapter. Instead, Section 2 will review some key topics while stressing the quantum-mechanical ideas we have used to analyse them. The realization that a few basic ideas can be used again and again is very important in physics. A sort of creative borrowing exists throughout the subject with, for example, ideas developed by particle physicists being adapted for use by solid-state physicists, and concepts invented by solid-state physicists being used by nuclear physicists, and vice versa. The fact that this is possible and valuable helps to bind physics together as a single subject, with a common set of concepts and techniques.

The second aim of this chapter is to develop skills needed to revise the course and sit the final examination, a subject that may be quite close to your heart at the moment! These skills will be discussed in Section 3. From time to time, we will refer to examinations taken in the Open University context, concentrating on the S207 exam in particular. Nevertheless, most of our comments are quite general and will apply to other exams taken elsewhere, so you should find this section useful, even if you are not an Open University physics student.

Finally, Section 4 is a collection of short questions designed to give you the chance to consolidate your understanding of the material covered in this book. Section 5 extends this consolidation by means of the interactive questions package for *Quantum physics of matter*.

There are no *Physica* questions associated with this book.

## 2   Overview of Chapters 1 to 4

This book has given you a second look at the physics of matter. Earlier in the course, we looked at matter from an entirely classical perspective, and introduced the subjects of classical statistical mechanics, thermodynamics and fluid mechanics, all of which could be described without any hint of quantum mechanics. However, classical physics is not an adequate theory of Nature. At the beginning of the book *Quantum physics: an introduction* you saw many reasons for rejecting classical physics and starting afresh with quantum mechanics. The present book has been concerned with developing an understanding of the properties of matter based on quantum mechanics.

In order to provide a fresh overview of the subject matter of this book, we shall look at some of the distinctive features of quantum physics, and discuss how they are reflected in the properties of matter.

The most obvious arena in which to look for quantum-mechanical effects is that of tiny objects, such as atoms or atomic nuclei. The main reason for this is the extremely small value of Planck's constant. For example, a component of angular momentum is quantized in units of $\hbar = 1.055 \times 10^{-34}\,\text{kg}\,\text{m}^2\,\text{s}^{-1}$, whereas a typical angular momentum encountered in ordinary life is likely to be billions of billions of billions of times larger. By going down to the scale of atoms, nuclei and fundamental particles, we invariably encounter angular momenta that are comparable to $\hbar$. The last two chapters of this book are concerned with such microscopic systems — nuclei and fundamental particles — where quantum mechanics is bound to be important.

Quantum-mechanical effects also make their presence felt in much larger systems, including macroscopic samples of gases and solids. For example, the quantum-mechanical effects that underlie the behaviour of electrons in a copper wire affect the conductivity of the wire, which is a macroscopic property. Some quantum phenomena are inherently macroscopic in scale. This is the case for Bose–Einstein condensation, for example, where the atoms congregate in a single quantum state and become so correlated that they can be described by a single wavefunction.

Before embarking on our survey of key topics in the physics of matter, it may be helpful to make a list of features of quantum mechanics that distinguish it from classical physics. For example,

1   Quantization
2   Uncertainty
3   Indeterminacy
4   Tunnelling
5   Wave-like behaviour of particles
6   Particle-like behaviour of waves
7   Indistinguishability of identical particles
8   Fermions and the Pauli exclusion principle
9   Bosons and Bose–Einstein condensation

We merely log these features now, but will revisit them in the context of reviewing the physics of matter. A series of marginal notes will indicate where these concepts are used.

## 2.1  Quantum gases

The first chapter of the book concentrated on quantum gases. Another way of describing its contents would be to say that it developed the subject of quantum statistical mechanics.

uncertainty

Quantum mechanics certainly has a profound effect on the way we describe gases. A classical description of the configuration of a gas relies on saying which particles occupy which phase cells, but the quantum-mechanical concept of *uncertainty*, embodied by the Heisenberg uncertainty principle, seriously compromises the concept of a phase cell, and places limits on how small a phase cell can be. This is because simultaneous knowledge of position and momentum is beyond our reach.

quantization

Fortunately, another aspect of quantum mechanics — *quantization* — comes to the rescue. Instead of specifying the configuration of a gas in terms of phase cells, we do so in terms of quantum states. The translational energy of a particle of mass $m$ (such as a molecule or an electron) in a cubic box with sides of length $L$ is

$$E = \frac{h^2}{8mL^2}\,(n_1^2 + n_2^2 + n_3^2). \tag{5.1}$$

The quantum numbers, $n_1$, $n_2$ and $n_3$ have possible values 1, 2, 3,…and serve to label a specific quantum state. The configuration of a gas is then defined by saying how the various quantum states are occupied by particles.

Although configurations, specified in terms of quantum states, are important from a conceptual point of view, they are not directly observed in most gases. We are usually interested in gases that are in a state of thermal equilibrium, and generally require much less information than is provided by a configuration at a given instant. Often we just want to know the average number of particles with translational energies in a small range, from $E$ to $E + \Delta E$. In most circumstances, this can be calculated using the classical continuum approximation, where the discrete nature of the energy levels is ignored. It turns out that the classical continuum approximation is valid, provided that the typical de Broglie wavelength of a particle is much less than the size of the container

$$\lambda_{dB} \ll L. \tag{5.2}$$

This criterion is, of course, phrased in terms of another striking quantum-mechanical property — the ability of particles to behave as *waves*.

wave-like behaviour of particles

We can represent the distribution function for translational energy as the product of two factors. When the criterion of Equation 5.2 is satisfied, the distribution function is:

$$G(E) = D(E) \times F(E). \tag{5.3}$$

The first of these factors, $D(E)$, is known as the density of states function. It has the significance that $D(E)\,\Delta E$ is the number of quantum states with translational energies between $E$ and $E + \Delta E$. The second factor, $F(E)$ is called the occupation factor. It represents the average number of particles in each quantum state.

When the density of states function is calculated for particles like molecules that have mass, it turns out that in the classical continuum approximation $D(E) \propto \sqrt{E}$. For particles like photons that have no mass, it has the form $D_p(E) \propto E^2$. It is not unreasonable for different functions to apply in these cases because a photon trapped in a box has an entirely different set of energy levels to a molecule. Certainly, the energy levels for a photon cannot be described by Equation 5.1, since putting $m = 0$ in that equation would lead to infinity, which is impossible. (The corresponding equation for the allowed energy levels of a photon in a box was not given in the text.)

A major quantum-mechanical issue arises when we consider the form of the occupation factor, $F(E)$. The issue is that of indistinguishability. In quantum mechanics, identical particles are completely indistinguishable from one another. This fact fits in perfectly with other aspects of quantum mechanics. For example, uncertainty makes it impossible to keep track of particles by means of their trajectories (because the trajectories don't exist or, at least, are fuzzy). Moreover, the wave-like property of particles implies that waves can overlap so, when particles approach one another, we can no longer tell which is which. Even quantization is relevant. Two electrons have precisely the same charge because charge itself is quantized — you will not find an electron that is distinguished from others by having 99.9% of the normal electronic charge.

identical particles

Even granted all these things, you might still wonder why complete indistinguishability should matter. At first sight, the distinction between 'almost the same' and 'exactly the same' may seem to be academic. It isn't! The reason is that indistinguishability destroys any possibility of labelling the particles and therefore affects what we can mean by a configuration. In classical statistical mechanics, a configuration is specified by stating *which* particle is in which phase cell. In quantum physics, we cannot presume such detailed knowledge, so a configuration is specified by saying *how many* particles are in each quantum state. We continue to assume that all configurations of the required total energy are equally likely but, now configurations are counted in a totally different way, and this affects the average numbers of particles in any given quantum state. It directly affects the function occupation factor.

According to quantum mechanics, particles can be divided into two broad categories. Fermions (such as electrons, protons and neutrons) obey the Pauli exclusion principle — there can be no more than one identical fermion in a given quantum state. Bosons (such as photons and $^4_2\text{He}$ atoms) do not obey the Pauli exclusion principle, and occupy quantum states without restriction. Corresponding to these categories, there are two types of occupation factor in quantum mechanics:

For identical fermions in equilibrium at temperature $T$,

$$F_\text{F}(E) = \frac{1}{e^{(E-\mu_\text{F})/kT} + 1} \qquad \text{(Fermi occupation factor).} \qquad (5.4)$$

For identical bosons in equilibrium at temperature $T$,

$$F_\text{B}(E) = \frac{1}{e^{(E-\mu_\text{B})/kT} - 1} \qquad \text{(Bose occupation factor).} \qquad (5.5)$$

The quantities $\mu_\text{F}$ and $\mu_\text{B}$ are characteristic energies for the gas generally called the *chemical potentials*. In general, they depend on the number density of particles, their mass and the temperature.

Under certain circumstances, both $F_\text{F}(E)$ and $F_\text{B}(E)$ become proportional to the Boltzmann factor, $e^{-E/kT}$. In this limiting case, it does not matter whether the particles are fermions or bosons — they both behave in the same way and occupy states as if they were distinguishable particles. For this to be possible, the quantum states must be occupied very sparsely. More exactly, we require that

$$\lambda_\text{dB} << d, \qquad (5.6)$$

where $\lambda_\text{dB}$ is a typical de Broglie wavelength of a particle, and $d$ is the typical spacing between particles. (For a gas of photons we interpret $\lambda_\text{dB}$ as the typical wavelength $\lambda$ of the electromagnetic radiation.) If this condition is met, the gas can be treated as if it were a collection of distinguishable particles, obeying the Maxwell–Boltzmann distribution.

In pursuing the influence of quantum mechanics we are interested in cases where Equation 5.6 is not satisfied, and indistinguishability is important. The photon gas of thermal radiation provides a striking example. This is the radiation inside a cavity, which has come into thermal equilibrium with the cavity walls. For photons, $\mu_\text{B} = 0$, and the Bose occupation factor becomes

$$F_\text{B}(E) = \frac{1}{e^{E/kT} - 1}. \qquad (5.7)$$

The distribution function for photons in thermal radiation is then given by Planck's radiation law:

$$G_\text{p}(E) = \frac{CE^2}{e^{E/kT} - 1}, \qquad (5.8)$$

where $C$ is a constant.

Planck's radiation law is quite unlike the Maxwell–Boltzmann energy distribution. The number of photons (given by the total area under the distribution curve) is not constant, but increases as $T^3$, and the total energy of the photon gas increases very rapidly, as $T^4$. A photon gas never behaves like a classical gas because, in all circumstances, Equation 5.8 implies that

$$\lambda = \lambda_{dB} \approx d, \tag{5.9}$$

and the quantum states are not sparsely occupied (this is demonstrated in Question 1.13).

Gases of boson *atoms* behave differently from photon gases because the number of atoms in a given sample is fixed. Spectacular effects are observed when a gas of boson atoms is cooled sufficiently for the typical de Broglie wavelength of an atom to be comparable with the interatomic spacing, i.e. Equation 5.6 is *not* satisfied; this is a regime where quantum-mechanical effects are certainly important. Around this temperature, Bose–Einstein condensation occurs — the gas makes a transition into a phase where a significant number of atoms share the same, low-energy quantum state. These atoms are said to form the Bose–Einstein condensate and can be described by a single wavefunction. Similar effects are observed in liquid helium-four, although this case is complicated by the existence of large interatomic forces.

The most famous example of a fermion gas is the free-electron gas model used to understand the behaviour of metals. The typical de Broglie wavelength of electrons is comparable with atomic dimensions, so Equation 5.6 is not satisfied and this gas must be treated quantum-mechanically. Strictly speaking, electrons in solids interact powerfully, but the free-electron gas model neglects these interactions, and treats the electrons as a gas of non-interacting fermions. Over the typical range of temperatures of interest in metals (say from 0 K to 1000 K) the characteristic energy $\mu_F$ of the electron gas changes hardly at all with temperature. We define the Fermi energy, $E_F$ to be the value of $\mu_F$ at 0 K. To an excellent approximation, we can then replace $\mu_F$, by the constant $E_F$, and write the occupation factor for electrons as

$$F_F(E) = \frac{1}{e^{(E-E_F)/kT} + 1}. \tag{5.10}$$

The corresponding distribution function for the electron gas is

$$G_e(E) = \frac{B'\sqrt{E}}{e^{(E-E_F)/kT} + 1}, \tag{5.11}$$

where $B'$ is a constant. This is Pauli's distribution for the electron gas.

Pauli's distribution incorporates the Pauli exclusion principle — no quantum state accommodates more than one electron. The low-energy quantum states are almost completely full but at a few $kT$ below the Fermi energy, the average number of particles per state starts to drop perceptibly below 1. At a few $kT$ above the Fermi energy, the quantum states are almost completely empty. Because of the restriction implied by the Pauli exclusion principle, and the need to find quantum states to accommodate all the electrons, the Fermi energy is high — around $10\,eV$ for most metals. This is much larger than the average energy per particle in a Maxwell–Boltzmann gas at room temperature, and leads to a correspondingly large pressure, which partly accounts for the hardness of metals. Pauli's distribution function, Equation 5.11, varies slowly with temperature — the changes involve mainly the occupation of states around the Fermi energy — and this helps to explain why the electron gas contributes little to the heat capacity of a metal (an unanswerable mystery in classical physics).

## 2.2 Solid-state physics

Many of the quantum properties of solids are related to the behaviour of electrons. For example, a covalent bond is formed when a pair of electrons is shared between two atoms. The wavefunctions of these electrons can interfere constructively or destructively. Constructive interference leads to a concentration of electrons between the positive ions, and binds the atoms together. This is an intrinsically quantum-mechanical effect, since it can be shown that an entirely classical system of positive and negative charges would be unstable, and could not bind together in this way. Metallic bonding also relies heavily on quantum-mechanical effects. Electrons become dissociated from their original atoms and free to roam over the whole metal. This leads to a decrease in their kinetic energy (since, effectively, $L$ in Equation 5.1 has increased from an atomic dimension to the size of the whole metal.) While this is offset, to some extent, by the requirements of the Pauli exclusion principle, the result is still a lowering in energy which binds the whole metal together.

Drude's classical free-electron model of electrical conduction, ran into a number of serious difficulties, including an incorrect prediction for the temperature dependence of resistivity and the need to assume that electrons have implausibly long mean free paths as they navigate their way through the lattice of positive ions.

wave-like behaviour of particles

Pauli's quantum free-electron model provides considerably more insight into electrical conduction in metals. In quantum mechanics, electrons behave as waves, so it is not surprising that they can pass through the lattice of positive ions without being scattered. If the lattice is perfect, and the very special conditions needed for Bragg reflection are avoided, most electrons pass through the metal without hindrance. Electrical resistance occurs because *imperfections* in the lattice scatter electrons. Defects, such as missing atoms or dislocations, give a temperature-independent contribution to the resistivity. In addition, the positive ions vibrate as a result of their thermal energy, and this departure from perfect order gives a contribution to the resistivity that is proportional to the absolute temperature.

wave-like behaviour of particles

quantization

Pauli exclusion principle

In spite of its merits, Pauli's quantum free-electron model has its failings. Crucially, it fails to predict whether a given material will be a conductor, an insulator or a semiconductor. Much better insight into electrical conduction is provided by the band theory of solids, which recognizes that the electron energy levels are organized into distinct bands, separated by energy gaps. The origin of the energy gaps can be understood in different ways, either as a consequence of Bragg reflection, which prevents electrons of certain energies from travelling through the crystal, or as a remnant of the gaps that occur between atomic energy levels. Either way, the Pauli exclusion principle means that it is possible to have an energy band that is completely full, so that no more electrons can be added to it.

A completely full band cannot conduct electricity. At a finite temperature, some electrons might be thermally excited across the energy gap between full and empty bands and so allow electricity to be conducted. Whether this happens or not depends on the size of the energy gap and on the temperature. At room temperature, a gap of more than about 3.0 eV corresponds to an insulator, a gap of between about 0.1 eV and 3 eV corresponds to a semiconductor, and a gap of less than about 0.1 eV to a conductor. Most conductors have a partly-filled energy band and so are able to conduct electricity even at 0 K.

In a semiconductor, conduction can be by means of electrons in the conduction band, or holes in the valence band. The concept of a hole arises because a nearly-full band can be regarded as a completely full band, plus a few positively-charged particles — the holes. Because the completely full band does not produce any current, it is ignored, and we need only consider the holes. In an intrinsic semiconductor, such as pure silicon, equal numbers of electrons and holes are generated by thermal excitation of electrons from the valence band to the conduction band. In an extrinsic semiconductor, the majority of charge carriers are produced by impurity atoms. Donor atoms, such as arsenic, can donate electrons to the silicon conduction band, producing an n-type semiconductor. Acceptor atoms, such as boron, can accept electrons from the silicon valence band, effectively creating holes in the valence band producing a p-type semiconductor.

The electrical behaviour of insulators, conductors and semiconductors is dominated by the fact that electrons and holes are fermions, and so obey the Pauli exclusion principle. However, if two fermions bind together, the composite particle produced is a boson. This is what happens in a superconductor. Pairs of electrons, with opposite momenta and spins, form composite particles called Cooper pairs. This has a profound effect on the electrical properties of a superconductor since the Cooper pairs undergo a condensation analogous to Bose–Einstein condensation, and enter a highly-ordered state in which currents can flow without electrical resistance. Special devices known as SQUIDS, can be constructed from superconductors. These devices rely on another quantum-mechanical effect — the ability of Cooper pairs to tunnel from one superconductor into another, passing through a thin insulating barrier. They are used to provide fast switches, and to measure tiny changes in magnetic fields.

*bosons*

*quantum-mechanical tunnelling*

## 2.3 Nuclear physics

Sometimes, it is worth standing back to note with wonder what has been discovered. It is remarkable enough that humans have discovered the existence and properties of atoms each with a volume of order $10^{-29}$ that of a person, but to go further and discover the detailed properties of nuclei, which have volumes $10^{-12}$ or so smaller still, is a truly impressive feat. The reliability of quantum physics and relativity has certainly been a factor in this success, allowing physicists to link together the clues provided by Nature and carefully designed experiments, in order to obtain a rounded and satisfying theory.

The sizes of nuclei can be explored by a number of experimental techniques. Atomic electrons, especially those in an s-state, can penetrate inside the atomic nucleus and this has a small influence on the energy levels of the electrons, allowing nuclear sizes to be inferred from close analysis of spectral lines. More direct measurements involve deliberately scattering beams of particles such as α-particles or electrons from nuclei. The patterns that emerge show that all nuclei have approximately the same density, so nuclei grow steadily in size as the number of nucleons increases. The constant density arises because the strong nuclear force, which binds nucleons together, has a very short range, so nucleons interact only with their immediate neighbours. The force is also very strongly repulsive at short distances, so nuclear matter cannot be compressed.

Nuclei are capable of extremely energetic reactions. Since nucleons are bound within the very small volume of the nucleus, their wavelengths must be very small. By de Broglie's relation, short wavelengths imply high momenta and hence high kinetic energies. Put another way, the positions of nucleons are known extremely well — they are within the tiny volume of the nucleus. By the uncertainty principle, their momenta must be very uncertain and this can only be achieved if the nucleons have high kinetic energies. The high kinetic energies of the nucleons are more than offset by the negative potential energy due to the strong nuclear force, which binds the nucleus together. Because such large energies are involved, a nucleus has noticeably less mass than the sum of the masses of its constituent nucleons.

*wave-like properties of particle*

*uncertainty principle*

The binding energy per nucleon varies in a systematic way for different nuclei, starting from a low value for very light nuclei, reaching a peak around iron ($Z = 56$) and then gradually tailing off for heavier nuclei. This pattern can be explained by two very different types of model. The semi-empirical model treats the nucleus as if it were a charged drop of water. The nuclear shell model is a fully quantum-mechanical approach, based on Schrödinger's equation for a nucleon moving through a potential due to the average influence of all the other nucleons. As in an atom, wavefunctions and their corresponding energies can be calculated, and it turns out that the quantum states can be grouped into sets of well-defined energies. Because neutrons and protons obey the Pauli exclusion principle, a shell-like structure is formed, reminiscent of that found in atoms. When a particular shell is filled (either for protons, or neutrons, or both) the nucleus is especially stable. The special numbers of protons and neutrons needed to achieve this are known as *magic numbers*.

The Pauli exclusion principle has another important effect. Imagine constructing a nucleus by adding protons and neutrons one at a time. If there were far more protons than neutrons, the final proton would go into an extremely high energy state, but the final neutron would go into a lower energy state. This ensures that the most stable nuclei have approximately equal numbers of protons and neutrons. Usually, there are slightly more neutrons than protons, because protons experience Coulomb repulsion which increases their energies. If there are too many neutrons for stability, the nucleus may undergo $\beta^-$-decay, converting a neutron into a proton, an electron and an antineutrino. If there are too many protons for stability, the nucleus may undergo $\beta^+$-decay, converting a proton into a neutron, a positron and a neutrino, or it will perform electron capture of an atomic electron and emit a neutrino.

indeterminacy

Radioactive decay occurs spontaneously. The probability that a nucleus will decay in a given time interval can be predicted using quantum mechanics, but the actual time of decay of any particular nucleus cannot be predicted. This apparently inescapable feature of quantum mechanics is known as indeterminacy. Consequently the radioactive decay law (Equation 3.5) predicts only the *average* number of radioactive nuclei that are present in a sample at any time.

In addition to $\beta$-decay, other types of radioactive decay are possible, Some nuclei emit $\alpha$-particles ($\alpha$-decay). Others, such as uranium, split into two fragments of similar mass. This process is called fission and is used as a source of power in nuclear reactors and atomic bombs. Spontaneous fission is rare, but the absorption of a neutron by a uranium nucleus can trigger break-up of the nucleus, leading to the production of more neutrons. Under the right conditions, a growing chain reaction takes place. Some light nuclei can fuse together, producing more tightly-bound nuclei and releasing considerable energy. This is the fusion process, used in hydrogen bombs but not yet tamed in commercial reactors.

tunnelling

Many nuclear reactions could never take place classically, because there is not enough energy available to get them started. In quantum mechanics, however, these reactions can take place, thanks to quantum-mechanical tunnelling. Tunnelling occurs in $\alpha$-decay, allowing $\alpha$-particles to tunnel out of unstable nuclei. It also occurs in the fission of uranium nuclei, and in the fusion of two protons that is the key step in a chain of nuclear reactions that provide the Sun's power.

## 2.4  Particle physics

uncertainty principle

Knowledge about particle physics has been even more hard-won than that about nuclei. One difficulty is that many of these particles are extremely short-lived, with lifetimes of the order $10^{-24}$ s. Such a particle leaves no measurable track in a particle detector, but its presence can be inferred from a bump (or resonance) in a graph of cross-section against beam energy. The energy–time uncertainty principle relates the mean lifetime of the corresponding composite particle to the width of the resonance.

The simplest aspects of particle physics are those to do with listing fundamental particles, classifying them into families, identifying their interactions and properties, and any conservation laws they must obey. The broad picture that emerges is relatively simple.

We can distinguish four types of interaction. In order of decreasing strength they are: the strong interaction, the electromagnetic interaction, the weak interaction and gravity.

Next, we can classify elementary particles into two main groups: hadrons and leptons. A third group consists of exchange particles or gauge bosons; they have a particular role in mediating interactions, a point we will come back to soon. Hadrons are made up of quarks. They can be either mesons (zero spin or integer spin bosons containing two quarks) or baryons (half-odd integer spin fermions containing three quarks). Irrespective of any other interactions they may feel, all hadrons, and no leptons, experience the strong interaction. All of these particles can be assigned quantum numbers, such as baryon number, lepton number and charge, which are conserved in all known processes. Other quantum numbers, such as strangeness number, are conserved only in certain types of process. These quantum numbers are, of course, an example of quantization.

quantization

The physics of fundamental particles is entirely quantum-mechanical. However, the quantum mechanics of particles, as exemplified by Schrödinger's equation, does not provide a suitable starting point. What is needed is a quantum theory that is fully consistent with special relativity. The details of such quantum field theories, as they are known, lie far beyond the scope of undergraduate physics, but we can outline some of their main features.

First, each type of particle is associated with a field — that is, a quantity that has a definite value at each point in space at a given time. For example, an electron will be described by an electron field. The field is not an immediately observable quantity (like an electric field or a magnetic field) but is more like a wavefunction — something that exists in the theory and is related to other quantities that can be measured. The field has various modes, which can be thought of as standing waves. Particles (such as electrons, quarks and photons) emerge as quanta of excitation of modes of these fields, with attributes of the particle (such as momentum, energy etc.) derived from attributes of the mode. One of the consequences of describing particles in this way is that they can appear or disappear as the excitation of different field modes changes.

The interaction between particles is associated with the emission and absorption of particles known as exchange particles, or gauge bosons. Each type of gauge boson is responsible for conveying a certain type of interaction. For example, photons convey electromagnetic forces, so when two electrons interact electromagnetically, one electron emits a photon and another electron absorbs it. In analysing the probability that a certain process will take place, it is necessary to list all possible sub-processes that contribute to that process. The enumeration of these sub-processes is greatly simplified by drawing diagrams, called Feynman diagrams, each of which corresponds to a particular sub-process. The contributions are added together, and then finally squared to find the probability of the process. The fact that Feynman diagrams are added *before squaring* leads to the possibility of interference between

interference

different contributions. Sometimes, different sub-processes will reinforce one another, and sometimes they will cancel out. This is directly analogous to the phenomenon of interference observed with waves.

In the standard model there are three distinct generations of fundamental particles. Each generation consists of two quarks, a lepton and a neutrino, giving 12 fundamental particles in all, plus their 12 antiparticles. In addition, the gauge bosons are classified as eight gluons (associated with the strong interaction), the $W^+$, $W^-$ and $Z^0$ particles (associated with the weak interaction) and the photon (associated with the electromagnetic interaction).

The standard model already unifies the electromagnetic and weak interactions into the electroweak interaction. It is hoped that grand unified theories will go further by unifying the electroweak interaction and strong interaction. Ultimately, superunified theories may unify all four interactions.

# 3 Preparing for and taking exams

Few people enjoy exams. At best, they are regarded as obstacles to be overcome. For Open University students, the situation is alleviated by the existence of continuous assessment, with the final grade on a course being determined from the exam mark *and* the continuous assessment mark. Nevertheless, exams still have to be sat, passed, and preferably passed well, so it is worth considering how you are going to approach them.

No form of assessment — whether essay writing, problem-solving, practical tests, oral or poster presentations, interviews or exams — can be a neutral test of pure knowledge. Sitting an exam involves a number of skills, such as those concerned with written communication and time-management. The good news is that, by approaching exams and revision in an organized way, you will be able to maximize your potential, and are likely to score higher than others who know as much as you, but cannot communicate their knowledge as effectively to the examiners.

From the outset, it is important to realize that the examiners will not try to catch you out with inconsequential questions about minor details of the course. Rather, they want to see if you have grasped the meaning and significance of the most important principles in the course. The issue is not whether you know the course by heart. In one sense, this would be far too much. For example, the exam paper of S207 contains a detailed list of equations, and you should always be able to recognize any results that you half-remember. But, in another sense, learning the course by rote would be far too little. You will need to understand the significance of key principles, so that you can choose which ones are relevant in a given situation, and can apply them to make specific deductions.

## 3.1 Before the exam

About four weeks before the exam you should start to revise. The first decision to make is what to revise — the part of the course you are most familiar with, the part you are least familiar with, or the whole lot?

In the S207 exam, more than half the marks come from short questions that cover the whole course and offer no choice. Less than half the marks come from longer questions where you do have a choice. In these circumstances, we advise you to make sure you have a basic knowledge of material from all books of the course.

Assuming that you have read the books at some stage, this need not be too daunting. There are several ways of doing this:

- Read the summaries and achievements of each chapter. Underline anything that is not clear and, if necessary, read that part of the chapter again.

- Look up entries in the *Physica* Browser.

- Look at a selection of worked examples, including those obtained in *Physica* by clicking the *Do it* button.

- Look through TMA questions you have done, not worrying about mistakes but concentrating on methods to use.

- Read through the advice on problem-solving and writing definitions, given in Consolidation and skills chapters.

- Do not bother to commit a large number of equations to memory, but take careful note of the situations where different equations are used, and the precise wording of any definitions.

- The specimen exam paper provides invaluable information about the real exam. Familiarize yourself with the overall format of the exam:

    How many parts it is split into?

    How long you are advised to spend on each part?

    What choice will you have in selecting questions?

    Notice which forms to fill in, which answer books to use, etc. This can save five minutes of reading time in the real exam, and this is time which can be better used in other ways.

A very important part of your preparation is to attempt the specimen exam. The ideal would be to set aside a three-hour period and sit the exam under realistic conditions. If this is not possible, attempt one section at a time, allocating the time suggested on the exam paper. It is essential to tackle questions against the clock because this is one of the few chances you will have to gauge your speed of working. In the context of continuous assessment, you may have decided to work slowly and surely, checking your algebra line-by-line. Under exam conditions, you may not be able to afford the luxury of working in this way. The specimen exam, especially the problems in Part C, will indicate the sort of fluency that is expected, if you are not to lose marks through running out of time.

When answering questions on the specimen exam, use the list of equations provided with it. It is important to get some practice in using this list, so that you will know how to find specific equations on it. Also, before the exam, make sure you are familiar with your calculator. You will be at a disadvantage if you are not fluent with using the basic function keys. For example, you should be able to evaluate

$$6.778 \times 10^{-23} \times \log_e(4.671^{5/3} + \sqrt{3.567})$$

without too much difficulty and, ideally, without writing down any intermediate steps. You should also be able to use functions such as $\sin(x)$ and $\arcsin(x)$, with your calculator in degree mode or radian mode. It is a shame to have to repeat calculations several times in an exam, just because you are insecure with using your calculator.

## 3.2 In the exam room

An old Chinese exam, with no time limit, is said to have consisted of a single question:

'Write everything you know.

(Use the paper, ink and food provided.)'

Nowadays, exams are utterly different. Exam questions are highly specific, and you will have a limited time to construct your answers. This implies that you should answer the questions as directly as possible. If you cannot answer a specific question, you might be tempted to write a few paragraphs about the general subject area, just to show that you know something about it. This lapse into 'writing everything you know' is a very bad tactic. You will be unlikely to pick up any marks this way, because the marking scheme will focus very directly on the specific question that is asked. Even worse, you will be using up time that you might have used more profitably to answer other questions.

The words in the question often give an important clue about the level of detail expected. If a question asks you to 'state whether such and such is true', you need only answer 'true' or 'false'. If it asks you to describe or explain something in a couple of sentences, you would not be expected to write a mini-essay spanning a page or more.

Your aim should be to tackle all the questions required by the rubric (but no more). In the case of multiple-choice questions in S207, no marks are deducted for wrong answers, so there is nothing to be gained by leaving an answer blank. If necessary, guess — after all, this will not be a random guess, but will be informed by your feeling that some options look more reasonable than others. For other questions it is worth bearing in mind that half the marks can often be gained rather quickly, though it may take much longer to make sure of full marks. So, if you find that shortage of time leads you to tackling fewer questions than you are asked to, it is likely that you have won your marks the hard way, and have dropped some marks you could easily have gained by at least making a start on a full set of questions. If you run short of time, keep your answers very brief by using bullet points or just writing down equations and saying what you would do with them if you had time.

Another point to be aware of is that examiners are not allowed to mark anything you have crossed-out. Even if you know your answer is wrong, there may be some merit in it, and some marks might be allocated. The best advice is not to cross anything out until you have replaced it by something better. Crossing-out is then important, to avoid presenting two contradictory answers.

Finally, remember that the examiners are on your side. They want you to succeed and, within the freedom provided by the marking scheme, will be looking for opportunities to give you marks, rather than take them away.

# 4   Revision questions

**Question 5.1**   A gas of helium molecules at 5 K, has a number density of $10^{28}\,\mathrm{m^{-3}}$, while a gas of bromine molecules at 300 K, has a number density of $10^{24}\,\mathrm{m^{-3}}$. Explain why the Maxwell–Boltzmann energy distribution would be expected to be more valid for the sample of bromine gas than for the sample of helium gas.

**Question 5.2**   Helium $^4_2$He atoms and photons are both examples of bosons. Would you expect a gas of $^4_2$He atoms and a gas of photons to obey the same energy distribution law at the same temperature? Give reasons for your answer.

**Question 5.3**   Show that the pressure of thermal radiation is proportional to $NkT/V$, where $N$ is the number of photons contained in a volume $V$ and $T$ is the absolute temperature. Is it valid to claim that the photon gas obeys the ideal gas equation of state?

**Question 5.4**   The measured heat capacities of metals show that the electrons make a much smaller contribution than that predicted by the Drude model and the equipartition of energy theorem. Explain in one or two sentences why the Pauli model solves this problem.

**Question 5.5**   A hypothetical gas consists of two identical particles. Each particle has five quantum states, of energy 0 J, $\varepsilon$, $2\varepsilon$, $3\varepsilon$ and $4\varepsilon$. The total energy of the gas is fixed at $4\varepsilon$. What is the probability of finding the two particles in the same quantum state if the particles are (i) indistinguishable fermions (ii) indistinguishable bosons (iii) distinguishable particles?

**Question 5.6**   Assume that the following gases all have the same number density, $10^{26}$ m$^{-3}$. Ignoring any potential energy contributions, which gas exerts the greatest pressure close to 0 K? (Both electrons and neutrons are spin 1/2 particles.)

(i) a gas of electrons

(ii) a gas of neutrons

(iii) a gas of helium $^4_2$He atoms

**Question 5.7**   For each of the following types of bonding in solids, write a sentence or two briefly describing its characteristics and explaining its origins: (i) covalent bonding (ii) ionic bonding (iii) metallic bonding.

**Question 5.8**   Briefly explain the role of cracks and dislocations in determining the mechanical strengths of solids.

**Question 5.9**   Contrast the explanations of electrical resistance given by Drude's free-electron model and Pauli's free-electron model, describing the extent to which these explanations are successful.

**Question 5.10**   Describe, in terms of energy levels and the band theory of solids, the main features of (i) a conductor (ii) an insulator (iii) an intrinsic semiconductor (iv) an n-type semiconductor (v) a p-type semiconductor.

**Question 5.11**   (a) Explain what is meant by the depletion region in a p–n junction. (b) Explain what is meant by the diffusion current and the pair current in a p–n junction. What are the directions and relative magnitudes of the diffusion current and the pair current (i) in an isolated p–n junction (ii) in a reverse-biased p–n junction (iii) in a forward-biased p–n junction?

**Question 5.12**   What is a Cooper pair and how does the formation of Cooper pairs help to explain the phenomenon of superconductivity?

**Question 5.13**   The following process is important in certain stars: $^4_2\text{He} + {}^{12}_6\text{C} \rightarrow {}^{16}_8\text{O} + \gamma$. Write down the atomic number and the number of neutrons for each of the three nuclei involved.

**Question 5.14**   The binding energy of a deuteron is 2.22 MeV. Could a $\gamma$-ray of energy 1.9 MeV photodisintegrate it? Could a photon of energy 3.2 MeV photodisintegrate it? In any case where photodisintegration occurs, what is the combined kinetic energy of the proton and neutron?

**Question 5.15**   When high-energy electrons are used to measure nuclear charge densities, the diffraction minima are not deep. What aspect of nuclear structure does this reflect?

**Question 5.16**   What method can be used to measure the size of unstable nuclei, that is nuclei which do not live long enough to be targets in an electron scattering experiment?

**Question 5.17**   Why, in a few words, does the half-life of an $\alpha$-particle-emitting nucleus depend so strongly on the energy of the $\alpha$-particle?

**Question 5.18**   What are the main contributions to the binding energy of a nucleus in the semi-empirical model? Name one important feature of nuclear stability which signals the limits of the semi-empirical model.

**Question 5.19**   For each of the classes of particles named in the following list, say whether the members are (a) all bosons (b) all fermions (c) a mix of bosons and fermions or (d) neither bosons nor fermions.

Quarks, leptons, exchange particles, hadrons, baryons, mesons.

**Question 5.20**   Which of the following quantities will be conserved in a collision process that involves only the strong interaction?

(a) total relativistic energy (b) relativistic kinetic energy (c) relativistic momentum (d) electric charge (e) baryon number (f) strangeness.

**Question 5.21**   State, in general terms, how hadron resonances were first discovered.

**Question 5.22**   What are the exchange particles that mediate a fundamental force in each of the following quantum field theories?

(a) QED (b) weak interaction theory (c) QCD.

**Question 5.23**   What, in broad terms, is the standard model of particle physics?

# 5 Interactive questions

Open University students should leave the text at this point and use the interactive questions package for this book. When you have completed these questions, you should return to the text. You should not spend more than 1 hour on this task.

There are no *Physica* questions associated with this book.

# 6 Postscript

So, there it is. Our circumnavigation of *The Physical World* is complete. If you have read all the books in this course, you will have a fairly detailed knowledge of the fundamental laws that bind this world together and make it a unified whole. We have illustrated these laws in different ways. Sometimes we have looked at simple but artificial examples — particles, ideal springs, perfect conductors and the like. Sometimes we have looked at experimental results — the Michelson–Morley experiment, Reynolds' experiments on turbulence or the Aspect experiment. Applications in everyday life and in industry have appeared in many parts of the course. But, through all this, we hope you will see that certain concepts and principles form a backbone of the subject.

These concepts, and the skills needed to apply them, lie at the core of physics. And because the central ideas of physics are so universal and broad-ranging, it is fair to say that physics has no firm subject boundaries. Ideas developed in quantum physics may be used in biology or chemistry, or may be used to develop a new electronic device. You should certainly not think of physics as that part of science that is left over after the biologists, chemists, astronomers etc. have marked off their patches. One of our colleagues in the Department of Physics and Astronomy at the Open University has worked on entirely different research problems in chemistry, biology, and Earth sciences, as well as in physics. This is not unusual, and indicates one reason why physicists are so valuable. Many interesting fields are opening up at the interfaces between different disciplines, and physicists will certainly take a prime role in developing these new areas.

At the same time, physics itself will develop and expand. Certainly, a great deal of effort is being spent trying to unify the laws of physics into a more coherent whole. This is part of a long historical development. Newton started the process by proposing a law of gravitation that would work throughout the Universe, for planets in the heavens as well as for apples on Earth. Electricity, magnetism and light were unified by Maxwell. Einstein insisted that the laws of physics should not depend on the observer's state of uniform motion, and so on. Nowadays, physicists are hoping to unify all four of the fundamental forces. The aim is always to simplify, to explain a wider set of phenomena with a smaller number of principles, and to show that apparently disparate facts are, after all, connected.

Another main concern of physicists is to understand complex phenomena. You have seen that situations that seem to be of hopeless complexity can often be tackled with a clever insight. Statistical mechanics provides one example of this, where accurate predictions can be made in spite of a lack of precise knowledge. In some cases, a well-chosen model or approximation can make all the difference. The simple gas model, the electron gas model and the nuclear shell model are all examples of this art.

Finally, at opposite ends of the spectrum, physics is related to technology and philosophy. In many ways, the physics of today will be the technology of tomorrow. There is often a considerable time-lag between a fundamental new physics discovery and its major impact on society. 30 to 40 years after the discovery of radio waves, broadcasts became widespread, and 30 to 40 years after the invention of solid-state electronics, society started to be transformed by the personal computer. But one thing is certain: without physics, neither of these revolutions in communications and information technology would have taken place. In the midst of these worldly concerns, physics also gives us a way of addressing questions of philosophical interest. Is the world really indeterminate? Is free will an illusion? Will it ever be possible to travel backwards in time? We do not claim that physics provides clear answers to these questions, but it casts light on them, and gives a framework within which they can be discussed.

In brief, we hope you have enjoyed the trip, and that you will be inspired to look deeper into some of the questions raised by this course.

John Bolton    Alan Durrant    Robert Lambourne    Joy Manners    Andrew Norton

Academic editors of *The Physical World*

# Answers and comments

**Q1.1** Refer to Equation 1.7. The typical spacing between translational energy levels is

$$\frac{h^2}{8mL^2} = \frac{(6.63 \times 10^{-34}\,\text{J s})^2}{8 \times (4.65 \times 10^{-26}\,\text{kg}) \times (10^{-1}\,\text{m})^2}.$$

$$\approx 1.2 \times 10^{-40}\,\text{J}.$$

At room temperature

$$kT \approx (1.381 \times 10^{-23}\,\text{J K}^{-1}) \times 300\,\text{K} \approx 4 \times 10^{-21}\,\text{J}.$$

Hence the criterion of Equation 1.7 is satisfied by 19 orders of magnitude (i.e. 19 powers of ten) and so the classical continuum approximation is extremely good in these circumstances.

**Q1.2** The continuum approximation can fail if the left-hand side of Equation 1.7 becomes large or the right-hand side becomes small. We note that $m$ and $L^2$ are in the denominator on the left-hand side of Equation 1.7. Hence small values of $m$, and especially $L$ (which is squared) help to make the left-hand side large. The temperature $T$ is in the numerator on the right-hand side and so low values of $T$ help it to fail also. Thus Equation 1.7 shows that the continuum approximation may fail for very small values of particle mass, container size or temperature.

**Q1.3** (a) The eight configurations are shown in Figure 1.33. The configurations do not all have the same total energy. For example, the total energy is zero when all three particles are in the state of lower energy, and $3\varepsilon$ when they are all in the upper state.

(b) There are twenty particles, each of which can be in a lower state or an upper state, and so the number of configurations is

$$2 \times 2 \times \ldots \times 2 = 2^{20} = 1\,048\,576$$

where there are twenty factors of 2 on the left-hand side of the immediately preceding equation.

***Comment:*** *A macroscopic sample of matter might contain $10^{23}$ particles, each of which has many possible quantum states. Clearly, the number of configurations of such a system is astronomically large.*

**Q1.4** (a) Three configurations correspond to particle A having energy $5\varepsilon$ with total particle energy $E_T = 9\varepsilon$ (Figure 1.34).

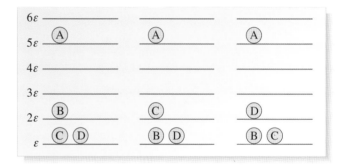

**Figure 1.34** Answer to Question 1.4(a).

(b) 15 configurations correspond to particle A having energy $2\varepsilon$ with total particle energy $E_T = 9\varepsilon$. They are shown in Figure 1.35.

(c) A value of $2\varepsilon$ for the energy of particle A is much more likely than $5\varepsilon$. This is because there are 15 configurations with particle A in the state of energy $2\varepsilon$, whereas there are only 3 configurations with particle A in the state of energy $5\varepsilon$. Since all configurations are equally likely, particle A is $15/3 = 5$ times more likely to have energy $2\varepsilon$ than energy $5\varepsilon$.

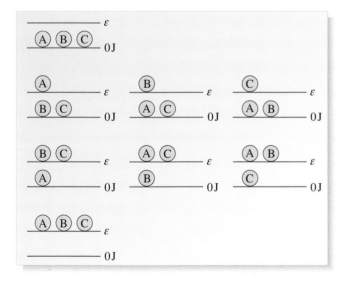

**Figure 1.33** Answer to Q1.3(a).

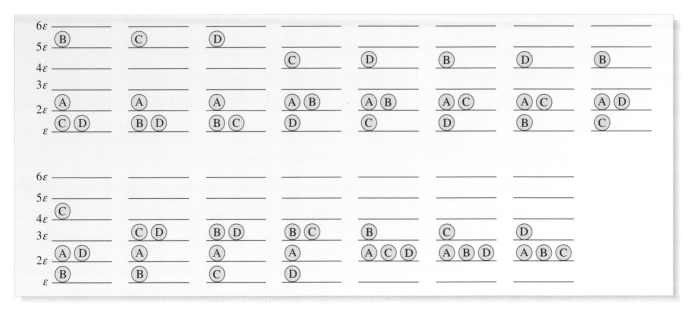

**Figure 1.35** Answer to Question 1.4(b).

**Q1.5** (a) If the particles are distinguishable, they can be labelled A, B and C. There are seven configurations with total energy $E_T = 3\varepsilon$, and one of these has three particles in the same state (see Figure 1.36). Since all seven configurations are equally likely, the probability of finding all three particles in the same state is 1/7.

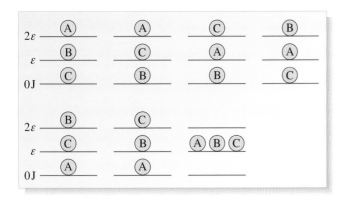

**Figure 1.36** Answer to Question 1.5(a).

(b) If the particles are identical bosons, they cannot be labelled so there are only two configurations with energy $3\varepsilon$ (see Figure 1.37). Each of these configurations is equally likely so the probability of finding all three particles in the same state is 1/2.

(c) There is no possibility of finding three identical fermions in the same state because this would contravene the exclusion principle.

**Figure 1.37** Answer to Question 1.5(b).

**Q1.6** (a) A hydrogen atom $^1_1\text{H}$ contains two fermions (one electron and one proton). It is therefore a boson.

(b) The symbol $^2_1\text{H}$ shows that a deuterium nucleus contains two particles, a neutron as well as a proton, each of which is a fermion. A deuterium atom therefore contains three fermions (one electron, one proton and one neutron) and so the atom $^2_1\text{H}$ is itself a fermion.

(c) A nitrogen atom ($^{14}_7\text{H}$) contains 21 fermions (7 electrons, 7 protons and 7 neutrons). The atom is therefore a fermion.

(d) A nitrogen molecule $N_2$ contains two nitrogen atoms, which are fermions (see part (c)). It follows that a nitrogen molecule is a boson.

(e) A rubidium ($^{87}_{37}\text{Rb}$) atom consists of 37 protons, 50 neutrons and 37 electrons. That's an even number (124) of fermions. Therefore the rubidium atom is a boson.

**Q1.7** Use $V = Nd^3$. Then $d = (V/N)^{1/3}$
$= (1.0 \times 10^{-14}\,\text{m}^3/(2 \times 10^4))^{1/3} = 7.9 \times 10^{-7}\,\text{m}.$

Now $\lambda_{dB} = h/(mv)$ and we can estimate the speed $v$ by taking $v = v_{rms} = (3kT/m)^{1/2}$. Hence

$$\lambda_{dB} = \frac{h}{m \times \sqrt{3kT/m}} = \frac{h}{\sqrt{3mkT}}$$
$$= \frac{6.626 \times 10^{-34}}{\sqrt{3 \times 1.45 \times 10^{-25} \times 1.381 \times 10^{-23} \times 1.0 \times 10^{-7}}}\,\text{m}$$
$$\approx 8.5 \times 10^{-7}\,\text{m}.$$

Hence $\lambda_{dB} \approx d$ and so the criterion of Equation 1.16 is not satisfied and indistinguishability may not be ignored.

**Q1.8** (a) It may surprise you to know that there is electromagnetic radiation inside a cave, but the rocks in the walls of the cave emit radiation, mostly in the infrared part of the spectrum, beyond the range of human vision. If the walls of the cave have a uniform and constant temperature and the interior of the cave is well enclosed, the radiation will be in thermal equilibrium with the rocks around it, so it can be described as thermal radiation.

(b) The radiation emitted by a glow-worm has nothing to do with thermal radiation. If the glow-worm were emitting visible light by virtue of being very hot, it would not be alive! The glow-worm's light is due to a completely different process, involving enzyme-catalysed photochemical reactions.

(c) The radiation emitted by the electric fire is not thermal radiation because the emitted radiation is not in thermal equilibrium — the bars of the fire emit many more photons than they absorb. (However, Planck's radiation law can often be used to give a good approximation to the spectrum of the radiation emitted by a hot object.)

(d) The radiation inside a hot oven is a good example of thermal radiation provided the temperature of the oven walls is constant and uniform, and the oven door is closed!

**Q1.9** The average number of photons with energies in the narrow range from $E$ to $E + \Delta E$ is $G_p(E)\,\Delta E$. From the data given in the question, $E = 1.00 \times 10^{-20}\,\text{J}$ and $E + \Delta E = 1.01 \times 10^{-20}\,\text{J}$, so that $\Delta E = 0.01 \times 10^{-20}\,\text{J} = 10^{-22}\,\text{J}$. The value of $G_p(10^{-20}\,\text{J})$, read off the 300 K graph in Figure 1.21, is $3.1 \times 10^{34}\,\text{J}^{-1}$. But the graph refers to a volume of $1\,\text{m}^3$. Since the cavity in this question has volume $V = 1\,\text{litre} = 10^{-3}\,\text{m}^3$, we have $G_p(10^{-20}\,\text{J}) = 3.1 \times 10^{31}\,\text{J}^{-1}$. Thus, at 300 K, the average number of photons in the given energy range is

$$G_p(E)\,\Delta E = G_p(10^{-20}\,\text{J}) \times 10^{-22}\,\text{J} = 3.1 \times 10^{31}\,\text{J}^{-1} \times 10^{-22}\,\text{J}$$
$$= 3.1 \times 10^9.$$

(This is so large that statistical fluctuations can be neglected and the word 'average' omitted.)

Repeating this calculation at 400 K, Figure 1.21 shows that

$$G_p(E)\,\Delta E = 6.2 \times 10^{34}\,\text{J}^{-1} \times 10^{-3} \times 10^{-22}\,\text{J} = 6.2 \times 10^9.$$

**Q1.10** The glass should be transparent to visible light, but opaque to thermal radiation at the typical ambient temperatures. Figure 1.21 shows the photon energy distribution function for thermal radiation at 300 K. At this temperature, most of the photons have energies less than $3 \times 10^{-20}\,\text{J}$. An energy of $3 \times 10^{-20}\,\text{J}$ corresponds to a wavelength of

$$\lambda = \frac{hc}{E} = \frac{6.63 \times 10^{-34}\,\text{J s} \times 3 \times 10^8\,\text{m s}^{-1}}{3 \times 10^{-20}\,\text{J}}$$
$$\approx 7 \times 10^{-6}\,\text{m}.$$

Most of the photons we wish to trap are of longer wavelength than this. Visible light, on the other hand, corresponds to wavelengths below $1 \times 10^{-6}\,\text{m}$. Glass for greenhouses should therefore be as opaque as possible at wavelengths above $7 \times 10^{-6}\,\text{m}$, and as transparent as possible at wavelengths below $1 \times 10^{-6}\,\text{m}$. Figure 1.38 illustrates this idealized behaviour.

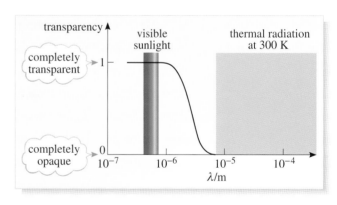

**Figure 1.38** Ideal properties of glass for greenhouses. The black curve shows the transparency of the glass as a function of optical wavelength.

**Q1.11** Use Equation 1.21 with $V = 1\,\text{m}^3$.

(a) At $T = 300\,\text{K}$,
$N = (2.0 \times 10^7\,\text{m}^{-3}\,\text{K}^{-3}) \times 1\,\text{m}^3 \times (300\,\text{K})^3 = 5.4 \times 10^{14}.$

(b) At $T = 380\,\text{K}$,
$N = (2.0 \times 10^7\,\text{m}^{-3}\,\text{K}^{-3}) \times 1\,\text{m}^3 \times (380\,\text{K})^3 = 1.1 \times 10^{15}.$

**Q1.12**  The average energy per photon is the total energy of the photon gas divided by the number of photons. Thus, to two significant figures,

$$\langle E \rangle = \frac{U}{N} = \frac{(\pi^4/15)C(kT)^4}{2.4C(kT)^3}$$

$$= \frac{\pi^4}{36}kT \approx 2.7kT.$$

Visible light corresponds to wavelengths in the range from about 700 nm (red) to about 400 nm (violet). Taking a typical visible wavelength to be, say, 600 nm, the photon energy is $hf = hc/\lambda$

$= (6.646 \times 10^{-34} \text{ J s} \times 2.998 \times 10^8 \text{ m s}^{-1})/600 \times 10^{-9} \text{ m}$
$\approx 3 \times 10^{-19} \text{ J}.$

In order for this to be the average photon energy, the temperature must be

$$T = \frac{\langle E \rangle}{2.7k} \approx \frac{3 \times 10^{-19} \text{ J}}{2.7 \times 1.38 \times 10^{-23} \text{ J K}^{-1}}$$

$$\approx 8000 \text{ K}.$$

**Comment**: *The temperature near the surface of the Sun is about 5800 K, so it is not surprising that the Sun emits a great deal of visible light.*

**Q1.13**  Let $d$ be the typical spacing between photons. The volume occupied by $N$ photons can be estimated as $V = Nd^3$. Therefore $d = (V/N)^{1/3}$.

The total number of photons is, from Equation 1.21,

$$N = 2.4C(kT)^3$$

$$= 2.4 \times \frac{8\pi V}{h^3 c^3} \times (kT)^3$$

$$= 60V\left(\frac{kT}{hc}\right)^3.$$

This gives a typical photon spacing of

$$d = \left(\frac{V}{N}\right)^{1/3} = \left(\frac{1}{60}\right)^{1/3} \frac{hc}{kT}.$$

You can compare this with the wavelength of a typical photon. Consider a photon with the average energy $\langle E \rangle = 2.7kT$. Its momentum is $p = \langle E \rangle/c = 2.7kT/c$ and the corresponding wavelength is

$$\lambda = \frac{h}{p} = \frac{1}{2.7}\frac{hc}{kT}.$$

Hence

$$\frac{d}{\lambda} = \frac{(1/60)^{1/3}}{(1/2.7)} = \frac{0.26}{0.37} = 0.70.$$

Thus $d \approx 0.7\lambda$, so, speaking very roughly, the typical spacing between photons is comparable to $\lambda$.

**Comment**: *The fact that Equation 1.16 (with $\lambda_{dB} = \lambda$) is not satisfied for a photon gas confirms that the indistinguishability of identical photons cannot be neglected.*

**Q1.14**  You can see that the total number of photons is larger than the total number of states of energies up to $kT$, and so the states are *not* sparsely occupied. This is true for all volumes $V$ and at all temperatures $T$ for any photon gas in thermal equilibrium. Hence there are no conditions under which the indistinguishability of identical photons can be neglected. (This conclusion is confirmed by Question 1.13.)

**Q1.15**  (a) We have valency $z = 1$, $M_r = 23$ and $\rho = 970 \text{ kg m}^{-3}$ for sodium. Substituting these values in Equation 1.26 gives the number density of electrons as

$$n = \frac{1 \times 970 \text{ kg m}^{-3} \times 6.022 \times 10^{23} \text{ mol}^{-1}}{(23 \times 10^{-3} \text{ kg mol}^{-1})}$$

$$= 2.54 \times 10^{28} \text{ m}^{-3}.$$

(b) Treating air as an ideal gas, we have $PV = NkT$. The number density of molecules is given by $n = N/V = P/kT$. Under normal conditions we put $P = 10^5 \text{ N m}^{-2}$ and $T = 300 \text{ K}$, so the number density of molecules is

$$n = \frac{10^5 \text{ N m}^{-2}}{1.381 \times 10^{-23} \text{ J K}^{-1} \times 300 \text{ K}}$$

$$= 2.41 \times 10^{25} \text{ m}^{-3}.$$

**Comment**: *Notice that the number density of free electrons in sodium metal is at least a thousand times greater than the number density of molecules in the atmosphere.*

**Q1.16**  The average speed is

$$\langle v \rangle = \sqrt{\frac{8kT}{\pi m_e}}.$$

At $T = 300 \text{ K}$, this gives

$$\langle v \rangle = \sqrt{\frac{8 \times 1.381 \times 10^{-23} \text{ J kg}^{-1} \times 300 \text{ K}}{\pi \times 9.109 \times 10^{-31} \text{ kg}}}$$

$$= 1.1 \times 10^5 \text{ m s}^{-1}.$$

Thus, the mean speed of the free electrons is predicted to be about $10^5\,\mathrm{m\,s^{-1}}$. Although this speed is very much larger than the average speeds in molecular gases at room temperature, it is still only a small fraction of the speed of light ($c = 2.998 \times 10^8\,\mathrm{m\,s^{-1}}$ and so $v^2/c^2 \approx 10^{-6}$). Therefore relativistic effects can be safely neglected.

**Q1.17**    The energy level $E = 11h^2/8m_eL^2$ arises from three different combinations of $n_1$, $n_2$ and $n_3$: namely, (3, 1, 1), (1, 3, 1) and (1, 1, 3). Each of these combinations corresponds to *two* quantum states, one for each of the two allowed orientations of the electron's spin. Thus, there are six different quantum states with energy $E = 11h^2/8m_eL^2$. According to the exclusion principle, each of these states contains a maximum of one electron. The maximum number of free electrons with the given energy is therefore six.

**Q1.18**    We use $\lambda_{\mathrm{dB}} = h/m_e v$ and $v = v_{\mathrm{rms}} = \sqrt{3kT/m_e}$ where the rms speed $v$ is a typical speed. Hence, at room temperature the typical de Broglie wavelength is This is to be compared with $d \approx 2 \times 10^{-10}\,\mathrm{m}$. Thus $\lambda_{\mathrm{dB}} \approx 30d$ at room temperature, which does not satisfy the criterion of Equation 1.16 for a classical treatment.

$$\lambda_{\mathrm{dB}} = \frac{h}{m_e\sqrt{3kT/m_e}} = \frac{h}{\sqrt{3m_e kT}}$$

$$= \frac{6.626 \times 10^{-34}\,\mathrm{J\,s}}{\sqrt{3 \times 9.109 \times 10^{-31}\,\mathrm{kg} \times 1.381 \times 10^{-23}\,\mathrm{J\,K^{-1}} \times 300\,\mathrm{K}}}$$

$$\approx 6.2 \times 10^{-9}\,\mathrm{m}.$$

(You can see that $\lambda_{\mathrm{dB}} \propto 1/\sqrt{T}$, and so even at a temperature of $3 \times 10^5\,\mathrm{K}$ the de Broglie wavelength would still be about the same as $d$. Of course any metal would have melted and vapourized before such a high temperature could be reached.)

**Q1.19**    For sodium, $n = 2.54 \times 10^{28}\,\mathrm{m^{-3}}$, and so

Equation 1.31 gives

$$E_F = \frac{(6.626 \times 10^{-34}\,\mathrm{J\,s})^2}{8 \times 9.109 \times 10^{-31}\,\mathrm{kg}} \left( \frac{3 \times 2.54 \times 10^{28}\,\mathrm{m^{-3}}}{\pi} \right)^{2/3}$$

$$= (6.025 \times 10^{-38} \times 8.376 \times 10^{18})\,\mathrm{J}$$

$$= 5.05 \times 10^{-19}\,\mathrm{J} = 3.15\,\mathrm{eV}.$$

For magnesium, $n = 8.62 \times 10^{28}\,\mathrm{m^{-3}}$. Hence

$$E_F = \frac{(6.626 \times 10^{-34}\,\mathrm{J\,s})^2}{8 \times 9.109 \times 10^{-31}\,\mathrm{kg}} \left( \frac{3 \times 8.62 \times 10^{28}\,\mathrm{m^{-3}}}{\pi} \right)^{2/3}$$

$$= (6.025 \times 10^{-38} \times 1.89 \times 10^{19})\,\mathrm{J}$$

$$= 1.14 \times 10^{-18}\,\mathrm{J} = 7.12\,\mathrm{eV}.$$

These answers agree well with the widths of the energy distributions in Figure 1.30

**Q1.20**    Only those electrons that can move into higher energy levels when the metal is heated can contribute to the heat capacity. The graphs show that only those electrons of energies within a few $kT$ of the Fermi energy can move into higher vacant energy levels when the temperature increases. For the lower lying electrons there are no empty levels immediately above them and the exclusion principle forbids them from moving into levels already occupied, and so they do not contribute to the heat capacity. Quantitatively, we know that at 1000 K, $kT = 1.381 \times 10^{-23} \times 1000\,\mathrm{J} \approx 0.09\,\mathrm{eV}$. This is very small compared to the spread of electron energies, given by the Fermi energy of several eV in most metals. Thus only a few percent of the electrons are excited.

**Q1.21**    At $T = 0\,\mathrm{K}$, the total energy of the gas is, from Equation 1.32, $U = 3NE_F/5$. The average energy per electron is therefore $U/N = 3E_F/5$. For zinc this is $3 \times 9.5\,\mathrm{eV}/5 = 5.7\,\mathrm{eV}$. This is also a good approximation at room temperature since very few electrons can move into vacant higher energy levels when the metal is heated (see answer to Question 1.20).

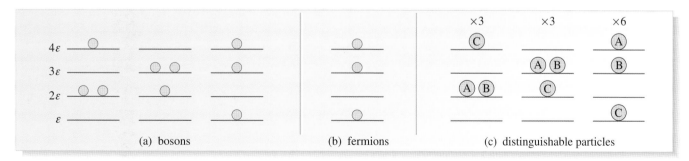

**Figure 1.39**    (a) 3 configurations for the identical bosons. (b) 1 configuration for identical fermions. (c) 12 configurations for distinguishable particles (the factors 3 and 6 arise from interchanging labelled particles in different quantum states).

**Q1.22** We can use Equation 1.33 which gives a good approximation at room temperature. Using values for $n$ and $E_F$ from Tables 1.3 and 1.4, we have $P = 2nE_F/5$ $= 2 \times 8.45 \times 10^{28}\,\text{m}^{-3} \times 7.0\,\text{eV} \times (1.609 \times 10^{-19}\,\text{J eV}^{-1})/5$ $= 3.81 \times 10^{10}\,\text{Pa}$. Knowing that 1 atmosphere = $10^5\,\text{Pa}$, we find $P = 3.81 \times 10^5$ atmospheres — a very high pressure indeed!

**Q1.23** The configurations are shown in Figure 1.39.

**Q1.24** The criterion for a classical continuum of energy levels is given by Equation 1.8: $\lambda_{dB} \ll L$. The criterion for neglecting the effects of indistinguishability and the uncertainty principle is usually much more stringent and is given by Equation 1.16: $\lambda_{dB} \ll d$, where $d$ is the typical distance between particles. The typical de Broglie wavelength is $\lambda_{dB} = h/mv = h/(3mkT)^{1/2}$, where we have taken the typical speed of a molecule to be

$v = v_{rms} = \sqrt{3kT/m}$. Thus for the given gas of hydrogen molecules we have

$\lambda_{dB} = (6.626 \times 10^{-34}\,\text{J s})/[(3 \times 2 \times 1.673 \times 10^{-27}\,\text{kg}) \times (1.381 \times 10^{-23}\,\text{J K}^{-1}) \times 300\,\text{K}]^{1/2} \approx 1.0 \times 10^{-10}\,\text{m}$.

We are given $L^3 = 0.2\,\text{m}^3$ and so $L = 0.6\,\text{m}$. We also have

$d = (V/N)^{1/3} = (0.2\,\text{m}^3/10^{23})^{1/3} = 1.3 \times 10^{-8}\,\text{m}$.

Thus both criteria are well satisfied and so the gas can be treated as a classical gas.

**Q1.25** We can use Equations 1.21a, 1.22 and 1.24 (or 1.25). Thus the number of photons in one cubic metre is

$N = (2.0 \times 10^7\,\text{m}^{-3}\,\text{K}^{-3}) \times (1\,\text{m}^3) \times (3 \times 10^6\,\text{K})^3 = 5.4 \times 10^{26}$.

The energy in one cubic metre is

$U = (7.5 \times 10^{-16}\,\text{J m}^{-3}\,\text{K}^{-4}) \times (1\,\text{m}^3) \times (3 \times 10^6\,\text{K})^4$
$= 6.1 \times 10^{10}\,\text{J}$,

and the pressure is

$P = U/3V = (6.1 \times 10^{10}\,\text{J})/(3 \times 1\,\text{m}^3) = 2 \times 10^{10}\,\text{Pa}$.

**Q1.26** (a) The $_2^3\text{He}$ nucleus contains two protons and one neutron, i.e. three fermions, and so it is itself a fermion. (b) The neutral atom is a helium nucleus plus two orbiting electrons. Thus it consists of five fermions and is also a fermion. (c) The ion has lost one electron and therefore consists of four fermions. It is therefore a boson.

**Q1.27** For silver, the Fermi energy is

$$E_F = \frac{h^2}{8m_e}\left(\frac{3n}{\pi}\right)^{2/3}$$

$$= \frac{(6.626 \times 10^{-34}\,\text{J s})^2}{8 \times 9.109 \times 10^{-31}\,\text{kg}}\left(\frac{3 \times 5.86 \times 10^{28}\,\text{m}^{-3}}{\pi}\right)^{2/3}$$

$$= 8.82 \times 10^{-19}\,\text{J} = 5.50\,\text{eV}.$$

The total translational energy of the electrons in a 1.0 cm cube is

$U = 3NE_F/5$
$= (3 \times 5.86 \times 10^{28}\,\text{m}^{-3} \times 10^{-6}\,\text{m}^3 \times 8.82 \times 10^{-19}\,\text{J})/5$
$= 3.10 \times 10^4\,\text{J}$.

The pressure is $P = 2U/5V$
$= 2 \times 3.10 \times 10^4\,\text{J}/(5 \times 1.0 \times 10^{-6}\,\text{m}^3) = 1.2 \times 10^{10}\,\text{Pa}$.

**Q2.1** The equilibrium separation of the two hydrogen atoms in the molecule is at $r = r_0$ where the total energy is a minimum. Reading from the graph you can estimate this minimum energy to be approximately $-6\,\text{eV}$. Thus the energy required to separate the two atoms by a large distance ($r \to \infty$) is approximately $6\,\text{eV}$. This is the binding energy of the molecule.

**Q2.2** From Bragg's law, $\sin\theta = n\lambda/2d = n \times 0.5520 \times 10^{-9}\,\text{m}/(2 \times 0.4250 \times 10^{-9}\,\text{m}) = n \times 0.6494$. Remembering that the magnitude of $\sin\theta$ cannot exceed 1, the only possible value of the integer $n$ in this case is 1, and so a reflection will occur when $\sin\theta = 0.6494$, i.e. when $\theta = 40.5°$.

**Q2.3** (a) $E_{pot} = e(-e)/4\pi\varepsilon_0 d$
$= -(8.988 \times 10^9\,\text{N m}^2\,\text{C}^{-2}) \times (1.602 \times 10^{-19}\,\text{C})^2/(0.282 \times 10^{-9}\,\text{m})$
$= -8.18 \times 10^{-19}\,\text{J} = -5.11\,\text{eV}$.

(b) From part (a) we have $d \times E_{pot} = -e^2/4\pi\varepsilon_0$
$= -2.31 \times 10^{-28}\,\text{N m}^2 = 1.44 \times 10^{-9}\,\text{eV m}$, which is a constant regardless of the value of $d$.

**Q2.4** The current is given by $i = nev_dA$, and so the drift speed is

$v_d = i/(neA) = 1\,\text{A}/(8.45 \times 10^{28}\,\text{m}^{-3} \times 1.602 \times 10^{-19}\,\text{C} \times \pi \times (5 \times 10^{-5}\,\text{m})^2)$

$= 9.4 \times 10^{-3}\,\text{m s}^{-1} = 9.4\,\text{mm s}^{-1}$.

**Q2.5** (a) (i) At 10 K we are on the flat portion of the curve where the resistivity is due to the scattering of electron waves by defects in the crystal structure. (ii) At 100 K the resistivity is increasing uniformly with temperature and the resistivity is due to thermal scattering as well as defect scattering. (b) Reading from the graph, the resistivity at 300 K is approximately $\rho = 2.4 \times 10^{-8}\,\Omega\,\text{m}$, and so the conductivity is approximately $\sigma = 1/\rho = 4.2 \times 10^7\,\Omega^{-1}\,\text{m}^{-1}$.

**Q2.6** A, B, and F are metals because there are partially filled bands. E is a metal because neither of the two bands shown is completely filled. These four materials will be metals at all temperatures. For the other materials you will find it helpful to refer to Figure 2.26. C has an energy gap between filled and empty bands of about 7 eV and so it will be an insulator at 300 K and an even better insulator at 10 K. D has a small energy gap, about 1 eV, and will therefore be a semiconductor at 300 K. At 10 K however, very few electrons will be excited across the gap, so it will be a good insulator.

**Q2.7** The correctly filled table is shown below. Points to notice are:

- The electron and hole densities in pure silicon (an intrinsic semiconductor) are the same since for each electron thermally excited from the valence band into the conduction band there is a hole left in the valence band.

- The doping levels are very dilute so that the number density of silicon is not significantly reduced by the replacement of some silicon atoms by donor impurities in n-type material or acceptor impurities in p-type materials.

- Doping vastly increases the electron density in the conduction band of an n-type material and the hole density in the valence band of a p-type material.

- Because of the above point, the recombination rate in doped materials is very much greater than that in the pure (intrinsic) material, and so the hole density in the valence band of an n-type material and the electron density in the valence band of a p-material are very much lower than in a pure (intrinsic) material.

Table 2.5  For Q2.7.

| Material | Electron number density in conduction band/m$^{-3}$ | Hole number density in valence band/m$^{-3}$ | Number density of silicon atoms/m$^{-3}$ |
|---|---|---|---|
| pure silicon | $10^{16}$ | $10^{16}$ | $5 \times 10^{28}$ |
| n-type silicon | $10^{22}$ | $10^{10}$ | $5 \times 10^{28}$ |
| p-type silicon | $10^{10}$ | $10^{22}$ | $5 \times 10^{28}$ |

**Q2.8**  (a) In a metal such as sodium, the electrons in the conduction band conduct electricity by essentially the same mechanism as in Pauli's quantum free-electron model. Electrical resistance is caused by the scattering of electron waves by defects and thermal agitation of the lattice. Thus the conductivity decreases with temperature (the resistivity of course increases with temperature) due to increased thermal scattering. In a semiconductor such as silicon, the current is carried by electrons that are thermally excited from the valence band across the energy gap into the conduction band and by the holes left behind in the valence band. The electrical conductivity increases with temperature because the thermal excitation rate increases with temperature. (b) The thermal excitation of electron–hole pairs described in part (a) increases rapidly

with temperature and so the conductivity of the pure material increases rapidly with temperature. In a doped material however, the current is carried mainly by the large number of donor electrons in the conduction band of an n-type material or donor holes in the valence band of a p-material. Thus, over a wide temperature range, the conductivity is determined mainly by the doping density and is little affected by temperature.

**Q2.9**  The change in the electrical potential energy of an electric charge $q$ when it undergoes a displacement $\Delta x$ in a region where there is a uniform electric field component $\mathscr{E}_x$ is (from *SFP*)

$$\Delta E_{el} = -q\mathscr{E}_x \Delta x.$$

For the given p–n junction, $\mathscr{E}_x = -1.5 \times 10^6$ V m$^{-1}$.

(a) For an electron, $q = -e$. (i) When an electron moves from n to p, $\Delta x = -1 \times 10^{-6}$ m, and so

$\Delta E_{el} = -(-e)(-1.5 \times 10^6 \text{ V m}^{-1})(-1.0 \times 10^{-6} \text{ m}) = 1.5 \text{ eV} = 2.4 \times 10^{-19}$ J.

(ii) When an electron moves from p to n, $\Delta x = +1.0 \times 10^{-6}$ m and so $\Delta E_{el} = -1.5$ eV $= -2.4 \times 10^{-19}$ J.

(b) For a hole, $q = +e$. (i) Thus for a hole moving from n to p,

$\Delta E_{el} = -(+e)(-1.5 \times 10^6 \text{ V m}^{-1})(-1.0 \times 10^{-6} \text{ m}) = -1.5 \text{ eV} = -2.4 \times 10^{-19}$ J.

(ii) For a hole moving from p to n, $\Delta x = +1.0 \times 10^{-6}$ m and so $\Delta E_{el} = 1.5$ eV $= 2.4 \times 10^{-19}$ J.

(c) (i) The force on an electron is in the direction opposite to that of the electric field, i.e. the force is in the positive $x$-direction. (ii) For a positive hole, the force is in the same direction as the electric field, i.e. in the negative $x$-direction.

**Q2.10**  Photons with energy less than the energy gap *cannot* excite electrons from the valence band to the conduction band and such photons will be ineffective for generating electricity with a solar cell. In silicon the energy gap is 1.12 eV. This corresponds to an optical wavelength $\lambda$ given by $E_{gap} = hf = hc/\lambda$. Thus $\lambda = hc/E_{gap} = (6.626 \times 10^{-34}$ J s$) \times (2.998 \times 10^8$ m s$^{-1})/(1.12 \times 1.602 \times 10^{-19}$ J$) = 1110$ nm. This wavelength is in the infrared part of the spectrum. The region of the solar spectrum with wavelengths greater than 1110 nm (i.e. energy less than 1.12 eV) is ineffective in a silicon cell. As you can see from Figure 2.39, a significant portion of the solar spectrum falls in this ineffective range.

**Q2.11**  No. The energy gap of GaAs corresponds to a wavelength of 867 nm which is in the infrared region of the spectrum and is not visible. The relationship between wavelength and energy gap is $\lambda = ch/E_{gap}$. You could work the wavelength out directly for GaP, but it is easier to notice that wavelength is inversely proportional to energy gap, so for GaP the wavelength is 867 nm × (1.43 eV/ 2.27 eV) = 546 nm, which is visible (green), and so GaP would be suitable.

**Q2.12** We are given $B_C = 0.20$ T. The maximum current density is given by $J_C = \dfrac{2B_C}{\mu_0 a}$, and the maximum current is $i_C = \pi a^2 J_C$.

(a) For a wire with $2a = 1.0 \times 10^{-4}$ m,

$J_C = 2 \times (0.20\,\text{T})/((4\pi \times 10^{-7}\,\text{T m A}^{-1}) \times (0.5 \times 10^{-4}\,\text{m})) = 6.4 \times 10^{9}\,\text{A m}^{-2}$.

The maximum current is $i_C = \pi(0.5 \times 10^{-4}\,\text{m})^2 \times 6.4 \times 10^{9}\,\text{A m}^{-2} = 50$ A.

(b) For a filament with $2a = 1.0 \times 10^{-6}$ m,

$J_C = 2 \times (0.20\,\text{T})/((4\pi \times 10^{-7}\,\text{T m A}^{-1}) \times (0.5 \times 10^{-6}\,\text{m})) = 6.4 \times 10^{11}\,\text{A m}^{-2}$.

The maximum current is $i_C = \pi(0.5 \times 10^{-6}\,\text{m})^2 \times 6.4 \times 10^{11}\,\text{A m}^{-2} = 0.5$ A.

**Q2.13** (a) Diamond and grey tin are both covalent solids. Covalent bond strengths fall off rapidly with distance between atoms. The tin atom is larger than the carbon atom and so the bond length in grey tin is greater and hence the bond strength is weaker. (b) The diamond crystal is held together entirely by covalent bonds. These are much stronger than the van der Waals bonds that hold the plane sheets of graphite together. (c) Metal crystals have dislocation defects in them which can move under stress and so metals are ductile. Glass is an amorphous solid in which stress concentrates at the ends of small cracks until the material fails catastrophically.

**Q2.14** The units of electrical conductivity $\sigma$ can be found by putting the units into Pauli's expression $ne^2\lambda_F/ m_e v_F$. Thus the units of $\sigma$ are: $(\text{m}^{-3})(\text{C}^2)(\text{m})/((\text{kg})(\text{m s}^{-1}))$ which simplifies to $\text{C}^2\,\text{s}/(\text{kg m}^3)$. The tricky bit is to bring the ohm ($\Omega$) into this. One approach is to use Ohm's law, which tells us that $1\,\Omega = 1\,\text{V}/1\,\text{A}$. Knowing also that $1\,\text{A} = 1\,\text{C s}^{-1}$ and $1\,\text{V} = 1\,\text{J C}^{-1} = 1\,\text{kg m}^2\,\text{s}^{-2}\,\text{C}^{-1}$, we have that

$$1\,\Omega = (1\,\text{kg m}^2\,\text{s}^{-2}\,\text{C}^{-1})/\text{C s}^{-1} = 1\,\text{kg m}^2\,\text{s}^{-1}\,\text{C}^{-2}.$$

Thus we can replace $\text{C}^2\,\text{s/kg}$ in the unit expression for $\sigma$ by $\text{m}^2/\Omega$. Hence the units of $\sigma$ are $(\text{m}^2/\Omega)/\text{m}^3 = \Omega^{-1}\,\text{m}^{-1}$.

**Q2.15** (a) The current stays the same. The current is independent of the length, if the electric field is constant. To see this, remember that electric field is measured in $\text{V m}^{-1}$, i.e. electric field is the potential difference per metre of wire. By Ohm's law the current flowing through each metre of wire is constant when the potential difference per metre, i.e. the electric field, stays the same.

(b) The current falls when the conductivity is reduced since the resistivity of the material increases (resistivity is the reciprocal of the conductivity) and so the resistance of the piece of wire increases.

(c) The current increases when the mean free path is increased. This is because the average time interval between scattering events increases with mean free path, and this allows the acceleration caused by the electric field between scattering events to increase, and hence the drift speed increases. Formally, Pauli's expression (Equation 2.8) for the electrical conductivity $\sigma$ has the mean free path $\lambda_F$ in the numerator showing that the conductivity increases in proportion to mean free path.

(d) The current will decrease if the temperature is increased because the increased thermal scattering increases the resistivity of the material and hence the resistance of the wire.

(e) The current will increase if the free-electron density is increased simply because this increases the density of current carriers in Pauli's model. Formally, the free-electron density $n$ appears in the numerator of the expression for electrical conductivity $\sigma$, showing that conductivity increases in proportion to $n$.

**Q2.16** The tight-binding model starts with the energy levels of identical atoms a long way apart. As the atoms are brought together to form the crystal, the wavefunctions of the electrons, especially the valence electrons, of different atoms begin to overlap, and the atoms begin to interact with one another. The result of this is that each individual level splits into a large number of closely separated levels thereby forming bands of energy levels separated by gaps. The lower atomic energy levels, belonging to the inner shells of the atoms, form very narrow bands, but those of the outer valence electrons expand into very wide bands which may overlap with one another. The atoms reach equilibrium at the separation and structure corresponding to minimum energy for the whole system.

**Q2.17** In some materials the highest occupied energy band, the valence band, is completely full of electrons, with a wide energy gap before the next band of allowed states, the conduction band, which is empty. Strictly the above is true only at absolute zero temperature, but for a wide gap the number of electrons excited into the upper band at room temperature is extremely small. These materials are insulators.

When the gap between these two bands is reduced to about 3 eV, enough electrons are excited into the conduction band to give significant effects, and such materials are classed as semiconductors. There is no qualitative difference between insulators and semiconductors. A metal is obtained when the two bands overlap in energy and some of the filling of the lower band is transferred to states in the lowest part of the upper band, so that both bands are only partly filled.

Alternatively the lowest occupied band may be only partly filled due to a lack of free electrons to occupy its upper states. In both these metallic cases, electrons at the highest energy can easily transfer to a wide variety of other states at essentially the same energy, allowing flow of electricity and heat by this high density of free electrons.

**Q2.18**  In equilibrium a small diffusion current flows from the p-region to the n-region due to high-energy electrons in the n-region and high-energy holes in the p-region climbing the potential energy hill and crossing into the p- and n-regions respectively where they subsequently recombine. At the same time there is a small current of the same magnitude, the pair current, flowing in the opposite direction, from the n- to the p-region, due to thermally excited electrons in the p-region and holes in the n-region that slide down the potential energy hill (or, if you like, are swept across the junction by the electric field in the depletion region).

In reverse bias, an external voltage source raises the electrostatic potential on the n-side of the junction relative to the p-side. The height of the potential energy hill is therefore increased (i.e. the magnitude of the electric field in the depletion region is increased), and this stops the diffusion current, allowing the small pair current only to flow from n to p. In forward bias, the potential hill is lower (the magnitude of the electric field is reduced), allowing a large diffusion current to flow without changing the small pair current. Hence the p–n junction acts as a rectifier, allowing a large current to flow in one direction (from p to n) only.

**Q3.1**  (a) $^1_1H_0$, $^2_1H_1$, $^3_1H_2$.

(b) $^{12}_8O_4$, $^{26}_8O_{18}$.

(c) $^1H$, $^2H$, $^{16}O$, $^{17}O$ and $^{18}O$.

**Q3.2**  (a) Take the mass of a nucleon to be $1.67 \times 10^{-27}$ kg. Then the nuclear mass density is $0.17 \times (1.67 \times 10^{-27}\,\text{kg})/(10^{-15}\text{m})^3 = 2.8 \times 10^{17}\,\text{kg m}^{-3}$. A 5 ml spoonful would have mass $(2.8 \times 10^{17}\,\text{kg m}^{-3}) \times (5 \times 10^{-6}\,\text{m}^3) = 1.4 \times 10^{12}$ kg.

(b) $A = 0.17\,\text{fm}^{-3} \times 4\pi r^3/3$. Hence $r = (1.1\,\text{fm})A^{1/3}$.

**Q3.3**  $A$ decreases by 4 and $N$ decreases by 2.

**Q3.4**  $^{226}_{88}Ra \rightarrow \, ^{222}_{86}Rn + \alpha$.

**Q3.5**  $^{15}_8O \rightarrow \, ^{15}_7N + e^+ + \nu_e$ and $^{11}_6C \rightarrow \, ^{11}_5B + e^+ + \nu_e$. The parent nuclei are $^{15}_8O$ and $^{11}_6C$, and the daughter nuclei are $^{15}_7N$ and $^{11}_5B$.

**Q3.6**  The nuclear energy is reduced by

$$E_2 - E_1 = (0.437 - 0.081)\,\text{MeV} = 0.356\,\text{MeV}$$
$$= (0.356\,\text{MeV}) \times (1.602 \times 10^{-13}\text{J MeV}^{-1})$$
$$= 5.70 \times 10^{-14}\,\text{J}.$$

The γ-ray photon frequency is therefore

$$f = (E_2 - E_1)/h = 5.70 \times 10^{-14}\,\text{J}/(6.626 \times 10^{-34}\,\text{J s})$$
$$= 8.61 \times 10^{19}\,\text{Hz},$$

and the wavelength is

$$\lambda = 2.998 \times 10^8\,\text{m s}^{-1}/(8.61 \times 10^{19}\,\text{Hz})$$
$$= 3.48 \times 10^{-12}\,\text{m}.$$

**Q3.7**  (a) The half-life of $^{238}U$ is, to the accuracy given in the question, equal to the age of the Earth and so there is about 50% of this isotope left. On the other hand, the remaining fraction of $^{235}U$ is

$$\frac{N(\text{now})}{N_0} = \exp\left(\frac{-0.6931 \times 4.5 \times 10^9 \text{ years}}{7.04 \times 10^8 \text{ years}}\right)$$
$$= 0.0119,$$

so about 1.2% of the original $^{235}U$ remains.

(b) We have found that there was originally twice as much $^{238}U$ as at present and $1/0.0119 = 84$ times as much $^{235}U$ as at present. Given that the present proportion of $^{235}U$ to $^{238}U$ is 0.0072 to 1, the primordial proportion was $0.0072 \times 84/2 = 0.30$ to 1.

**Q3.8**  Figure 3.11 shows that there are 4 naturally abundant isotopes of lead Pb ($Z = 82$); they have $N = 122$, 124, 125 and 126, i.e they are $^{204}_{82}Pb_{122}$, $^{206}_{82}Pb_{124}$, $^{207}_{82}Pb_{125}$ and $^{208}_{82}Pb_{126}$. The isotope $^{205}_{82}Pb_{123}$ is not one of them. (It was discovered in 1940 and has a half-life of $1.5 \times 10^7$years.)

**Q3.9**  (a) The available kinetic energy is $(8.00 - 2.22)\,\text{MeV} = 5.78\,\text{MeV}$.

(b) An 8.00 MeV photon has insufficient energy to break up a triton, binding energy 8.48 MeV, into its constituent nucleons. However, only $(8.48 - 2.22)\,\text{MeV} = 6.26\,\text{MeV}$ is required to break it up into a neutron and a deuteron, so the 8.00 MeV photon has sufficient energy to do this.

**Q3.10**  For a triton, $A = 3$ and so $B/A = 8.48\,\text{MeV}/3 = 2.83\,\text{MeV}$ per nucleon.

**Q3.11**  Total binding energy of $^{238}U = 238 \times 7.6\,\text{MeV}$.

Total binding energy of $^{234}Th$ and $^4He$ is $234 \times (7.6 + 0.027)\,\text{MeV} + 28.3\,\text{MeV}$.

The total kinetic energy available (almost all of which is carried away by the α-particle) is equal to the increase in total binding energy, i.e. $(234 \times 0.027 - 4 \times 7.6 + 28.3)\,\text{MeV} = 4.2\,\text{MeV}$.

**Q3.12**  The energy released in the proton capture reaction is $(3 \times 2.57 - 2 \times 1.11)\,\text{MeV} = 5.5\,\text{MeV}$.

**Q3.13**  Essential points are:

(i) Nucleons in the surface are within the strong nuclear force range of fewer other nucleons than are nucleons well inside the nucleus, and so the surface nucleons are less tightly bound. Thus the $B/A$ values fall with increasing surface to volume ratio, i.e. decreasing $A$.

(ii) The Coulomb energy is a positive contribution to the total energy, and therefore a negative contribution to the binding energy, representing the cumulative effect of the long-range Coulomb potential energy between all proton pairs. It therefore reduces $B/A$ for large $Z$.

(iii) The symmetry energy follows from the larger nucleon–nucleon binding energy for unlike nucleons and the Pauli exclusion principle. For $N > Z$ the proportion of unlike-nucleon pairs to like-nucleon pairs is less, and so $B/A$ is reduced.

**Q3.14**  The surface contribution would tend to reduce the binding energy per nucleon $B/A$ for the smaller nucleus $^{124}$Sn because of the larger surface to volume ratio of the smaller nucleus. The Coulomb contribution also would tend to reduce the $B/A$ of the smaller nucleus because the smaller separations of the protons increase the Coulomb potential energy. (Note that each isotope has the same number of protons.)

However, we are told that the larger nucleus has the smaller $B/A$. Hence we conclude that this must be an asymmetry effect. (This conclusion is also supported by the fact that the ratio of unlike nucleon pairs to like nucleon pairs is greater in $^{130}_{50}$Sn than in $^{124}_{50}$Sn.)

**Q3.15**  The next neutron closed shell (i.e. magic number) after 50 is 82, and so the tin isotope $^{132}_{50}$Sn$_{82}$ is doubly magic. It is also neutron rich, i.e. it lies well below the path of stability in the $Z$–$N$ plane, and so would be expected to decay by $\beta^-$-decay. (In fact it does so with a half-life of 39.7 s. Still heavier isotopes of tin exist with rapidly decreasing half-lives, e.g. about 150 ns for $^{136}$Sn.)

**Q3.16**  The path of stability curves downwards away from the line $Z = N$ as $Z$ increases, indicating that the $N/Z$ ratio for stable nuclei increases with increasing nuclear size. The fission products of the relatively stable $^{238}_{92}$U$_{146}$ ($Z = 92$) will have $Z$ values in the region 46 and $N$ values in the region 73 which gives $N/Z$ in the region 73/46 which is too large for stability, i.e. the fission products lie below the path of stability. Thus the fission products are likely to decay by $\beta^-$-decay which reduces the $N/Z$ ratio. (This is one reason why nuclear reactors produce such unpleasant waste.)

**Q3.17**  The ratio of lifetimes is

$$\frac{1.405 \times 10^{10} \times 3.16 \times 10^7}{1.09 \times 10^{-7}} = 4.07 \times 10^{24}.$$

The prediction, with $Z = 90 - 2$ is

$$\frac{\exp(88a/\sqrt{4.08\,\text{MeV}})}{\exp(88a/\sqrt{9.85\,\text{MeV}})} = \exp\left[88\left(\frac{3.97}{\sqrt{4.08}} - \frac{3.97}{\sqrt{9.85}}\right)\right]$$
$$= \exp(61.7) = 6 \times 10^{26}.$$

Although this result differs by two powers of ten from the 24 powers of ten observed, the agreement is really quite good when you consider that the calculation involved the exponential function of a large number ($\exp(61.7)$) which is very sensitive to small errors. To illustrate the point, if the energy of 4.08 MeV is replaced by 4.00 MeV and the 9.85 MeV is replaced by 10.0 MeV, changes of just a few per cent, we obtain the result $8 \times 10^{27}$, which is more than ten times larger. Furthermore, approximations were made in deriving the formula that we have used for $T_{1/2}$.

**Q3.18**  We use the ideal gas relationship:

average kinetic energy $= \dfrac{3kT}{2}$

$$= \frac{3 \times 1.381 \times 10^{-23}\,\text{J K}^{-1} \times 2 \times 10^{10}\,\text{K}}{2}$$
$$= 4.14 \times 10^{-13}\,\text{J} = \frac{4.14 \times 10^{-13}\,\text{J}}{1.602 \times 10^{-13}\,\text{J MeV}^{-1}}$$
$$= 2.58\,\text{MeV}$$

which is near enough to 2.22 MeV in view of the statement of the temperature to one significant figure. (A useful rule of thumb is that a temperature of 10 000 K corresponds to about 1 eV.)

**Q3.19**  The reaction is $^2$H + $^2$H = $^4$He. The binding energy of the two deuterons is $2 \times 2.22$ MeV and that of the $\alpha$-particle is 28.3 MeV, and so the energy released is $(28.3 - 4.44)$ MeV = 23.9 MeV.

**Q3.20**  (a) First calculate the nuclear radii given by $1.2 A^{1/3}$ fm. For the rubidium nucleus $^{93}$Rb, this is $1.2 \times 93^{1/3} = 5.437$ fm and for caesium nucleus $^{141}$Cs it is $1.2 \times 141^{1/3} = 6.245$ fm. When the nuclei are 'just touching' the distance between their centres is just the sum of these radii, which we shall call $R = 11.682$ fm. Using this, the Coulomb potential energy is

$$\frac{\text{product of charges}}{4\pi\varepsilon_0 R} = \frac{Z_1 Z_2 e^2}{4\pi\varepsilon_0 R}$$
$$= \frac{(8.988 \times 10^9\,\text{N m}^2\,\text{C}^{-2}) \times 37 \times 55 \times (1.602 \times 10^{-19}\,\text{C})^2}{11.682 \times 10^{-15}\,\text{m}}$$

$$= 4.0 \times 10^{-11}\,\text{J} = 250\,\text{MeV}.$$

This is the potential energy that the fission products have at the instant of separation. As they move apart under the

Coulomb repulsion this potential energy is converted into kinetic energy. Our result agrees quite well with the total energy released by fission, about 200 MeV mostly in the form of kinetic energy of the fission products.

(b) We know that 1 mole of $^{235}$U has a mass of $235 \times 10^{-3}$ kg and contains Avogadro's number of atoms. Thus the number of atoms in 1 kg of $^{235}$U is approximately $(1/(235 \times 10^{-3})) \times$ Avogadro's number = $(1/235) \times 6.022 \times 10^{26}$ atoms = $2.563 \times 10^{24}$ atoms (and the same number of nuclei). Hence the total energy released by fission is approximately $2.563 \times 10^{24} \times 200$ MeV = $8.21 \times 10^{13}$ J. This is sufficient to run a 1000 MW at 20% efficiency for about $0.20 \times 8 \times 10^4$ s, i.e. about 4 hours.

**Q3.21** $^{233}$U is expected to be fissile for the same reason as $^{235}$U is fissile, i.e. the nucleus $^{234}_{92}U_{142}$ produced when a thermal neutron is absorbed has $Z$ and $N$ even. Thus, because of the pairing energy, sufficient binding energy is released on absorbing a neutron to put the $^{234}$U nucleus at the top of its barrier. $^{240}$Pu is not expected to be fissile for the same reason as $^{238}$U is not fissile: the resulting $^{241}$Pu nucleus has $N$ odd and therefore a relatively small $B/A$ due to the pairing energy. As a result the excited $^{241}$Pu nucleus is still too far below the top of its barrier for tunnelling to be likely.

**Q3.22** (a) Knowing that $Z$ decreases by 2 in α-decay you can consult the Periodic Table, Figure 3.2, to see that the resulting nucleus is radium. Thus

$$^{224}_{90}\text{Th} \rightarrow \,^{220}_{88}\text{Ra} + \,^{4}_{2}\text{He}.$$

You will need to consult the Periodic Table again for the β-decays, remembering that $A$ remains the same in β-decay, $Z$ increases by one in β⁻-decay, while $Z$ decreases by one in electron capture. Thus we find

$$^{197}_{78}\text{Pt} \rightarrow \,^{197}_{79}\text{Au} + e^- + \bar{\nu}_e$$

and

$$^{197}_{80}\text{Hg} + e^- \rightarrow \,^{197}_{79}\text{Au} + \nu_e.$$

(b) We know that $^{197}_{80}$Hg undergoes electron capture and therefore has too many protons, or too few neutrons, for stability. This suggests that $^{197}_{81}$Tl and $^{197}_{82}$Pb will also undergo electron capture (or positron emission) since they have lower neutron–proton ratios. We are also told that $^{197}_{78}$Pt is β⁻-active, i.e. it has too many neutrons and lies on the other side of the valley of stability. This leaves $^{197}_{79}$Au as the likely stable nucleus, lying on the valley floor. (In fact $^{197}_{79}$Au is the only stable one and is the only stable gold isotope.)

**Q3.23** Use the exponential decay law. We want the fraction remaining after 5 minutes, i.e. 300 s. Thus

$$N(300\,\text{s}) = N_0 \exp(-0.6931 \times 300\,\text{s}/124\,\text{s}),$$

And so $N(300\,\text{s})/N_0 = \exp(-1.677) = 0.1870$. Hence there is 18.7% left.

The isotope has to be made in-house and used within minutes because of its short half-life; if brought from elsewhere it would decay away before it could be transported to the hospital and used.

**Q3.24** The highest energy γ-ray has energy $E_2 - E_1 = (880 - 0)$ keV. The γ-ray frequency $f$ is found from Plank's law, $E_2 - E_1 = hf$, and so the wavelength is

$hc/(E_2 - E_1) = 6.626 \times 10^{-34}$ J s $\times 2.998$ $\times 10^8$ m s$^{-1}$/$(880 \times 10^3$ eV $\times 1.602 \times 10^{-19}$ J eV$^{-1})$

$$= 1.41 \times 10^{-12}\,\text{m}.$$

**Q3.25** The binding energy changes from $5.33 \times 6$ MeV to $(2.83 \times 3 + 7.07 \times 4)$ MeV. Thus the binding energy increases by 32.0 MeV − 36.7 MeV = 4.8 MeV, and so the reaction is exothermic releasing 4.8 MeV.

**Q3.26** (a) Total power output from the Sun is $P = (4\pi(1.5 \times 10^{11})^2 \times 1.4 \times 10^3)$ W $= 4.0 \times 10^{26}$ W.

(b) Rate of fusion is

$4.0 \times 10^{26}$ W$/(26.7 \times 10^6 \times 1.60 \times 10^{-19}$ J$) = 9.36 \times 10^{37}$ s$^{-1}$.

Hence rate of consumption of $^1$H is

$4 \times 1.7 \times 10^{-27}$ kg $\times 9.36 \times 10^{37}$ s$^{-1} = 6.4 \times 10^{11}$ kg s$^{-1}$,

and time taken for all $^1$H to be consumed is

$\frac{1}{3} \times 2.0 \times 10^{30}$ kg$/(6.4 \times 10^{11}$ kg s$^{-1}) = 1.0 \times 10^{18}$ s $= 3.3 \times 10^{10}$ years.

(c) Solar power per unit mass of Sun = $4.0 \times 10^{26}$ W$/2.0 \times 10^{30}$ kg $= 2 \times 10^{-4}$ W kg$^{-1}$. Your body's power output per unit mass is about 70 W/50 kg = 1.4 W kg$^{-1}$. Hence, weight for weight, your body generates over a thousand times more power than the Sun! You may wonder then why the Sun is so hot and you are so cool. This can be understood in terms of *surface to volume ratio*. In the steady state, the rate of heat production is equal to the rate of heat loss. Heat production occurs throughout the volume at a constant rate whereas heat loss occurs only at the surface and at a rate that increases with temperature. Thus the Sun compensates for its tiny surface to volume ratio by being very hot.

(The surface to volume ratio for a sphere of radius $R$ is

$$\frac{4\pi R^2}{4\pi R^3/3} = \frac{3}{R}$$ which decreases with $R$. Hence a huge

sphere like the Sun has a tiny surface to volume ratio.)

**Q4.1**   Since $1\,\text{eV} = 1.602 \times 10^{-19}\,\text{J}$, it follows that

$$1\,\text{MeV}/c^2 = \frac{1.602 \times 10^{-13}\ \text{J}}{(2.997 \times 10^8\ \text{m s}^{-1})^2}$$
$$= 1.784 \times 10^{-30}\,\text{kg}.$$

Hence, we obtain

$$m_e = 9.109 \times 10^{-31}\,\text{kg} = \frac{9.109 \times 10^{-31}}{1.784 \times 10^{-30}}\ \text{MeV}/c^2$$
$$= 0.511\ \text{MeV}/c^2$$

$$m_p = 1.673 \times 10^{-27}\,\text{kg} = \frac{1.673 \times 10^{-27}}{1.784 \times 10^{-30}}\ \text{MeV}/c^2$$
$$= 938\,\text{MeV}/c^2$$

$$m_n = 1.675 \times 10^{-27}\,\text{kg} = \frac{1.675 \times 10^{-27}}{1.784 \times 10^{-30}}\ \text{MeV}/c^2$$
$$= 939\,\text{MeV}/c^2\,.$$

**Q4.2**   A collision in which two electrons become three is forbidden by the conservation of electric charge. Initially, the total charge is $2 \times (-e) = -2e$, while the total charge of three electrons is $3 \times (-e) = -3e$. Conservation of electric charge prohibits any process in which the initial total charge is different from the final total charge.

**Q4.3**   The mass and spin of the positron are the same as those of the electron. The charge of the positron is equal in magnitude, but opposite in sign, to that of the electron. Hence, $m_{pos} = m_e = 0.511\ \text{MeV}/c^2$ $(= 9.11 \times 10^{-31}\,\text{kg})$ the spin is 1/2 and the charge is $+e = 1.602 \times 10^{-19}\,\text{C}$. The energy required to create an electron and a positron is $2m_e c^2 = 2 \times 0.511\,\text{MeV} = 1.02\,\text{MeV}$ (or $1.63 \times 10^{-13}\,\text{J}$).

**Q4.4**   A magnetic field exerts a force perpendicular to the direction of motion of the charge and so does no work on it. A magnetic field cannot therefore change the particle's kinetic energy.

**Q4.5**   According to Coulomb's law of electrostatics, the electrical force between two protons separated by $10^{-14}\,\text{m}$ is repulsive and is of magnitude

$$F_{el} = \frac{e^2}{4\pi\varepsilon_0(10^{-14}\ \text{m})^2}.$$

According to Newton's law of gravitation, the gravitational force between two protons separated by $10^{-14}\,\text{m}$ is attractive and is of magnitude

$$F_{grav} = \frac{Gm_p^2}{(10^{-14}\ \text{m})^2}.$$

It follows that

$$\frac{F_{grav}}{F_{el}} = \frac{Gm_p^2 4\pi\varepsilon_0}{e^2}$$

$$= \frac{6.67 \times 10^{-11} \times (1.67 \times 10^{-27})^2 \times 4\pi \times 8.85 \times 10^{-12}}{(1.60 \times 10^{-19})^2}$$

$$= 8.08 \times 10^{-37}.$$

The gravitational force is clearly very weak in comparison to the electrostatic force. Note that this ratio is actually independent of the separation of the protons, since the gravitational and electrostatic forces are both inverse square laws, i.e. the magnitude of each force is proportional to $1/r^2$.

**Q4.6**   The proton is a baryon, i.e. a particle with baryon number $B = 1$. Since baryon number is a conserved quantity, it follows that the proton may only decay if there is a lighter baryon that can carry away the baryon number. However, the proton is the lightest baryon, so, unless processes that violate baryon number conservation are eventually observed, it appears that protons must be stable.

**Q4.7**   From Table 4.4, the width for the $\Delta^{++}(1232)$ resonance is:

$$\Gamma \approx (115 \times 10^6\,\text{eV}) \times (1.602 \times 10^{-19}\,\text{J}) = 1.84 \times 10^{-11}\,\text{J}.$$

Moreover, according to the uncertainty principle (*QPI* Eqn 2.8), $\Delta E\,\Delta t \geq \hbar/2$

where $\Delta t$ represents the mean lifetime $\tau$ and $\Delta E$ is the width $\Gamma$. Hence

$$\tau \approx \frac{\hbar}{2\Gamma} \approx \frac{1.05 \times 10^{-34}\,\text{J s}}{2 \times 1.84 \times 10^{-11}\,\text{J}}$$
$$\approx 3 \times 10^{-24}\,\text{s}.$$

**Q4.8**   (a) The reaction $n \to e^+ + e^- + \nu_e + \bar{\nu}_e$ is forbidden. The neutron on the left-hand side has $B = 1$ but all the particles on the right-hand side have $B = 0$. (b) The reaction $p \to n + e^+ + \bar{\nu}_e$ is forbidden. The left-hand side has $L_e = 0$. On the right-hand side, $e^+$ has $L_e = -1$ and $\bar{\nu}_e$ also has $L_e = -1$ and so the total $L_e$ on the right-hand side is $-2$.

**Q4.9**   (a) The answers are given in Table 4.9.

**Table 4.9** For Q4.9.

| Particle | Quarks | $Q$ | $B$ | $S$ | $C$ |
|---|---|---|---|---|---|
| $\Sigma^0$ | uds | 0 | 1 | −1 | 0 |
| $\Delta^{++}$ | uuu | 2 | 1 | 0 | 0 |
| $\Lambda_c^+$ | udc | 1 | 1 | 0 | 1 |
| $\overline{K}^0$ | $\overline{s}d$ | 0 | 0 | −1 | 0 |
| $\psi$ | $c\overline{c}$ | 0 | 0 | 0 | 0 |

In each case, the quark assignment is taken from Table 4.7 and the total values of $Q$, $B$, $S$ and $C$ are found by adding the corresponding values for each quark as given in Table 4.6.

(b) For the antiparticle of $\Lambda_c^+$, we have $Q = -1$, $B = -1$, $S = 0$, $C = -1$.

(c) A meson may be its own antiparticle if it is electrically neutral and is a combination of quark–antiquark pairs such as $u\overline{u}$, $d\overline{d}$, $s\overline{s}$, etc. For example, $\overline{d\overline{d}}$ is the same combination as $d\overline{d}$.

Since each baryon consists of three quarks, it is not possible for a baryon to be its own antiparticle. For example, the neutron is a udd combination but the antineutron is $\overline{u}\overline{d}\overline{d}$.

**Q4.10** The strong interaction is mediated by gluons which, according to QCD, couple to coloured particles (i.e. quarks). Since electrons (and other leptons) are not composed of quarks and have no colour, they cannot interact directly with gluons. Hence, electrons do not 'feel' the strong force.

**Q4.11** A linac is a linear particle accelerator, where the accelerator cavities are arranged in a straight line and the particles being accelerated make a single pass through each of the cavities in the line. A synchrotron is a cyclic particle accelerator in which the accelerator elements are arranged in a circle and the particles are made to go from one element to the next by means of a magnetic field (usually within a 'bending magnet'). As a result of its circular structure, particles can travel around a synchrotron many times, being repeatedly accelerated by the same elements. Keeping the particles moving along a path of fixed radius, despite their increasing speed, requires a carefully synchronized variation in the electric and magnetic fields that are responsible for the acceleration and the bending (and focusing) of the particles — hence the name synchrotron.

Cyclic particle accelerators have the advantage of compactness (and hence cheapness) compared with linacs of the same outgoing particle energy. However, they have the disadvantages of greater complexity. Also the bending of particle trajectories causes the particles concerned to emit electromagnetic radiation — so-called *synchrotron radiation* — and hence to lose energy.

**Q4.12** The four fundamental forces, or interactions, are the strong, electromagnetic and weak gravitational interactions. Their basic properties are listed in Table 4.2.

**Q4.13** Values of charge, baryon number and strangeness are written below each of the strong interaction processes. As can be seen, (a) and (b) are forbidden.

(a)

|   | $\pi^-$ | + | p | → | n | + | p |   |
|---|---|---|---|---|---|---|---|---|
| $Q$ | −1 |  | +1 | ≠ | 0 |  | +1 | forbidden |
| $B$ | 0 |  | +1 | ≠ | +1 |  | +1 | forbidden |
| $S$ | 0 |  | 0 | = | 0 |  | 0 | allowed |

(b)

|   | $\Sigma^-$ | + | p | → | $K^0$ | + | n |   |
|---|---|---|---|---|---|---|---|---|
| $Q$ | −1 |  | +1 | = | 0 |  | 0 | allowed |
| $B$ | 1 |  | +1 | ≠ | 0 |  | +1 | forbidden |
| $S$ | −1 |  | 0 | ≠ | +1 |  | 0 | forbidden |

(c)

|   | $K^-$ | + | p | → | $\overline{K}^0$ | + | n |   |
|---|---|---|---|---|---|---|---|---|
| $Q$ | −1 |  | +1 | = | 0 |  | 0 | allowed |
| $B$ | 0 |  | +1 | = | 0 |  | +1 | allowed |
| $S$ | −1 |  | 0 | = | −1 |  | 0 | allowed |

**Q4.14** (a) The muon is a lepton. (b) A gluon is an exchange particle for the strong force. (c) The pion is a meson. (d) The photon is the exchange particle for the electromagnetic force. (e) The $Z^0$ is an exchange particle for the electroweak interaction.

**Q4.15** The $D^+$ is a meson, so it must be a quark–antiquark combination. This eliminates options (i), (ii) and (iii). Moreover it has charm $C = 1$ which means that it must contain a charm quark and therefore eliminates options (vi) and (vii). Considering quark charges, with $Q = 2/3$ for c, $Q = 1/3$ for $\overline{d}$ and $Q = -2/3$ for $\overline{u}$, it is clear that only option (iv) can provide the necessary total charge ($Q = +1$). By supposing that the spins of the quarks are oppositely orientated, the necessary total spin can also be explained. Thus, the only plausible quark structure for the $D^+$ from the given list is $c\overline{d}$ (option (iv)).

**Q5.1** The Maxwell–Boltzmann energy distribution is based on the assumption that the molecules are indistinguishable. Quantum mechanics tells us that this assumption is false, but that the effects of indistinguishability are negligible provided the typical de Broglie wavelength of the molecules is much less than their typical separation. At temperature $T$, the typical speed of a molecule of mass $m$ is

$$v_{rms} = \sqrt{\frac{3kT}{m}}$$

and the typical de Broglie wavelength is

$$\lambda_{dB} = \frac{h}{mv_{rms}} = \frac{h}{\sqrt{3mkT}}.$$

Because the helium gas is at a lower temperature than the bromine gas, and because helium molecules have less mass than bromine molecules, the typical de Broglie wavelength of the helium molecules is greater than that of the bromine molecules. The typical separation of particles is given by

$$d = n^{-1/3}.$$

This is smaller for the sample of helium gas than for the sample of bromine gas. For all these reasons, $\lambda_{dB}/d$ is greater for the sample of helium gas than for the sample of bromine gas. We conclude that the effects of indistinguishability are greater in the sample of helium gas, and the Maxwell–Boltzmann energy distribution is less valid for the helium gas than for the bromine gas.

(In fact, inserting values given in the question, $\lambda_{dB}/d = 1.2$ for the helium gas and $\lambda_{dB}/d = 0.001$ for the bromine gas. This shows that the Maxwell–Boltzmann distribution would not give a good description of the helium sample but would give a good approximation for the bromine sample.)

**Q5.2** No. The two gases would not be expected to obey the same distribution law. The distribution law for photons is in fact Planck's radiation law (Equation 1.20). A gas of $^4_2$He atoms is described by the Maxwell–Boltzmann distribution law (provided $\lambda_{dB} \ll d$). Looking more deeply, we can identify two reasons why the two gases obey different energy distribution laws:

(i) The density of states function for photons is proportional to $E^2$, while the density of states function for helium atoms is proportional to $E^{1/2}$. This reflects the fact that photons are massless and their energy $E$ and momentum $p$ are related by $E = pc$, whereas helium atoms have a finite mass and $E = p^2/2m$. Consequently, photons confined in a box have a completely different set of quantum states and energy levels to helium atoms, and so have a very different density of states. (The energy levels for a photon confined in a box are not given in the text.)

(ii) A second major distinction between photons and $^4_2$He atoms is that the number of helium atoms is conserved, while the number of photons is not. This means that the total area under the translational energy distribution function for helium atoms, which is equal to the number of helium atoms, is independent of temperature. By contrast, the total area under the energy distribution function (Planck's radiation law) for photons increases rapidly with temperature (Equation 1.21 shows that it is proportional to $T^3$.)

**Q5.3** From Equations 1.21, 1.22 and 1.24, we have

$$U = 3PV = \frac{\pi^4}{15} C(kT)^4$$

$$= \frac{\pi^4}{15} \frac{NkT}{2.4}.$$

Thus, for a photon gas,

$$PV = \frac{\pi^4}{15 \times 2.4 \times 3} NkT$$

$$\approx 0.9NkT.$$

This shows that the pressure of thermal radiation is proportional to $NkT/V$, but the ideal gas equation of state, $PV = NkT$, is not satisfied because the constant of proportionality is not equal to 1. The similarity is rather deceptive — in a molecular gas the number of particles, $N$, is fixed but in a photon gas $N$ increases rapidly with temperature, in fact $N \propto T^3$.

**Q5.4** Electron states well below the Fermi energy are fully occupied at 0 K, and continue to be fully occupied at 1000 K. It is only electron states within a narrow range around the Fermi energy that have their occupations changed as the temperature rises. Only a small fraction of the electrons contributes to changes in the internal energy, and hence to the heat capacity.

**Q5.5** (i) If the particles are indistinguishable fermions, they will obey the Pauli exclusion principle, so the probability of finding both particles in the same quantum state is zero. (ii) If the particles are indistinguishable bosons, there are three different configurations that have total energy $4\varepsilon$ (see Figure 5.1a). Each of these configurations is equally likely, and only one has both particles in the same quantum state, so the probability of both particles being in the same quantum state is 1/3. (iii) If the particles are distinguishable, there are five different configurations that have energy $4\varepsilon$. (see Figure 5.1b). Each of these configurations is equally likely, and only one has both particles in the same quantum state, so the probability of both particles being in the same quantum state is 1/5.

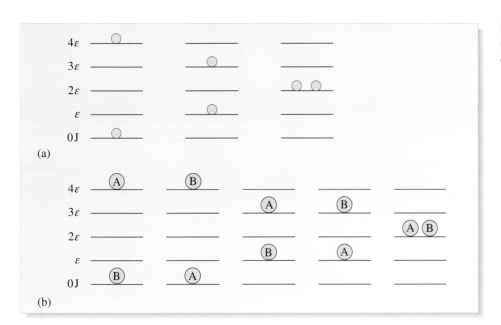

**Figure 5.1** Configurations for (a) indistinguishable bosons, (b) distinguishable particles. For Q5.5.

***Comment****: The fact that the answer for case (i) is zero, and the answer for case (ii) is greater than that for case (iii) illustrates the slogan that identical fermions are unsociable and identical bosons are sociable.*

**Q5.6** Electrons and neutrons are both fermions, so both obey the Pauli exclusion principle. Close to 0 K, the pressure due to these gases is given by Equation 1.33

$$P = \tfrac{2}{5} n E_{\mathrm{F}},$$

where $n$ is the number density of particles and $E_{\mathrm{F}}$ is the Fermi energy. So the gas with the greater Fermi energy will exert the greater pressure. The Fermi energy is given by Equation 1.31

$$E_{\mathrm{F}} = \frac{h^2}{8m}\left(\frac{3n}{\pi}\right)^{2/3}.$$

This expression was derived for electrons in the text, but it is clear that the derivation would be the same for neutrons which, like electrons, are spin 1/2 particles with two possible spin states. Because the mass of an electron is nearly 2000 times smaller than that of a neutron, while the number densities of the gases are assumed to be the same, the Fermi energy, and hence the pressure, of the electron gas will be by far the highest.

$_{2}^{4}$He atoms are bosons. Close to 0 K they will tend to congregate in quantum states of very low energy, and the corresponding pressure will be very low.

**Q5.7** (i) Covalent bonding occurs when pairs of electrons are shared between atoms, with wavefunctions that interfere constructively so as to produce a concentration of electrons between the two atoms, which opposes and overcomes the repulsion of the atomic nuclei. Covalent bonding saturates and is directional.

(ii) Ionic bonding occurs when one type of atom in the bond is strongly electronegative, i.e. it tends to attract electrons, and the other type tends to lose electrons. The resultant electrostatic attraction between the positive and negative ions binds the solid together.

(iii) Metallic bonding occurs in metals and is associated with the presence of a cloud of free electrons that are not associated with any particular atom or pair of atoms. A large part of the binding energy arises because these electrons are more delocalized than they would be if bound to individual atoms, leading to lower kinetic energies.

**Q5.8** Real materials, as opposed to perfect crystals, have tiny cracks. Failure occurs because stress is concentrated near the end of an existing crack, and this causes the crack to propagate through the material. Thus, fracture occurs under loads that are much smaller than would be needed to simultaneously break all the bonds between two atomic planes.

Dislocations are incomplete sheets of atoms which facilitate flow of the material by 'unstitching' the bonding one plane at a time. This again requires less force than simultaneously breaking all the bonds between two atomic planes.

**Q5.9** In the Drude model, electrical resistance arises because free electrons, accelerated by an applied electric field, lose energy by colliding with the fixed positive lattice ions. This simple model is able to explain Ohm's law at the expense of requiring implausibly long mean free paths (especially at low temperatures), and of predicting a resistivity proportional to $T^{1/2}$, rather than the linear temperature-dependence observed experimentally.

In the Pauli model, the electrons are treated as waves, which have the ability to propagate through a perfectly crystalline structure without being scattered. In practice, a crystal lattice is never perfect and the electron waves are scattered by defects, like vacancies and impurities, and by thermal vibrations of the lattice. The Pauli model is again able to reproduce Ohm's law, but now with a resistivity that correctly depends linearly on temperature.

**Q5.10** (i) A conductor has a partly-filled band, or overlapping bands that are partly filled, so that the electrons can move into unoccupied levels under the influence of an applied electric field. (ii) At 0 K, an insulator has a completely full valence band, separated from a completely empty conduction band by an energy gap. The gap is so wide that practically no electrons are thermally excited across it at room temperature (or at the temperature of interest). (iii) An intrinsic semiconductor is like an insulator, but with a smaller energy gap. The gap is small enough that, at the temperature of interest, a small, but not totally insignificant, number of electrons is thermally excited across the gap into the conduction band, producing the same number of holes in the valence band. (iv) In an n-type semiconductor, impurity atoms with energy levels just below the bottom of the conduction band donate electrons into the conduction band. The number of these electrons exceeds those produced by thermal excitation across the energy gap. (v) In a p-type semiconductor, impurity atoms with energy levels just above the top of the valence band accept electrons from the valence band, producing holes in it. The number of these holes exceeds those produced by thermal excitation across the energy gap.

**Q5.11** (a) The depletion region is a thin layer at the junction between n-type and p-type material where electrons diffusing from the n-region have recombined with holes diffusing from the p-region, producing a layer depleted of electrons and holes, with an electric field due to the donor and acceptor ions, directed from the now positively-charged n-region to the now negatively-charged p-region.

(b) The diffusion current is caused when a few electrons from the n-region and a few holes from the p-region have enough energy to surmount the potential energy barrier associated with the electric field, and drift across the boundary (before recombining). The pair current arises because electrons and holes can be thermally excited, as in an intrinsic semiconductor. Some of the electrons excited on the p-type side of the junction, and holes excited on the n-type side of the junction, are swept across the junction by the electric field: they constitute the pair current. In an isolated p–n junction, (i) the diffusion current (from p to n) and the pair current (from n to p) are equal in magnitude and opposite in direction. (ii) In a reverse-biased p–n junction, the diffusion current (from p to n) is reduced or stopped, and is less than the pair current (from n to p). (iii) In a forward-biased p–n junction, the diffusion current (from p to n) is increased, and is greater than the pair current (from n to p).

**Q5.12** A Cooper pair is a bound pair of electrons with opposite momenta and opposite spins. The electrons bind together by forces mediated by the lattice of positive ions. Cooper pairs help to explain superconductivity because they are bosons. They undergo a kind of Bose–Einstein condensation into a highly correlated state in which they all have the same momentum. It takes a significant amount of energy to scatter individual Cooper pairs out of this state, or to split a Cooper pair apart, so an electrical current can flow without resistance.

**Q5.13** The atomic numbers ($Z$ values) of He, C and O are 2, 6 and 8, respectively. The neutron numbers of the isotopes in the reaction are $N = 4 - 2 = 2$, $N = 12 - 6 = 6$ and $N = 16 - 8 = 8$ respectively.

*Comment*: *The oxygen in your body, in the air and in the seas was made in this reaction in massive stars going through an explosive phase.*

**Q5.14** The 1.9 MeV $\gamma$-photon does not have enough energy but the 3.2 MeV photon does, giving $3.2 - 2.22 = 0.98$ MeV of kinetic energy shared between the proton and the neutron.

**Q5.15** Electrons passing through a slit with sharp edges exhibit a diffraction pattern with deep minima, but the surface of a nucleus is not sharp but diffuse, with the density falling to zero over 1 fm or so. This leads to shallow diffraction minima.

**Q5.16** The wave functions of atomic electrons in s-states (i.e. electrons with $l = 0$) penetrate into the nuclear region. This affects the energies of these electrons and therefore the wavelengths of the corresponding atomic spectral lines. High-resolution optical techniques can be used to measure the wavelengths of the atomic spectral lines, so that nuclear sizes can be deduced.

*Comment*: *This method can be used for very short-lived nuclei which live only for the time they pass in a beam through the apparatus.*

**Q5.17** It is a characteristic of quantum tunnelling processes that the probability of passing through the barrier depends very sensitively indeed on how far the energy falls below the top of the barrier. A higher energy $\alpha$-particle will be closer to the top of the barrier and hence tunnel through much more rapidly.

**Q5.18**   The volume energy, the surface energy, the Coulomb energy and the symmetry energy. (There is also a smaller pairing energy that gives nuclei with even numbers of protons or neutrons extra binding energy.) Magic numbers fall outside the semi-empirical model, revealing a shell structure for the energy levels of protons and neutrons in the nucleus.

**Q5.19**   Quarks, leptons and baryons are all fermions. Exchange particles and mesons are all bosons. The class of hadrons includes baryons and mesons, so it represents a mix of fermions and bosons. There are no sub-atomic particles that are neither fermions nor bosons (though theorists have speculated about the possibility of discovering such particles).

**Q5.20**   The conserved quantities will be (a) total relativistic energy, (c) relativistic momentum, (d) electric charge, (e) baryon number, (f) strangeness.

**Q5.21**   Hadron resonances are too short-lived to leave measurable tracks in detectors. They were first detected because of the enhancements (bumps) they produced in certain cross-sections (e.g that for $\pi^+p$ elastic scattering) when those cross-sections were viewed as a function of energy.

**Q5.22**   (a) photons, (b)  $W^+$, $W^-$ and $Z^0$ particles, (c) gluons.

**Q5.23**   The standard model is a quantum field theory of elementary particles that is based on the 12 fundamental quarks and leptons (and their antiparticles), together with the exchange particles that feature in QCD and the electroweak theory. The standard model includes other ingredients, such as the Higgs boson.

# Acknowledgements

Grateful acknowledgement is made to the following sources for permission to reproduce material in this book:

*Front cover*: Courtesy of Don Eigler, IBM Research Division.

*Chapter 1*

*Figures p.7*: Mary Evans Picture Library; *Figures 1.1, 1.27*: Ketterle, W. et al, MIT 1995: *Figure 1.19b*: Photograph by Corus UK Limited; *Figure 1.24 left*: Science & Society Picture Library; *Figure 1.24 right*: Indian National Council of Science Museums, courtesy AIP Emilio Segrè Visual Archives. *Figure p.41*: AIP Emilio Segrè Visual Archives, Physics Today Collection.

*Chapter 3*

Figures 3.1, 3.8, 3.19 right, 3.33: Mary Evans Picture Library; Figure 3.3: National Portrait Gallery; Figures 3.5, 3.6, 3.13b, 3.14, 3.18, 3.22: Krane, K. S. (1988) Introductory Nuclear Physics. Copyright © 1988 by John Wiley and Sons, Inc. Reprinted by permission of John Wiley and Sons, Inc; Figure 3.7, 3.16: Heyde, K. (1999) Basic Ideas and Concepts in Nuclear Physics: An Introductory Approach. 2nd ed. Institute of Physics Publishing; Figure 3.11: Courtesy of Richard B. Firestone, Lawrence Berkeley National Laboratory, California; Figure 3.11 inset: Eisberg, R.M. (1961) Fundamentals of Modern Physics. Copyright © 1961 by John Wiley and Sons, Inc. Reprinted by permission of John Wiley and Sons, Inc; Figure 3.13c: Tim Beddow/Science Photo Library; Figures 3.19 left: AIP Emilio Segrè Visual Archives; Figure 3.23: Bohr, A/Mottelson, B.R. Nuclear Structure, vol. 1, Single-Particle Motion. Copyright © 1969 W.A. Benjamin, Inc; Figure 3.29: Heyde, K. (1994) Basic Ideas and Concepts in Nuclear Physics : An Introductory Approach. Institute of Physics Publishing.

*Chapter 4*

*Figures 4.5, 4.7, 4.9, 4.13, 4.22, 4.29 left:* Mary Evans Picture Library; *Figure 4.10:* AIP Emilio Segrè Visual Archives, Lindsay Collection; *Figure 4.11:* © SLAC & US Department of Energy; *Figures 4.14a, 4.14b, 4.17, 4.18:* © CERN; *Figure 4.14c;* © Manfred Schulze-Alex, Hamburg, Germany; *Figure 4.24:* Dr Donald Davis, University College London & the WA17 collaboration; *Figure 4.29 right:* AIP Emilio Segrè Visual Archives, Weber Collection.

Every effort has been made to trace all the copyright owners, but if any has been inadvertently overlooked, the publishers will be pleased to make the necessary arrangements at the first opportunity.

# Index

Entries and page numbers in **bold type** refer to key words which are printed in **bold** in the text and which are defined in the Glossary.